Climate Change

A Multidisciplinary Approach

2nd Edition

Climate Change: A Multidisciplinary Approach provides a concise, up-to-date presentation of current knowledge of climate change and its implications for society as a whole. This new edition has been thoroughly updated and extended to include the latest information.

The textbook describes the components of the global climate, and the physical principles of its behaviour on all timescales. It then considers how the many elements of climate combine to define its behaviour. The author reviews how climate change is measured. He stresses the importance of careful statistical analysis in seeking connections between observations and possible physical causes. The book discusses how the causes of climate change, including human activities, can be related to the evidence of change, and modelled to predict future changes. It then consider how these models inform the economic and political debate surrounding climate change, prevention and mitigation.

Climate Change: A Multidisciplinary Approach can be used on a wide range of introductory courses that consider the impact of a changing climate on the Earth and its inhabitants, within the departments of meteorology, oceanography, environmental science, earth science, geography, agriculture and social science. It will also appeal to a wider audience who wish to go beyond the standard 'coffee table' book and get to grips with the gritty issues of climate change.

William Burroughs is a professional science writer, and has published many books on weather and climate including *Climate Change in Prehistory, Does the Weather Really Matter?, Weather Cycles, The Climate Revealed, Watching the World's Weather*, and *Climate: Into the 21st Century* (all with Cambridge University Press). In 2005 he received the Michael Hunt Award from the Royal Meteorological Society for his work in popularising meteorology.

Reviews of the first edition

'... a recommended read for the informed layman and student seeking a wider background in this topical but complex field.'

Grant Bigg, *Weather*

'The book is well written, contains practically no mathematics and yet manages to explain, in a clear and attractive style, the subtleties of the subject ... I recommend it to everybody interested in the climate of our Earth.'

Michael Hantel, *Meteorologische Zeitschrift*

'... the book enthusiastically achieves its aims of not oversimplifying but explaining the complexities of what is well established and unknown about the climate system for a wider audience ...'

Claire Goodess, *International Journal of Climatology*

'Burroughs is to be congratulated for having written a serious and up-to-date book that competently surveys many highly technical aspects of modern climate science but manages to do so in a nonmathematical manner.'

American Meteorological Society

Climate Change

A Multidisciplinary Approach

2nd Edition

WILLIAM JAMES BURROUGHS

CAMBRIDGE
UNIVERSITY PRESS

CAMBRIDGE UNIVERSITY PRESS
Cambridge, New York, Melbourne, Madrid, Cape Town, Singapore, São Paulo, Delhi

Cambridge University Press
The Edinburgh Building, Cambridge CB2 8RU, UK

Published in the United States of America by Cambridge University Press, New York

www.cambridge.org
Information on this title: www.cambridge.org/9780521690331

First published 2007
Reprinted 2009

Printed in the United Kingdom at the University Press, Cambridge

A catalogue record for this publication is available from the British Library

ISBN 978-0-521-87015-3 hardback
ISBN 978-0-521-69033-1 paperback

Contents

Preface to the second edition

Since the first edition of this book was published in 2001, the subject of climate change has grown from just one of a number of pressing environmental issues to being seen as comparable with terrorism and nuclear proliferation as one of the greatest threats to humankind. Why has this dramatic change occurred? The principal reason is summed up in a quip by the British Prime Minister, in the late 1950s/early 1960s, Harold Macmillan. When asked by a young journalist after a long dinner what can most easily steer a government off course, he answered, "Events, dear boy. Events".

The climate has been anything but uneventful in recent years. The inundation of New Orleans, the European heat wave of 2003 and the accelerating melting of polar ice sheets are widely seen as examples of how climate change is a growing threat to the planet. It is not just the scale of these events but also the sense of inadequacy of our preparedness for managing adverse events and how this requires us to think in a multidisciplinary way in our preparedness. This wider thinking has inspired politicians to express far greater commitment to action on climate change. In part, this is the realisation that the first step had to be to sign up to and express commitment to the Kyoto Treaty of 1997. Then there was the remarkable impact of one of their number – Al Gore. Having come within a few hanging chads of becoming US President in 2000, he is now the international standard bearer for action on the climate.

Another feature of these changes is the increasing anthropomorphisation of the plight of creatures at threat from global warming. While this is a powerful method of publicising the potential consequences of melting sea ice for polar bears, penguins and walrus, it does need to be put in the context of past climate change. All these polar species survived the last interglacial warm spell around 125 000 years ago, when the peak global temperature was some 2 °C warmer than now and sea levels were about 6 metres higher. So, on the basis of this experience of these examples of polar species, they appear to have the capacity to survive for a while yet.

At the same time, the scientific community has reached a consensus on the nature of the scale of human contribution to recent climate change. These forecasts centred on the impact of the emission of greenhouse gases, which were seen as the dominant anthropogenic component in the recent warming. This confidence has, however, been qualified by a widespread recognition that there are considerable uncertainties in the handling of such factors as cloudiness, the associated consequences of dust and aerosols. Furthermore, some of the high hopes expressed as to progress with seasonal and longer-term forecasts, based on the growing understanding of the El Niño, have had to recognise that such forecasts cannot restrict the analysis to the Pacific alone, but must include the rhythms of the Atlantic and Indian Oceans.

There is, however, no doubt we are part of the recent changes and that global models of the climate are the only way to calculate future trends. What is less clear is the true scale of the natural variability. Paradoxically, while the majority are inclined to downgrade the impact of natural fluctuations in recent centuries, there has been a growing recognition that abrupt and unpredictable changes in the climate have been a feature of climatic history. Furthermore, it is clear that we have had the good fortune to live in a period of relative climatic stability during the last 10 000 years. Our hope has to be that neither natural changes nor any changes that will be precipitated by anthropogenic activities will push us too far from the conditions of this relatively stable period, which encompasses all the recorded history of humankind and more.

Confronted with this array of forbidding prospects, the calls for action have abounded. The most practical seek to show that a virtuous circle of improved efficiency and progressive tax policies can produce major reductions in greenhouse emissions. In practice, some of these will be politically acceptable, others will not. So, while progress will be made, handling political objections to more demanding cuts will increasingly be difficult to overcome.

It is at this stage that the soundness of the scientific arguments will come in for more aggressive scrutiny. Balanced arguments that recognise the strengths and weaknesses of the scientific case are more likely to prevail. The alternative is to insist that the only answer is to concentrate on carbon dioxide emissions and, that these need to be cut by, say, 80% by 2050 stands little chance of electoral support. Of course, events of sufficient magnitude may steam-roller all objections, but let us hope it does not come to that.

In the meantime, the objective of this book remains to build on the groundwork of pioneers such as Hubert Lamb and Murray Mitchell, who devoted so much effort to climate change when it was not regarded as an important aspect of meteorology. The pace of events does not alter the need to present the current state of climatic knowledge in a balanced manner. In so doing, it must provide a gateway to the intricacies of the climate system, that it properly integrates the analysis of how we have to face up to the challenge of changes that represent a serious threat to our current lifestyle. This can only be achieved by continued research into what are the most important features of climate change, how they will combine to produce imminent change, and what are the most realistic ways of averting these changes.

1 Introduction

There is always an easy solution to every human problem – neat, plausible and wrong. **H. L. Menken**

The climate has always been changing. On every timescale, since the Earth was first formed, its surface conditions have fluctuated. Past changes are etched on the landscape, have influenced the evolution of all life forms, and are a subtext of our economic and social history. Current climate changes are a central part of the debate about the consequences of human activities on the global environment, while the future course of the climate may well exert powerful constraints on economic development, especially in developing countries. So, for many physical and social sciences, climate change is an underlying factor that needs to be appreciated in understanding how these disciplines fit in to the wider picture. The aim of this book is to provide a balanced view to help the reader to give the right weight to the impact of climate change on their chosen disciplines. This will involve assessing how the climate can vary of its own accord and then adding in the question of how human activities may lead to further change.

The first thing to get straight is that there is nothing simple about how the climate changes. While the central objective of the book is to make the essence of the subject accessible, it is no help to you, the reader, to duck the issues. All too often people try to reduce the debate to what they see as the essential features. In so doing, either unwittingly or deliberately, they run the risk of squeezing the evidence down too much. So, from the outset it pays to appreciate that the behaviour of the Earth's climate is governed by a wide range of factors all of which are interlinked in an intricate web of physical processes. This means identifying the factors that matter most and when they come into play. To do this we have to define the meaning of climate change, because various factors assume different significance depending on the timescales under consideration.

1.1   Weather and climate

It is not easy to form a balanced view of the importance of climate change in our lives. We are bombarded with a continual stream of dramatic information about how the climate is becoming more extreme and threatening. This information is drawn from meteorological authorities around the world. While the process of translation is not always accurate, it sets the scene for interpreting the importance of climate change in our world. So, how are we to handle claims about the record-breaking nature of droughts, heatwaves and hurricane seasons?

The first stage in establishing how the climate varies is to discriminate between weather and climate. At the simplest level, weather is what we get, climate is what we expect. So, the weather is what is happening to the atmosphere at any given time, whereas climate is what the statistics tell us should occur at any given time of the year. Although climatic statistics concentrate on averages built up over many years, they also give an accurate picture of the incidence of extreme events, which are part of the normal for any part of the world. Here, the emphasis will be principally on the average conditions, but it is in the nature of statistics that giving proper weight to rare extreme events is difficult to handle.

Wherever possible the analysis will be in terms of well-established lengthy statistical series. These are more likely to provide reasonable evidence of change in the incidence of extremes. Where the changing frequency of extreme weather events exerts a major influence on the interpretation of changes in the climate, however, there may be some blurring of the distinction between weather and climate in considering specific events. The essential point is, in considering climate change, we are concerned about the statistics of the weather phenomena that provide evidence of longer term changes.

1.2   What do we mean by climate variability and climate change?

It follows from the definition of weather and climate, that changes in the climate constitute shifts in meteorological conditions lasting a few years or longer. These changes may involve a single parameter, such as temperature or rainfall, but usually accompany more general shifts in weather patterns that might result in a shift to, say, colder, wetter, cloudier and windier conditions. Due to the connection with global weather patterns these changes can result in compensating shifts in different parts of the world. More often they are, however, part of an overall warming or cooling of the global climate, but in terms of considering the implications of changes in the climate, it is the regional variations that provide the most interesting material, as long as they are properly set in the context of global change.

This leads into the question of defining the difference between climate variability and climate change. Given that we will be considering a continuum of variations across the timescale from a few years to a billion years, there is bound to be a degree of artificiality in this distinction and so it is important to spell out clearly how the two categories will be treated in this book. Figure 1.1a presents a typical set of meteorological observations; this example is a series of annual average temperatures, but it could equally well be rainfall or some other meteorological variable for which regular measurements have been made over the years. This series shows that over the period of the measurements the average value remains effectively constant (the series is said to be *stationary*) but fluctuates considerably from observation to observation. This fluctuation about the mean is a measure of *climate variability*. In Figures 1.1 b, c and d the same example of climatic variability is combined with examples of *climate change*. The combination of variability and a uniform cooling trend is shown in Figure 1.1b, whereas in curve c the variability is combined with a periodic change in the underlying climate, and in curve d the variability is combined with a sudden drop in temperature, which represents, during the period of observation, a once and for all change in the climate.

The implication of the forms of change shown in Figure 1.1 is that the level of variability remains constant while the climate changes. This need not be the case. Figure 1.2 presents the implications of variability changing as well. Curve (a) presents the combination of the amplitude of variability doubling over the period of observation, while the climate remains constant. Although this is not a likely scenario, the possibility of the variability increasing as, say, the climate cools (Fig. 1.2b) is much more likely. Similarly, the marked increase in variability following a sudden drop in temperature (Fig. 1.2c) is a possible consequence of climate change. The examples of climate variability and climate change presented on Figures 1.1 and 1.2 will be explored in this book, and so we will return to the concepts in these diagrams from time to time.

For most of these variations, a basic interpretation is that climate variability is being a matter of short-term fluctuations, while climate change is concerned about longer-term shifts. There are two reasons why this approach runs the risk of oversimplifying the issues. First, there is no reason

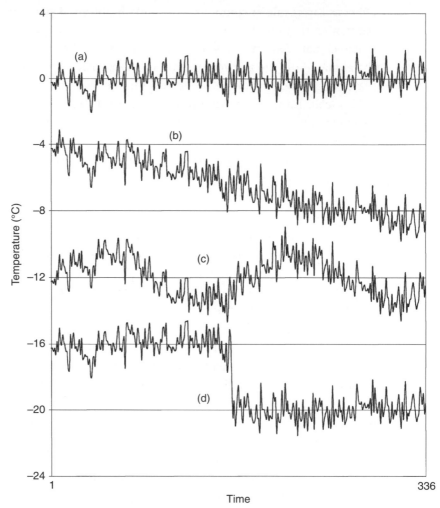

Fig. 1.1 The definition of climate variability and climate change is most easily presented by considering a typical set of temperature observations which show (a) climate variability without any underlying change in the climate, (b) the combination of the same climate variability with a linear decline in temperature of 4 °C over the period of the observations, (c) the combination of climate variability with a periodic variation in temperature of 3 °C, and (d) the combination of climate variability with a sudden drop in temperature of 4 °C during the record, with the average temperature otherwise remaining constant before and after the shift. Each record is displaced by 4 °C to enable a comparison to be made more easily. (From Burroughs, 2001, Fig.1.1.)

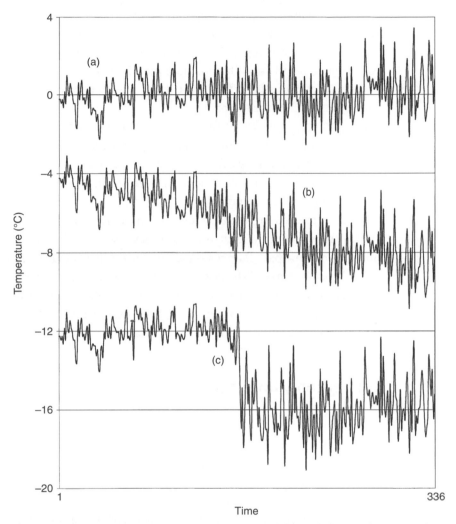

Fig. 1.2 The combination of increasing climate variability and climate change can be presented by considering a set of temperature observations similar to those in Figure 1.1 which show (a) climate variability doubling over the period of the record without any underlying change in the climate, (b) the combination of the same increasing climate variability with a linear decline in temperature of 4 °C over the period of the observations, (c) the combination of one level of climate variability before a sudden drop in temperature of 4 °C, which then doubles after the drop while with the average temperature remains constant before and after the shift. Each record is displaced by 4 °C to enable a comparison to be made more easily, and the example of periodic climate change in Figure 1.1 (curve c) is not reproduced here, as the nature of changing variability in such circumstances is likely to be more complicated. (From Burroughs, 2001, Fig. 1.2.)

why the climate should not fluctuate randomly on longer timescales, and recognising this form of variability is a major challenge in teasing out the causes of climate change. Secondly, as already discussed, climate change may occur abruptly and there is no doubt that in terms of the changes we will consider in this book this form of change has been an important factor in the Earth's climatic history.

Detecting fluctuations in the climate of the type described in Figures 1.1 and 1.2 involves measuring a range of past variations of meteorological parameters around the world over a wide variety of timescales. This poses major challenges that will be a central theme in this book. For now, the important point, is that when we talk about climate variability and climate change we are dealing with evidence from many parts of the world, which comes from a comprehensive range of sources and this varies greatly in quality. So, some past changes stand out with startling clarity, as in the case of the broad features of the last ice age, whereas others, such as the effect of solar activity on more recent changes, are shadowy and surrounded by controversy. It is this mixture of the well established and the unknown, which make the subject so hard to pin down and so fascinating.

1.3   Connections, timescales and uncertainties

From the outset the golden rule is: *do not oversimplify the workings of the climate*. In truth it is an immensely complicated subject. Part of the excitement of discovering the implications of climate change is, however, the fact it involves so many different processes linked together in an intricate web. In particular it requires us to understand the nature of *feedback processes* (see Box 1.1). The fact that a perturbation in one part of the system may produce effects elsewhere that bear no simple relation to the original stimulus provides new insights into how the world around us functions. It does require, however, a disciplined approach to the physical processes at work, otherwise the analysis is liable to be partial and misleading.

The challenge of discovering which processes matter most involves not only knowing how a given perturbation may disturb the climate but also of how different timescales affect the analysis of change. From the imperceptibly slow movement of the continents to the day-to-day flickering of the Sun, every aspect of the forces driving the Earth's climate is varying. Deciding which of these changes matters most, and when, requires evaluation of how they occur and how they are linked to one another. So while continental drift (*plate tectonics*) only comes into play when interpreting

Box 1.1 The nature of feedback

Understanding and quantifying various *feedback processes* is a central challenge of explaining and predicting climate change. These processes arise because, when one climatic variable changes, it alters another in a way that influences the initial variable that triggered the change. If this circular response leads to a reinforcement of the impact of the original stimulus then the whole system may move dramatically in a given direction. This runaway response is known as *positive feedback* and is best illustrated by the high-pitch whistle that a microphone system can produce if it picks up some of its own signal. In the case of the climate, an example of this behaviour might be the effect of a warming leading to a reduction in snow cover in winter. This, in turn, could lead to more sunlight being absorbed at the surface and yet more warming, and so on and on.

The reverse situation is when the circular response tends to damp down the impact of the initial stimulus and produce a steady state. This is known as *negative feedback*. An example of this type of climatic response is where a warming leads to more water vapour in the atmosphere, which produces more clouds. These reflect more sunlight into space thereby reducing the amount of heating of the surface and so tends to cancel out the initial warming. Depending on the complexity of the system involved, these types of chain reaction can lead to a variety of responses to any perturbations including sudden switches between different states, and reversals where a positive stimulus can produce a negative reaction by the system as a whole. Throughout this book the appreciation of feedback mechanisms will play an essential part in understanding how the climate responds to change.

geological records over millions of years, its more immediate consequences (e.g. volcanism) can have a sudden and dramatic impact on interannual climatic variability. Similarly, fluctuations in the output of the Sun may occur on every timescale. When it comes to looking into the future, longer-term variations are bound to be the subject of speculation and, as such, are of questionable value. In the case of predictions a few days to several decades ahead, however, the potential benefits of accurate forecasts are immense. No wonder this is one of the hottest topics in climate change.

This differing perspective, depending on whether the analysis of the past is relevant to forecasting the future, will influence how climatic change is presented in this book. To use an example, the oil industry exploits the growing information about past climates on geological timescales to assist in the search for hydrocarbon deposits. For instance, current research includes computer models of past ocean circulation that can be used to identify regions of high marine productivity in the geologic past, where large amounts of organic matter may have been laid down, but changing knowledge about the climate on these timescales will have no influence on the industry's view on how the climate in the future will affect its plans.

By way of contrast, as production extends further offshore, the oil industry needs to know whether changes in, say, the incidence of hurricanes in the Gulf of Mexico or deep depressions in the North Atlantic will make its deep-water operations more hazardous. Such forecasts are entirely dependent on both reliable measurements of how the climate has been changing in recent decades and on whether it is possible to predict future trends. This sets more demanding criteria on what we need to know and whether it is sufficient to make useful forecasts. This applies to all areas of climate variability and climate change and so the improved understanding of the long-term climatic processes that have shaped the world around us, while central to the making of more efficient use of natural resources, can be dealt with in general terms. The more immediate issues of how the climate could change in the near future and how it will shape our plans for weather-sensitive investments requires more detailed analysis.

1.4   The big picture

One further aspect of climate change must not be overlooked. This is the fact that everything in the system is connected to everything else and so we have to be exceedingly careful in trying to explain how things fits together. Inevitably it will be necessary to look at various aspects of the climate in isolation. In so doing, however, we must not lose sight of the big picture. The connections and feedback processes described in the preceding section link every process into the system as a whole. Adopting this wider perspective helps to achieve some balance. Indeed, if you think there is a simple answer to any issue associated with climate change then you have not read enough about the subject. So, there is no point in beating about the bush; the climate is fearfully complex and while the objective of this book is to make climate change as accessible as possible, it is of no benefit to the reader to understate the complexity of the connections.

This means that any analysis depends on how changes in every aspect of the Earth's physical conditions and extraterrestrial influences combine.

These range from the ever-changing motions of the atmosphere, through the variations in the land surface, including vegetation type, soil moisture levels and snow cover, to sea-surface temperatures, the extent of pack ice in polar regions, plus the stately motions of the deep-ocean currents, which may take over a thousand years to complete a single cycle, but which are also be capable of suddenly switching within a few years to entirely different patterns. Add to this that the prevailing climate, combined with the distribution of the continents controls the amount of nutrients washed into the oceans, which in turn affects the oceans productivity, and carbon dioxide levels in the atmosphere and you begin to get an idea of the complexity of climate change.

Some of these phenomena are predictable and some are not. For instance the features of the Earth's orbital motion, which govern the daily (diurnal) and annual cycles in the weather, can be predicted with great precision. Other gravitational effects, such as the longer-term cycles in the lunar tides, or the possible effects of the other planets in the solar system on both the Earth and the Sun, can be calculated with considerable precision, but their influence on the climate is much more speculative. The same applies to even longer-term changes in the Earth's orbital parameters, which alter the amount of sunlight falling at different latitudes throughout the year. But, as we will see later, this is the most plausible explanation of the periodic nature of the ice ages that have engulfed much of the planet during much of the last million years.

By comparison, many of the predictions of how the climate will respond to changes in various parts of the system are far less reliable. Indeed many of them expose the chaotic nature of the weather and the climate (see Box 1.1). So many aspects of climate change may well prove to be wholly unpredictable, but in studying how the various components of the climate interact with one another it may be possible to establish statistical rules about their behaviour. This might then lead to the making of useful forecasts of the probability of certain outcomes occurring as well as providing valuable insights into how the system works.

The consequence of the complexity is that from time to time we must enter the worlds of mathematics, statistics and physics. To many people studying other disciplines, which may be influenced by climate change, this may seem an unwelcome prospect but it is a necessary condition to understanding whether the issues raised by the claims of climatologists are of real consequence to your chosen discipline and so there has to be some mathematics and physics in this book. My objective will be, however, to keep the amount down to a minimum and to do my best to make it as user-friendly as possible.

Armed with these insights, my aim will be to steer a judicious course through the essential aspects of climate variability and climate change to show how many aspects of our lives are influenced by past and present fluctuations. This will then set the scene for considering the urgent issues of the potential impact of human activities on the climate in the future and what we can do about it. The first step in this process is to consider the Earth's energy balance.

FURTHER READING

A complete reference list is available at the end of the book but the following is a selection of the best books or articles to follow up particular topics within this chapter. Full details of each reference are to be found in the Bibliography.

> IPCC (1990), (1992), (1994), (1995), (2001) and (2007): In terms of obtaining a comprehensive picture of where the debate on the science of climate change has got to, these are the definitive statements of the consensus view. They are, however, both carefully balanced in their analysis, and exhaustive in their presentation of the competing arguments. As such, they are not an easy read and may appear evasive, if not confusing, until you have got a grip of the basic issues. So they are of greatest value once you have got your bearings clearly established. It follows that the greatest value will be in working with the latest review, but the earlier volumes provide an important historical perspective of the development of climate science in the last two decades.

2 Radiation and the Earth's energy balance

And God said, Let there be light: and there was light.   **Genesis 1,1.**

Although a wide range of factors governs the Earth's climate, the essential, driving process is the supply of energy from the Sun and what happens to this is energy when it hits the Earth. To understand how this works, we must consider the following processes:

(a) the properties of solar radiation and also how the Earth re-radiates energy to space;
(b) how the Earth's atmosphere and surface absorb or reflect solar energy and also re-radiate energy to space; and
(c) how all these parameters change throughout the year and on longer timescales.

The consideration of variations over timescales longer than a year has to take account of the fact that many of the most important changes may occur over widely differing periods (i.e. from a few years to hundreds of millions of years). The differing time horizons cannot be specified in advance and so we may find ourselves moving back and forth over this wide range as we progress through the book. The only way to deal with these shifts in time-scale is to identify clearly what we are talking about as we go along.

2.1   Solar and terrestrial radiation

At the simplest level the radiative balance of the Earth can be defined as follows: over time the amount of solar radiation absorbed by the atmosphere and the surface beneath it is equal to the amount of heat radiation emitted by the Earth to space. Global warming does, however, have the effect of retaining some solar energy in the climate system. It is reckoned that over the last 50 years the annual storage rate has been between 0.5 and 1.0 watts m^{-2}: a tiny amount compared with the fluxes shown in Figure 2.7 below, although it has significant implications for current climate change.

2.1.1   Radiation laws

It is a fundamental physical property of matter that any object not at a temperature of absolute zero (−273.16 °C) [for the purposes of discussing the radiative properties of bodies it is necessary to measure the temperature of the body in relation to absolute zero by defining how many degrees Celsius it is above absolute zero – this value is defined as the body's temperature in Kelvin (K)] transmits energy to its surroundings by radiation. This radiation is in the form of electromagnetic waves travelling at the speed of light and requiring no intervening medium. Electromagnetic radiation is characterised by its wavelength that can extend over a spectrum from very short gamma (γ) rays, through X-rays and ultraviolet, to the visible, and on to infrared, microwaves, and radio waves (Fig. 2.1). The wavelength of visible light is in the range 0.4–0.7 μm (1 μm = 10^{-6} m).

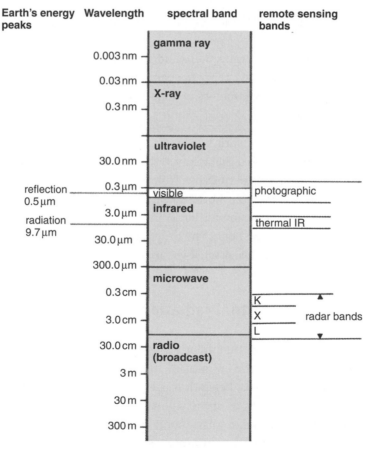

Fig. 2.1 The electromagnetic spectrum covers the complete range of wavelengths from the longest radio waves to the shortest γ-rays. (From Burroughs, 2001, Fig. 2.1.)

Throughout this book electromagnetic radiation will be identified in terms of its wavelength [generally in microns (μm) in the visible and infrared, and millimetres (mm) in the microwave region]. There are, however, times when it is more convenient to discuss the frequency of the radiation. The relationship between the wavelength and the frequency is

$$c = \lambda v,$$

where λ is the wavelength in metres, v is the frequency in cycles per second, and c is the velocity of light ($3 \times 10^8 \, \mathrm{m \, s^{-1}}$).

A body which absorbs all the radiation and which, at any temperature, emits the maximum possible amount of radiant energy is known as a 'black body'. In practice no actual substance is truly radiatively 'black'. Moreover, a substance does not have to be black in the visible light to behave like a black body at other wavelengths. For example, snow absorbs very little light but is a highly efficient emitter of infrared radiation. At any given wavelength, however, if a substance is a good absorber then it is a good emitter, whereas at wavelengths at which it absorbs weakly it also emits weakly.

The wavelength dependence of the absorptivity and emissivity of a gas, liquid, or solid is known as its spectrum. Each substance has its own unique spectrum that may have an elaborate wavelength dependence. Hence, the radiative properties of the Earth are made up of the spectral characteristics of the constituents of the atmosphere, the oceans, and the land surface. These combine in a way that is essential to explaining the behaviour of the global climate.

For a black body, which emits the maximum amount of energy at all wavelengths the intensity of radiation emitted and the wavelength distribution depend only on the absolute temperature. The expression for emitted radiation is defined by the Stefan–Boltzmann law, which states that the flux of radiation from a black body is directly proportional to absolute temperature. That is

$$F = \sigma T^4,$$

where F is the flux of radiation, T is the absolute temperature as measured in K from absolute zero ($-273.16\,^\circ$C), and σ is a constant ($5.670 \times 10^{-7}\,\mathrm{W\,m^{-2}\,(K)^4}$).

It can also be shown that the wavelength at which a black body emits most strongly is inversely proportional to the absolute temperature. Known as the Wien displacement law, this is expressed as

$$\lambda_\mathrm{m} = \alpha/T,$$

where λ_m is the wavelength of maximum energy emission in metres and α is a constant (2.898×10^{-3} m K).

So if the Earth were a black body and the Sun emitted radiation as a black body of temperature 6000 K, then a relatively simple calculation of the planet's radiation balance produces a figure for the average surface temperature of 270 K. This figure is somewhat lower than the observed value of about 287 K. Furthermore the Earth does not absorb all the radiation from the Sun and so, in principle, might be expected to be even cooler at around 254 K. The reason for this difference is, as we will see, the properties of the Earth's atmosphere. Widely known as the *Greenhouse Effect* (see Box 2.1), its impact is to alter how the Earth radiates energy from different levels in the atmosphere. To understand how this works we need to look more closely at the properties of both solar and terrestrial radiation.

Box 2.1 The Greenhouse Effect

Strictly speaking the term the *Greenhouse Effect* is a misnomer. The principal mechanism operating in a greenhouse is not the trapping of infrared radiation but the restriction of convective losses when air is warmed by contact with ground heated by solar radiation. The radiative properties of the glass in terms of preventing the transmission of terrestrial radiation are inconsequential, as R. W. Wood demonstrated with an elegant experiment in 1909. He showed that when the glass in a model greenhouse was replaced by rock salt, which is transparent to infrared radiation, it makes no difference to the temperature of the air inside. What matters is that the incoming solar radiation is not impeded by the glass, and once the ground and adjacent air is warmed up, turbulent movement of the air does not remove the heat too rapidly.

The fact that greenhouses do not *trap* terrestrial radiation is important not only in understanding how they work, but also in getting an accurate appreciation of how the build-up of radiatively active gases in the atmosphere alters the temperature. As the density of the atmosphere decreases rapidly with altitude any absorption of terrestrial radiation will take place principally near the surface (the exception is ozone, which occurs at higher concentrations in the stratosphere (see Box 2.2), and so will exert a different influence on radiative balance). Moreover, since the most important absorber is water vapour, which is concentrated in the lowest levels of the atmosphere, the greatest part of the absorption of terrestrial radiation emitted by the Earth's surface occurs at the bottom of the atmosphere.

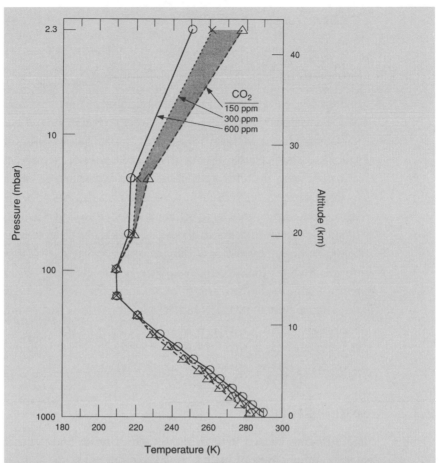

Fig. B2.1 The vertical temperature profile of the atmosphere showing how an increase in the concentration of carbon dioxide (CO_2), relative humidity and cloud cover remaining unchanged, alters the temperature distribution leading to a warming at low levels and a cooling at high levels. (From Trenberth, 1992, Fig. 20.4.)

If the concentration of radiatively active gases in the atmosphere increases and the amount of solar energy absorbed at the surface remains unchanged then the lowest levels of the atmosphere will be warmed up a little. This warming process is the result in the lower atmosphere absorbing and consequently emitting more infrared radiation, both upwards and downwards. In achieving balance between incoming and outgoing radiation this process leads to the surface and lower atmosphere warming and the upper atmosphere cooling (see Fig. B2.1). The consequence of this adjustment to additional absorbing gases is for the surface to warm up and for the atmosphere, as a whole, effectively to radiate from a higher

Box 2.1 (cont.)

level (i.e. the equivalent black-body temperature of the atmosphere required to maintain the Earth's energy balance remains unchanged but rises to a greater altitude). This change is usually expressed in terms of the change in the average net radiation at the top of the troposphere (known as the *tropopause*) and is defined as *radiative forcing*. It is changes in this parameter that is the real measure of the impact of alterations in the greenhouse effect, whether due to natural causes or human activities.

It may seem pedantic to quibble about whether this change can be described as the Greenhouse Effect. After all the net effect is warming where we all live, just as if we were trapped in a greenhouse with no scope to open ventilation windows, so does it matter? The answer has to be *yes*, because the essence of the process is the theoretical change in the temperature profile of the atmosphere. As we will see this is an essential feature in appreciating the arguments about the physical consequence of the atmospheric build-up of radiatively active gases due to human activities. So it is better not to regard these gases as putting the lid on the atmosphere, but rather of subtly altering how it distributes the heat absorbed from the Sun and reradiates it to space.

2.1.2 Solar radiation

The Sun does indeed behave rather like a black body with an effective surface temperature of 6000 K. Its maximum emission is in the region of 0.5 μm near the middle of the visible portion of the electromagnetic spectrum (Fig. 2.2). Almost 99% of the Sun's radiation is contained in the short wavelength range 0.15 to 4 μm. Some 9% fall in the ultraviolet, 45% in the visible and the remainder at longer wavelengths. In the atmosphere much of the shorter ultraviolet is absorbed by oxygen and ozone high in the atmosphere. Water vapour and carbon dioxide (CO_2) absorb much of the infrared radiation beyond 1.5 to 2 μm at lower levels in the atmosphere. So the spectrum of solar radiation reaching the Earth's surface is cut down to a band from around 0.3 to 2 μm (Fig. 2.2).

2.1.3 Terrestrial radiation

A black body with a temperature of 287 K emits its energy in the mid-infrared. Most of this energy is emitted in the range 4 to 50 μm (Fig. 2.3). With the Earth the radiation of heat is complicated by both emissivity of the surface, and,

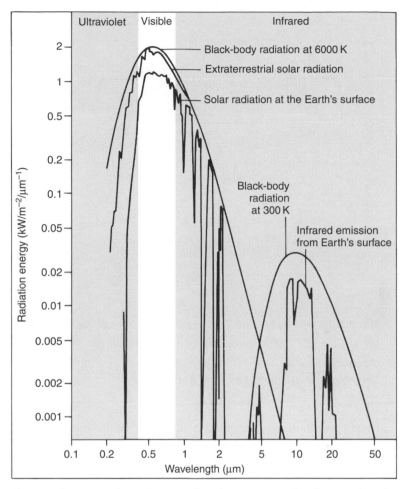

Fig. 2.2 The energy from the Sun is largely concentrated in the wavelength range 0.2 – 4 μm. (From Burroughs, 2001, Fig. 2.2.)

even more important, the spectral characteristics of the atmosphere. In practice most land surfaces and the oceans are efficient emitters of infrared radiation, and so can be regarded as radiatively 'black' at these wavelengths. This means that at, any given point, the amount of outgoing radiation from the Earth is proportional to the fourth power of the temperature, and the characteristics of the intervening atmosphere. As the principal atmospheric gases (oxygen and nitrogen) do not absorb appreciable amounts of infrared radiation, the radiative properties of the atmosphere are dominated by certain trace gases, notably water vapour, carbon dioxide and ozone. Each of these interacts with infrared radiation in its own way and so the surface radiation is modified by absorption and re-emission in the atmosphere by

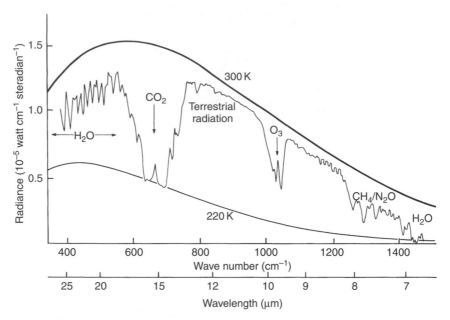

Fig. 2.3 The energy emitted by the Earth (terrestrial radiation), as measured by an orbiting satellite, is concentrated on the 4–50 μm region and the amount radiated at any wavelength depends on both the temperature and trace constituents in the atmosphere. (From Burroughs, 2001, Fig. 2.3.)

these trace constituents, whose temperatures will generally be different from that of the surface. As this upwelling energy comes from both the surface and the atmosphere it is defined as *terrestrial radiation*.

The effect of the radiatively active trace gases on the form of the terrestrial radiation is complex. Each species has a unique set of absorption and emission properties known as its *molecular spectrum*, which is the product of its molecular structure. This spectrum is made up of a large number of narrow features, referred to as *spectral lines*, which tend to be grouped in broader bands around certain wavelengths. For different molecules the spectral lines and bands are at different wavelengths. The intensity of individual spectral lines and broader bands varies in a way that is related to the physical properties of each gas. Therefore we need to know precisely how each gas absorbs and emits infrared radiation to understand how the climate is controlled by variations of the concentration of these gases.

One simple way to quantify the impact of the naturally occurring radiatively active gases is to estimate the contribution they each make to the warming of the Earth above the figure of 254 K. This calculation shows that water vapour contributes 21 K, carbon dioxide (CO_2) 7 K, and ozone (O_3) 2 K.

These figures show that water vapour is the most important greenhouse gas. Moreover, if the climate warms the amount of water vapour in the atmosphere will increase. On its own, this will have a positive feedback (see Box 1.1) on the warming process, whereas a cooling could be amplified by the reduction of water vapour in the atmosphere. The significant contribution of CO_2 and O_3 highlight the potential impact of human activities, which alter the concentrations of these gases in the atmosphere. In addition, other products of human activities [e.g. methane, oxides of nitrogen, sulphur dioxide and chlorofluorocarbons (CFCs)] will also modify the radiative properties of the atmosphere.

The impact of the most important radiatively active trace constituents shows up clearly in the spectrum of terrestrial radiation observed from space (Fig. 2.3). This spectrum holds the key to understanding how changes in these important trace gases exert control over the climate. Taking the example of the CO_2 band centred around 15 μm: the curve shows that, where this gas absorbs and emits most strongly, the radiation escaping to space comes from high in the atmosphere where the temperature is low (typically ~220 K). Where the atmosphere is largely transparent on either side of the CO_2 band (often termed *atmospheric window regions*), the radiation comes from the bottom of the atmosphere or the Earth's surface, which is warmer (typically ~287 K). Therefore, if the amount of CO_2 in the atmosphere increases due to, say, human activities, the band around 15 μm will become more opaque and effectively broaden. This will make no difference where the absorption is strong, but in other parts of the band the increase in absorption and emission will reduce the amount of terrestrial radiation emitted by CO_2 to space as more of it will come from the cold top of the atmosphere. For the Earth's energy budget to remain in balance, however, the amount of terrestrial radiation must remain constant and so the temperature of the lower atmosphere has to rise to compensate for the reduced emission in the CO_2 band. This is the basic physical process underlying the Greenhouse Effect (see Box 2.1). It applies to changes in the concentration of all radiatively active gases, and is central to how the temperature of the Earth adjusts to the amount of energy received from the Sun.

The analysis of the radiative impact of greenhouse gases depends on their distribution in the atmosphere. Most trace constituents are relatively uniformly spread both geographically and vertically. Two of the most important greenhouse gases (water vapour and O_3) do, however, have a much more complicated distribution. As the concentration of water vapour depends on temperature and the dynamic balance with clouds and precipitation, which makes up the *hydrological cycle*, the amount in the atmosphere varies substantially from place to place. Humidity levels are highest over the tropical oceans and lowest over continental interiors in winter. In addition, the concentration falls off rapidly with altitude. Therefore any analysis of the radiative impact of water vapour has to take full account of the geographical and seasonal nature of these variations.

There is an unresolved question about the extent to which global warming will alter the concentration of water vapour in the atmosphere. This is an essential part of the debate about how the climate will respond to human activities. As water vapour is the most important greenhouse gas and its concentration is dependent on the surface temperature of the Earth it is expected that it will contribute strongly to future warming (a positive feedback mechanism). This conclusion depends crucially on whether in a warmer world the increase in water vapour will occur throughout the troposphere. If it does, then its impact will be roughly to double the radiative effect of the increase due to the rising concentration of other greenhouse gases. If, however, the increase in water vapour is restricted to the lower levels of the troposphere then the feedback effect will be reduced. Thus far, the experimental evidence suggests that the latter may be the case, but that the feedback remains positive.

In addition, there is still some doubt about the absorption properties of water vapour. It has by far the most complicated absorption spectrum of the principal greenhouse gases. This affects not only how much incoming sunlight it absorbs, especially in the near infrared, but also how much terrestrial radiation it absorbs. These uncertainties add to those associated with the actual level of water vapour in the upper atmosphere in a warmer world (see Chapter 10).

With O_3, a different set of physical processes are at work. The majority of O_3 is created by the *photochemical* action of sunlight on oxygen in the upper atmosphere (see Box 2.2). This production process depends on the amount of sunlight and so influences the amount of sunlight absorbed by the atmosphere as well as having an impact on the outgoing terrestrial radiation. As such, O_3 is an important example of various photochemical processes in the atmosphere, which are an important factor in the radiation balance of the atmosphere. Owing to their dependence on the amount of sunlight present, these processes exhibit a marked annual cycle, especially at high latitudes. In the case of O_3, transport and photochemical processes then modify this annual cycle. In addition, pollution in urban areas, notably hydrocarbons and oxides of nitrogen from vehicles, can produce the right conditions for the photochemical production of O_3. This is a drawn-out process and so spreads far beyond urban areas. The result is significant widespread

Box 2.2 Photochemical processes

A number of minor constituents of the atmosphere are created by the effect of the absorption of short wavelength [ultraviolet (UV)] solar radiation. This high energy radiation breaks down certain molecules in the atmosphere to form highly reactive fragments, known as *free radicals*, which react with one another and with other molecules in the atmosphere to form new molecular species. The most important absorbers of UV radiation are oxygen (O_2) and ozone (O_3). The former absorbs photons with a wavelength shorter than 240 nm. This absorption process uses the photon's energy to break apart the bonds holding the oxygen atoms together:

$$O_2 + h\nu \rightarrow O + O,$$

where $h\nu$ is the energy of a photon having a frequency ν and h is Planck's constant. Oxygen atoms (O) produced are free radicals, which react with other oxygen molecules to form ozone,

$$O + O_2 + M \rightarrow O_3 + M,$$

where M denotes any air molecule, usually nitrogen or oxygen, that acquires the excess energy generated in this reaction, and dissipates to surrounding molecules by colliding with them, thereby denying the newly formed ozone molecule the excess energy that would cause it to fall apart to O and O_2. Ozone absorbs solar radiation between 240 and 310 nm to revert to O and O_2,

$$O_3 + h\nu \rightarrow O_2 + O.$$

The energy from this reaction heats up the adjacent air a little.

The consequence of this set of reactions is that solar radiation between 200 and 310 nm is absorbed in the upper stratosphere, at an altitude of around 25 to 30 km (shorter wavelength solar radiation is absorbed at higher levels). The dynamic balance between the creation and destruction of O_3 means that the concentration of O_3 is a maximum in the stratosphere. Furthermore, the photochemical nature of O_3 production is the reason why other chemical species can alter the balance between the creation and destruction of ozone. In particular, chlorine atoms formed by the breakdown of chlorofluorocarbons (CFCs) in the stratosphere can, in certain circumstances, participate in efficient catalytic reactions with atomic oxygen (O), which destroy ozone. These reactions are a central component of the creation of the *ozone hole*, which has appeared over Antarctica each October since the early 1980s (see Section 7.4).

Box 2.2 (cont.)

Other radiatively important photochemical processes are associated with the formation of other free radicals such as the hydroxyl radical (OH), the methyl radical (CH_3) and nitric oxide (NO) formed by the reactions between atomic oxygen and trace atmospheric constituents as follows:

$$O + H_2O \rightarrow OH + OH$$
$$O + CH_4 \rightarrow OH + CH_3$$
$$O + N_2O \rightarrow NO + NO.$$

These free radicals are involved in a wide range of chemical reactions with other atmospheric species, especially pollutants such as sulphur dioxide (SO_2), carbon monoxide (CO), oxides of nitrogen (NO_x) and unburnt hydrocarbons (HCs).

In the stratosphere the many reactions are part of the overall balance of ozone and other species. Of particular interest is the formation of nitric acid (HNO_3) that combines in the formation of reactive ice particles over Antarctica to provide the right conditions for destroying ozone more efficiently. By way of contrast, near ground level the cocktail of pollutants, free radicals and bright sunlight produces raised levels of ozone as part of a photochemical smog. Here again the formation of particulates is an integral feature of the complex chemistry involved.

increases in O_3 in the lower atmosphere over much of the more populous parts of the world. So the radiative impact of O_3 is concentrated in the stratosphere and close to the surface, and shows marked geographical and seasonal variations.

In considering how human activities may alter the radiative properties of the atmosphere it is normal to present the impact of a specific change. The commonly adopted measure is to calculate the effect of the equivalent of doubling the atmospheric concentration of CO_2 from pre-industrial levels [around 280 parts per million by volume (ppmv)] to, say, 560 ppmv. This change, which is defined as radiative forcing is estimated to be about $4\,W\,m^{-2}$, is likely to occur principally from the build-up from CO_2, but other gases will also make a contribution (Fig. 2.3). To put this radiative impact in context, the amount of energy received in the flux from the Sun is around $1366\,W\,m^{-2}$ (see Fig. 2.11), which equates to $342\,W\,m^{-2}$ when distributed uniformly over the Earth's surface, whereas the average amount of energy radiated by the Earth is about $240\,W\,m^{-2}$ after taking account of how much sunlight is reflected back into space (Section 2.1.4.).

2.1.4   The energy balance of the Earth

The manner in which changes in the atmosphere alter the climate can only be discussed in the context of the other factors that control the radiation balance of the Earth. First, there is the fundamental factor of how the Earth's orbit around the Sun and its own rotation about its tilted axis controls the amount of sunlight falling on any part of the globe (Fig. 2.4). These regular changes control both the daily (*diurnal*) cycle and the annual seasonal cycle and dominate the climatology of the Earth. The fact that the orbit is elliptical with the Sun at one focus of the ellipse means that the amount of solar radiation reaching the Earth varies throughout the year. At present, the Earth is closest to the Sun in December, so the northern winter is somewhat warmer than it would be if the winter solstice were at the opposite end of the orbit.

For most purposes, orbital variations in the amount of sunlight reaching the Earth during the year can be taken as read and do not need to be considered further. Longer-term variations in orbital parameters do, however, matter. The ratio of the distance of the Sun from the centre of ellipse forming the Earth's orbit and length of the semi-major axis (*eccentricity*) of the orbit changes with time, as does the time of year when the Earth is furthest from the Sun (*precession of the equinoxes*), while the inclination of the Earth's axis to the plane of the orbit (*obliquity*) rocks back and forth

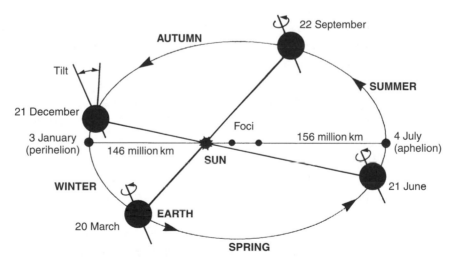

Fig. 2.4 The orbit of the Earth is an ellipse but the Sun is not one of the focal points. As a result the Earth is farther from the Sun (at aphelion) at one end of the long axis and closer (at perihelion) at the other. At present the Earth is closest to the Sun in December. (From Van Andel, 1994, Fig. 5.1.)

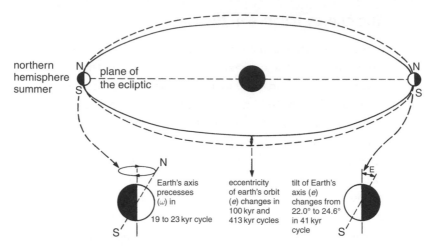

Fig. 2.5 The changes in the precession, tilt and shape of the Earth's orbit that are the underlying cause of longer term variations in the climate. (From Burroughs, 2003, Fig. 6.9.)

periodically. The combination of these effects means that over timescales from several thousand years to a million years the amount of solar energy received at different latitudes at different times of the year changes appreciably.

Having assumed that the output of the Sun is constant, the amount of solar radiation striking the top of the atmosphere at any given latitude and season is fixed by three elements: the eccentricity (e) of the Earth's orbit; the tilt of the Earth's axis to the plane of its orbit, obliquity of the ecliptic (ε); and the longitude of the perihelion of the orbit (ω) with respect to the moving vernal point as the equinoxes precess (Fig. 2.5). Integrated over all latitudes and over an entire year, the energy flux depends only on e. However, the geographic and seasonal pattern of irradiation essentially depends on ε and $e \sin \omega$. The latter is a parameter that describes how the precession of the equinoxes affects the seasonal configuration of Earth–Sun distances. For the purposes of computation the value of $e \sin \omega$ at AD 1950 is subtracted from the value at any other time to give the precession index δ ($e \sin \omega$). This is approximately equal to the deviation from 1950 value of the Earth–Sun distance in June, expressed as a fraction of invariant semi-major axis of the Earth's orbit.

Each of these orbital elements is a quasi-periodic function of time (Fig. 2.6). Although the curves have a large number of harmonic components, much of their variance (see Chapter 7) is dominated by a small number of features. The most important term in eccentricity (e) spectrum has a period of 413 kyr. Eight of the next 12 most significant terms lie in the

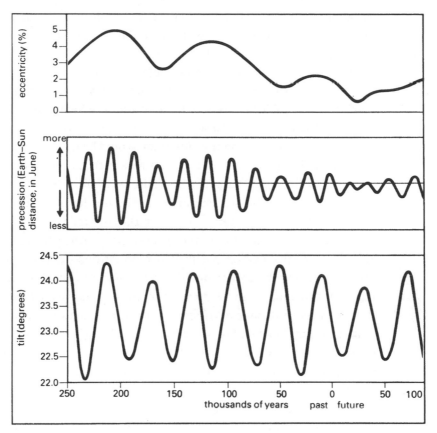

Fig. 2.6 Calculated changes in the Earth's eccentricity, precession and tilt. These changes reflect the fact that the Earth's orbit is affected by variations in the gravitational field due to planetary motion. (From Burroughs, 2003, Fig. 6.10.)

range from 95 to 136 kyr. In low-resolution spectra these terms contribute to a peak that is often loosely referred to as the 100-kyr-eccentricity cycle. In contrast, as can be seen in Figure 2.6, the variation of the obliquity (ε) is a much simpler function and its spectrum is dominated by components with periods near 41 kyr. The precession index (δ) is an intermediate case with its main components being periods near 19 and 23 kyr. In low-resolution spectra these act effectively as a single broad feature with a period of about 22 kyr.

Calculations of past and future orbits provide figures for the variation of the orbital elements. The present value of e is 0.017. Over the past million years it has ranged from 0.001 to 0.054. Over the same interval (ε), which is now 23.4°, has ranged from 22.0 to 24.5° and the precession index (δ, which is defined as zero at AD 1950) has ranged from -6.9 to 3.7%. As these variations alter the seasonal and latitudinal input of solar radiation to the top of the atmosphere they will affect the climate. Most obviously the changes in the obliquity will have seasonal effects. If the obliquity were reduced to zero, the seasonal cycle would effectively vanish and the pole-to-equator contrasts would sharpen. So low values of the obliquity should correlate with colder periods at high latitudes, which is indeed the case. The eccentricity also exerts a seasonal influence. If *e* was zero and the Earth had a circular orbit around the Sun, there would be no seasonal effect from this source. The precession of the orbit means that if the summer solstice were shifted towards the perihelion and away from its present position relatively far from the Sun, summers in the northern hemisphere would become warmer and winters colder than they are today.

The key to explaining how these variations can trigger ice ages is the amount of solar radiation received at high latitudes during the summer. This is critical to the growth and decay of ice sheets. At 65° N this quantity has varied by more than 9% during the last 600 000 years. Fluctuations of this order are sufficient to trigger significant changes in the climate that will be explored in detail in Section 8.7. At this stage, however, we need to concentrate on the fact that the total amount of solar energy reaching Earth each year is unaltered by these slow changes in the planet's orbital parameters.

Overall, the total incoming flux of solar radiation is, over time, balanced by the outgoing flux of both solar and terrestrial radiation. This depends on various processes (Fig. 2.7). Some solar radiation is either absorbed or scattered by the atmosphere and the particles and clouds in it. The remainder is either absorbed or reflected by the Earth's surface. The amount of energy absorbed or reflected is dependent on the surface properties. Snow reflects a high proportion of incident sunlight, whereas moist dark soil is an efficient absorber.

The amount of solar radiation reflected or scattered into space without any change in wavelength is defined as the albedo of the surface. The mean global albedo is about 30%. The albedo of different surfaces can vary from 90% to less than 5%. Examples of the albedo of different surfaces are given in Table 2.1. The striking feature of these figures is the differences of the albedo of snow-covered surfaces. These include values that have recently been obtained from satellite measurements showing that about twice as much sunlight is reflected back to space by snow-covered croplands and grasslands as is reflected by snow-covered forests. This has major implications for the climatic impact of deforestation at high latitudes in the northern hemisphere, especially in late winter and early spring (see Section 3.6).

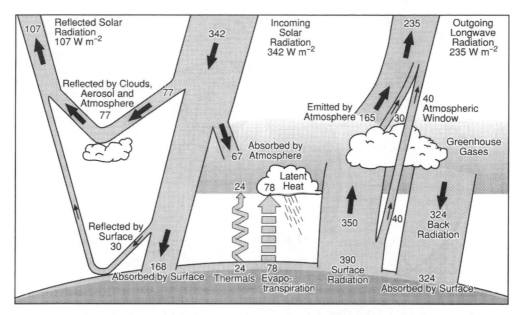

Fig. 2.7 The Earth's radiation and energy balance. The net incoming solar radiation of 342 W m^{-2} is partially reflected by clouds and the atmosphere, or at the surface, but 49% is absorbed by the surface. Some of that heat is returned to the atmosphere as sensible heating and considerably more as evapotranspiration that is released as latent heat in precipitation. The amount of energy emitted as thermal infrared radiation from the surface depends on how much is absorbed by the atmosphere, which in turn emits radiation both up and down, as part of the greenhouse effect. The net terrestrial radiation lost to space from the surface, from cloud tops and from throughout the atmosphere balances out the total amount of absorbed incoming solar radiation. (From IPCC, 1995, Fig. 1.3.)

Roughly two-thirds the Earth's surface is obscured by some form of clouds. The albedo of different types and levels of cloud is roughly proportional to cloud thickness. Values for various types of clouds are also given in Table 2.1. As about 30% of incoming solar radiation is reflected or scattered back into space, the remainder must be absorbed. Of this remaining flux, about three-quarters (i.e. about half of the total incoming flux) penetrates the atmosphere and is absorbed by the Earth's surface. The remainder (i.e. some 16% of the total incoming flux) is absorbed directly by the atmosphere. Both the atmosphere and the surface re-radiate this absorbed energy as long-wave radiation.

How the solar radiation is absorbed at the Earth's surface differs profoundly on land and at sea. On land most of the energy is absorbed close to the surface, which warms up rapidly, and this increases the amount of terrestrial radiation leaving the surface. At sea the solar radiation

Table 2.1 The proportion of sunlight reflected by different surfaces (albedo)

Type of surface	Albedo
Tropical forest	0.10–0.15
Woodland	
deciduous	0.15–0.20
coniferous	0.05–0.15
Farmland/natural grassland	0.16–0.26
Bare soil	0.05–0.40
Semi-desert/stony desert	0.20–0.30
Sandy desert	0.30–0.45
Tundra	0.18–0.25
Water (0°–60°)*	less than 0.08
Water (60°–90°)*	0.10–1.0
Fresh snow/snow-covered permanent snow	0.80–0.95
Snow-covered ice	0.75–0.85
Sea ice	0.25–0.60
Snow-covered evergreen forest	0.20–0.60
Snow-covered deciduous forest	0.25–0.50
Snow-covered farmland/natural grassland	0.55–0.85
Clouds	
low	0.60–0.70
middle	0.40–0.60
high (cirrus)	0.18–0.24
cumuliform	0.65–0.75

*The closer the sun is to the zenith the less sunlight is absorbed. The presence of whitecaps on the surface also increases the albedo.

penetrates much more deeply, with over 20% reaching a depth of 10 m or more. So the sea is heated to a much greater depth and the surface warms up much more slowly. This means more energy is stored in the top layer of the ocean and less is lost to space as terrestrial radiation. This absorptive capacity of the oceans plays an important part in the dynamics of the Earth's climate. In effect they act as a huge climatic flywheel damping down fluctuations in other parts of the climate.

Clouds play an equally important role. Their high albedo and extensive nature means they are responsible for effectively doubling the albedo of the Earth from the value if there were no clouds. Their transitory nature and variable properties, however, make them particularly difficult to represent accurately in models of the climate (see Chapter 9), so their role is not easy to quantify accurately. This uncertainty extends to the part they play in modifying the emission of terrestrial radiation. They are efficient absorbers and emitters of infrared radiation and their thickness and the temperature of their tops plays a great part in how much energy they radiate to space. Furthermore, because water vapour is also the major greenhouse gas, the transition between clouds and vapour poses additional problems in calculating their impact on the outgoing terrestrial radiation. Therefore it is no exaggeration to say that the physics of clouds is the greatest obstacle to improving predictions of climate change.

A related issue is the role of particulates. These may be the product of natural variations (e.g. dust from drought-prone areas) or human activities, including both agriculture and, more important, sulphate particulates from the combustion of fossil fuels. Unlike ice crystals and water droplets in natural clouds, these tiny particles are better reflectors of sunlight than they are absorbers of terrestrial radiation and so their impact on the Earth's energy balance is to reduce the net amount received, and hence to lead to a cooling effect.

This simple input/output analysis is not the whole story. As the atmosphere and oceans transport energy from one place to another this motion is an integral part of the Earth's radiative balance. On a global scale the most important feature is that, while most solar energy is absorbed at low latitudes, large amounts of energy is radiated to space, at high latitudes, despite lower temperatures (typically 240 K in polar regions in winter). This process is greatest in winter, when there is little or no solar radiation reaching these regions. This loss must be balanced by energy transport from lower latitudes (Fig. 2.8). Even though this energy transfer is reduced at other times of the year, it provides the energy to fuel the engine which drives the global atmospheric and oceanic circulation throughout the year (see Section 3.6).

The geographical and seasonal transport of energy polewards by the atmosphere and the oceans is complicated. Recent analysis of both satellite data and the re-analysis work of the ECMWF and NCEP (see Section 4.3) shows that the atmosphere transports the majority of energy. Although there is considerable uncertainty about the exact numbers, broadly speaking, the atmosphere transports 78% of the energy polewards at 35° N and 92% of the energy in the opposite direction at 35° S (Fig. 2.9). This energy is made up of approximately equal amount of sensible heat and latent heat, so the transport of water vapour, and hence fresh water, is a major part of this process. In the oceans, the Atlantic carries rather more heat northwards than the Pacific (Fig. 2.9). In the southern hemisphere the Pacific transports roughly the same amount of energy southwards as the Indian Ocean,

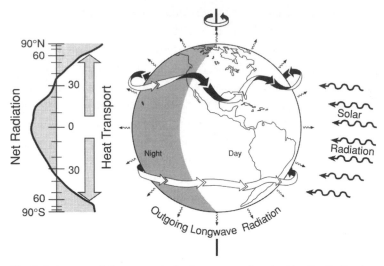

Fig. 2.8 In general the amount of solar energy absorbed by the Earth at each latitude differs from the amount of terrestrial radiation emitted at the same latitude, and so energy has to be transferred from equatorial to polar regions. (From IPCC, 1995, Fig. 1.2.)

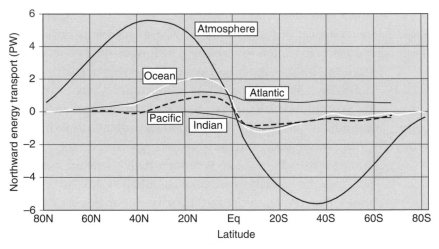

Fig. 2.9 The poleward atmosphere and ocean heat transports in each ocean basin, and summed over all oceans shows how the greatest amount of energy is transferred to mid-latitudes. For clarity the poleward transfer in the northern hemisphere is shown as positive, whereas in the southern hemisphere it is shown as negative. (Data from NCEP and ECMWF Oceanic and Atmospheric Transport Products website: http://www.cgd.ucar.edu/cas/catalog/ohts/ANNUAL_TRANSPORTS_1985_1989.ascii.)

whereas the Atlantic has a significant counterbalancing effect carrying energy northwards across the equator.

The role of the oceans in this process cannot be thought as being separate from the atmosphere. There is continual exchange of energy in the form of heat, momentum as winds stir up waves, and moisture in the form of both evaporation from the oceans to atmosphere and precipitation in the opposite direction. This energy flux is particularly great in the vicinity of the major currents, and where cold polar air moves to lower latitudes over warmer water. These regions are the breeding grounds of many of the storm systems that are a feature of mid-latitude weather patterns.

Maintaining an energy balance by transport from low to high latitudes has another fundamental implication. This relates to the mean temperatures of different latitude belts. Although these can fluctuate by a certain amount, at any time of the year, overall they will be close to the current climatic normal. Moreover, if one region is experiencing exceptional warmth, adjacent regions at the same latitudes will experience compensating cold conditions. So, in the shorter term, climatic fluctuations are principally a matter of regional patterns. These are often the product of the occurrence of certain extremes (e.g. cold winters, hot summers, droughts and excessive rainfall) becoming more frequent than usual, but, in the longer term, it is possible for mean latitudinal temperatures to move to different values while maintaining global energy balance. For the most part, however, these overall changes are small compared with the regional fluctuations that can occur on every timescale.

2.2   Solar variability

If the amount of energy emitted by the Sun varies over time it is bound to have an effect on the Earth's climate. Since the beginning of the seventeenth century, when Galileo first made telescope observations, it has been known that the Sun exhibits signs of varying activity in the form of sunspots. These darker areas, which are seen at lower latitudes between 30° N and 30° S crossing the face of the Sun as it rotates, are cooler than the surrounding chromosphere. Each sunspot consists of two regions: a dark central *umbra* at a temperature of around 4000 K and surrounding lighter *penumbra* at around 5000 K. The darkness is purely a matter of contrast. They appear dark in comparison with the general brightness of the Sun's brilliant 6000 K surface temperature.

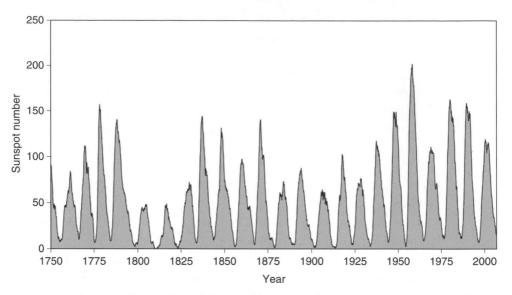

Fig. 2.10 The variation of the monthly number of sunspots from 1750 to 2006 (13-month smoothed data showing fluctuations of a year or longer. (Data from NOAA website: ftp://ftp.ngdc.noaa.gov/STP/SOLAR_DATA/SUNSPOT_NUMBERS/MONTHLY.PLT.)

Sunspots vary in their number, size and duration. There may be as many as 20 or 30 spots at any one time, and a single spot may be from 1×10^3 km to 2×10^5 km in diameter with a life cycle from hours to months. The average number of sunspots and their mean area fluctuate over time in a more or less regular manner with a mean period of 11.2 years (Fig. 2.10).

Since 1843, when Heinrich Schwabe discovered that the number of sunspots varied in a regular, predictable manner, the possibility that the energy output of the Sun could vary in a periodic manner has been the subject of great scientific debate. It was not until the late 1980s that satellites provided sufficiently accurate measurements to confirm that the output of the Sun did rise and fall during the sunspot cycle (Fig. 2.11). To the surprise of many scientists the output increases with the number of sunspots. It had been assumed that, because the spots are cooler areas, as they multiplied they would reduce the Sun's output. It is now known that the convection associated with the sunspots brings hotter gases from within the Sun to the surface in the surrounding areas known as *faculae*. These hotter areas outshine the cooling effect of the sunspots and result in more energy being emitted at the peak of sunspot cycles.

The change in the total amount of energy emitted by the Sun during the 11-year sunspot cycle is only about 0.1%. On the basis of energy balance

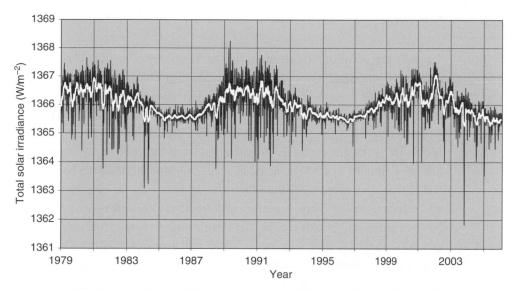

Fig. 2.11 Various satellite measurements of the total solar irradiance, showing how the output from the Sun varied between the peaks in sunspot numbers around 1980, 1990 and 2000, plus the minima in 1986 and 1996. (Data from PMOD/WRC, Davos, Switzerland, including unpublished data from the VIRGO Experiment on the cooperative ESA/NASA Mission SoHo.)

considerations this is not thought sufficient to produce significant climate change. This is, however, only the first part of the story. Much of the change in solar output is concentrated in the UV part of the spectrum. Much of this energy is absorbed high in the atmosphere by both oxygen and ozone molecules (see Box 2.2). If changes at these levels are capable of exerting a disproportionate effect on the weather patterns lower down, then it is possible for small variations in solar radiation to be the cause of larger fluctuations in the climate. So it is useful to know how solar UV energy affects the upper atmosphere, which is why the creation and destruction of ozone is discussed in Box 2.2. The physical arguments explaining how changes of ozone and other trace constituents in the upper atmosphere may play a part in climate variability and change are then explored in Section 8.5.

On geological timescales the output of Sun has risen steadily. It is estimated that four billion years ago the solar constant was 80% of the current value and it has risen monotonically since then. These changes are of interest in considering the long-term nature of the response of the global climate to fundamental changes in energy input, but are of no direct consequence for current climate studies.

2.3   Summary

The starting point in examining how the Earth's climate may change over time is to understand the radiative balance of our planet. All the energy that matters comes from the Sun. The incoming flux is matched over time by the outgoing terrestrial energy. The basic laws of physics define this balance. Anything that has the potential to influence this balance, either in global terms, or in respect of one part of the planet as opposed to another, could alter the climate. Therefore, at every stage of our analysis of fluctuations in the climate, we must anticipate how the amount of solar energy absorbed and reflected by the Earth, together with how much heat is radiated to space by the planet, will be affected by the changes we are exploring. In addition, anything that can alter how much solar radiation reaches the Earth, (e.g. solar activity, orbital variations, or even amounts of cosmic dust) could play a part in altering the climate.

So there are a lot of things that can change in the Earth's energy balance. What is more, they are connected in a variety of ways. These connections are best considered in the context of the various components of the Earth's climate. Now we must move on to how the balance between incoming and outgoing radiation establishes the global pattern of climatic regimes.

QUESTIONS

1. If the radiative forcing due to a doubling in the atmospheric concentration of CO_2 is $4\,W\,m^{-2}$, by what amount does the extent of cloud cover have to increase to offset this radiative effect by reflecting additional sunlight back into space? (To do this calculation use the albedo figures in Table 2.1 and assume that at any given time two-thirds of the Earth's surface is covered by clouds.) How do these figures vary with the mix of cloud types?

2. Why does the figure calculated in answer to Question 1 above overstate the impact of clouds on the radiative balance of the Earth?

3. Why is the radiation budget at any particular place seldom in balance?

FURTHER READING

A complete reference list is available at the end of the book but the following is a selection of the best books or articles to follow up particular topics within this chapter. Full details of each reference are to be found in the Bibliography.

Houghton (2002): This provides the basic physical principles of the behaviour of atmospheres: heavy on mathematics, and not an easy read, but does contain the essence of the physics.

Houghton (2004): A highly accessible discussion of the many issues surrounding the global warming, with a particularly clear section on the physics of the Greenhouse Effect.

McIlveen (1992): A readable and comprehensive presentation of many aspects of meteorology and climate change.

3   The elements of the climate

I am the daughter of Earth and Water,
        And the nursling of the Sky;
I pass through the pores of the ocean and shores;
        I change, but I cannot die.   **Percy Bysshe Shelley, 1792–1822**

In understanding what constitutes climate variability and climate change and predicting how each may evolve in the future we need to have a clear idea of which aspects of the climate matter. This is crucial as these fluctuations are not simply a matter of the Earth as a whole warming up or cooling down, or even of certain regional climates (e.g. deserts) expanding or contracting. The physical processes described in Chapter 2 may drive the mechanics of change, but the network of links within the system complicates how they combine with the different components of the climate. The objectives of this chapter are therefore to establish what are the most important components of the climate, and to identify the strongest links between them, so we can have a framework for examining the essential aspects of climate variability and climate change.

3.1   The atmosphere and oceans in motion

How the atmosphere and the oceans transfer energy around the globe is the key to climate studies. It is also central to the sciences of meteorology and climatology. There are a lot of more than adequate books on these subjects (see Bibliography) and so many of the physical fundamentals or details of the Earth's climate will not be regurgitated here. Instead we will concentrate on those latent aspects that are less frequently addressed but which are central to understanding the behaviour of the Earth's climate and how it may change of its own accord or be provoked into changing.

When viewed from space the atmosphere appears to consist principally of an endless succession of transient eddies swirling erratically across the face of the planet. If observed for long enough clear patterns emerge. The

Fig. 3.1 Viewed from a geostationary satellite over the equator many components of the Earth's climate system can be seen in a single image (with permission of the Japanese Meteorological Agency). (From Burroughs, 1991, Fig. 1.1.)

most obvious is the huge pulse of the annual cycle that moves from hemisphere to hemisphere. If watched with heat-sensing equipment it looks for all the world like a great blush spreading from one pole to the other – throughout the course of the year. At the more detailed level the essential features of the global circulation become obvious (Fig. 3.1). In the equatorial regions a region of intense convective activity – the intertropical convergence zone (ITCZ), girdles the planet. During the northern summer this develops into the more sustained expansion of the monsoon over the Indian subcontinent. Then, during late summer and early autumn activity in the ITCZ can erupt, now and then, into more organised outbursts as tropical cyclones emerge in the easterly flow and then, as they intensify, swerve northwards before merging into the high-latitude westerly flow.

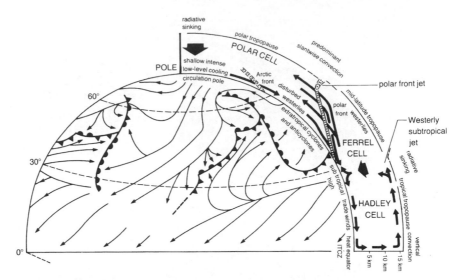

Fig. 3.2 A schematic model of the general circulation of the northern hemisphere (in cross-section), the locations of both the principal surface features, the vertical circulation and the polar front jet stream and the westerly subtropical jet. (From Musk, 1988, Fig. 12.5.)

To the north and south of this tropical hive of activity the continental deserts mark regions of subsiding air and low rainfall. Beyond this, the westerly conveyor belt of mid-latitude depressions swirl endlessly poleward. In the southern hemisphere this procession continues unabated throughout the year with only a modest shift to higher latitude in the austral summer. In the northern hemisphere there is a crescendo of activity in the winter half of the year, with a marked lull during the summer. Polar regions covered with ice and shrouded by clouds conceal much of their behaviour, but beneath this the arctic pack ice comes and goes (see Section 3.4), while much of the northern continents is blanketed by snow during the winter: the extent of which varies markedly from year to year.

Associated vertical motion of all these systems is central to understanding the climatic processes at work. These are shown in Figure 3.2. The rising air in equatorial regions generates the ITCZ. When it reaches an altitude of around 12 to 15 km virtually all the moisture has been wrung out of it and it spreads out and then descends to the north and south of this region creating zones of dry, hot air, which maintains the deserts in these parts of the world and also drives the Trade winds which flow back towards the equator. First interpreted by George Hadley in 1735, this circulation pattern now bears his name (*the Hadley Cell*). The mid-latitude depressions also raise huge quantities of air as they swirl polewards. This air descends over the polar regions and creates cold deserts in these regions.

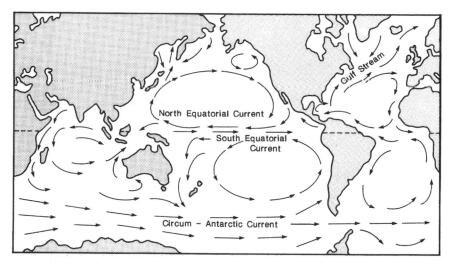

Fig. 3.3 The major ocean current systems of the world. (From Van Andel, 1994, Fig. 10.2.)

The movement of the oceans is not so easy to see but is just as important. Satellite measurements do, however, confirm the picture of ocean currents built up from centuries of maritime observations (Fig. 3.3). They also show that the flow is complicated to eddies and swirls (Fig. 3.4), which mirror the turbulent nature of the atmosphere. In addition, temperature measurements show that, on top of the annual pulse, there are major fluctuations in sea-surface temperatures (SSTs) on longer timescales, especially in the equatorial Pacific (see Section 3.5). Finally, completely hidden from view the deep waters of the oceans silently circulate, moving huge amounts of energy on the timescale of centuries.

All of these processes can interact with one another to produce variations on every possible timescale. In the short term they represent the fascinating extremes of our daily weather, but where they combine in a more sustained way then they produce the recipe for natural variations of the climate. So, the first step is to appreciate what drives these motions, and why they may fluctuate of their own accord. Then we can move on to how they can act in concert.

3.2   Atmospheric circulation patterns

The first issue to be addressed in finding out the causes of fluctuations in the climate is why large-scale global atmospheric circulation patterns

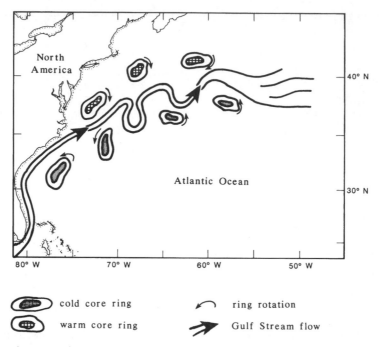

Fig. 3.4 Schematic representation of ring structures associated with the Gulf Stream, which are typical of the detailed structure that exists in the major ocean currents. (From Bryant, 1997, Fig. 4.6.)

shift from year to year. This is not a matter of looking at individual weather systems, but instead concentrating on their average positions and paths they take. If mid-latitude depressions take a different path during one winter, or for a number of winters, then some regions will get excessive rainfall while others suffer drought. Similarly, the repeated establishment of high pressure where more variable weather usually occurs can produce abnormally cold winters or hot summers and so understanding the nature of these patterns is the first stage of our search. In the case of patterns in the tropics, which are just as important, and in particular, the summer monsoon in southern Asia, these are best considered in the context of SSTs (see Section 3.5). Here, we will concentrate on mid-latitudes.

The key to long-lasting weather patterns is found in the middle levels of the atmosphere. By showing the average height of a given pressure surface (e.g. 500 mb), it is possible to strip away the complication of ground-level weather features and reduce the associated effects of landmasses. So the winter 500-mb northern hemisphere circulation pattern (see Fig. 3.5a) is

(a)   Geopot * 0.1020   at 500mb.
ECMWF ReAnalysis. 15-year mean. Period: DJF
contour interval: 40 gpm

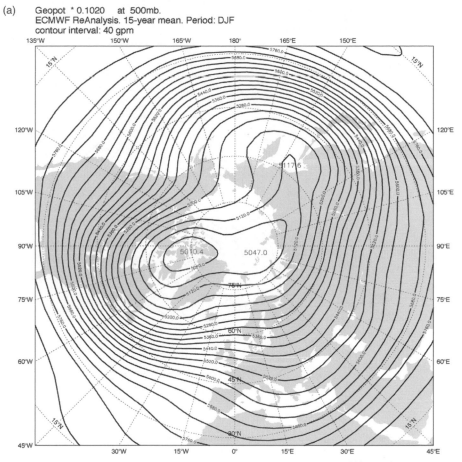

Fig. 3.5 The mean height of the 500 mb pressure surface (in decametres) in (a) winter and (b) summer for the northern hemisphere. (With permission from the ECMWF.)

dominated by an extensive asymmetric cyclonic vortex with the primary centre over the eastern Canadian arctic and a secondary one over eastern Siberia. In summer (see Fig. 3.5b) the pattern is similar, but the vortex is much less pronounced. These deviations from a purely circular pattern represent troughs and ridges in the circulation. At any time the number of troughs and ridges may vary as will their position and amplitude. These patterns are known as *long waves* (or *Rossby waves* – after the Swedish meteorologist Carl Gustav Rossby who first provided a physical explanation of their origin).

The basic physical explanation of the existence of long waves in the upper atmosphere is linked to the circulation of transient eddies at lower levels.

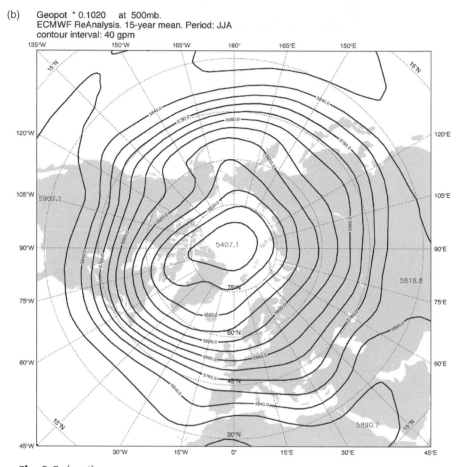

(b)   Geopot * 0.1020   at 500mb.
ECMWF ReAnalysis. 15-year mean. Period: JJA
contour interval: 40 gpm

Fig. 3.5 (cont).

The combined spin of one of these low-pressure systems and its rotation about the Earth's axis (known as *vorticity*) effectively remains constant. Therefore a system in a stream of air moving towards polar regions will tend to lose vorticity as the distance from the Earth's axis of rotation reduces. This is compensated by the flow adopting an anticyclonic curvature (clockwise in the northern hemisphere) relative to the surface at higher latitude. Eventually, the curved path takes the airstream back towards the equator. As it reaches lower latitudes the reverse effect will take place as the distance from the axis of rotation increases and the airflow will adopt a cyclonic curvature. So the airstream tends to swing back and forth as it circulates the Earth. The climatological mean of this flow has two major troughs at around 70° W and 150° W. Their position is linked to the impact

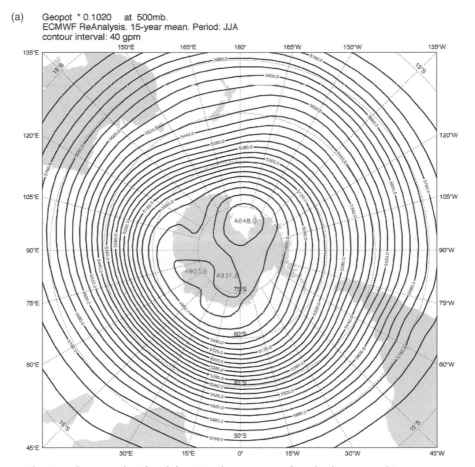

(a)   Geopot * 0.1020   at 500mb.
ECMWF ReAnalysis. 15-year mean. Period: JJA
contour interval: 40 gpm

Fig. 3.6 The mean height of the 500 mb pressure surface (in decametres) in
(a) winter and (b) summer for the southern hemisphere. (With permission from the
ECMWF.)

of the major mountain ranges on the flow, notably the Tibetan plateau and
the Rocky Mountains, together with the heat sources such as ocean currents
(in winter) and landmasses (in summer). As the southern hemisphere is
largely covered by water, the pattern is much more symmetrical and
shows less variation from winter to summer (see Fig. 3.6a and b).

What is not evident from these broad patterns is the nature of winds in
the upper atmosphere. The contour maps record the thickness of the atmosphere between sea level and the 700-mb surface, and this thickness is
proportional to the mean temperature – low thickness values correspond
to cold air and high thickness values to warm air. This means that the
circumpolar vortices (see Figs 3.5 and 3.6) reflect a poleward decrease in

(b)   Geopot * 0.1020   at 500mb.
        ECMWF ReAnalysis. 15-year mean. Period: DJF
        contour interval: 40 gpm

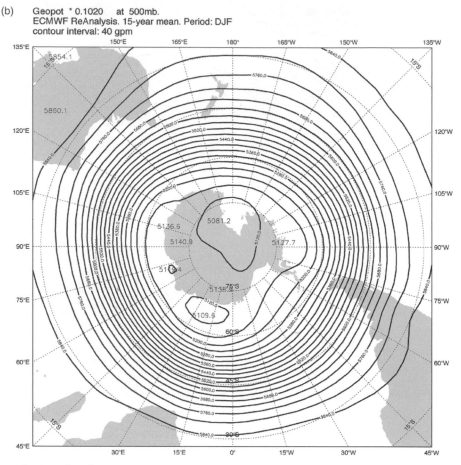

Fig. 3.6 (cont).

temperature and this drives strong westerly winds in the upper atmosphere. But, while the thickness decreases gradually with increasing latitude, the strongest winds are concentrated in a narrow region, often situated around 30° of latitude at an altitude between 9 and 14 km. The concentrated wind flow is known as the *jet stream*, and reaches maximum speeds of 160 to 240 kph, and can exceed 450 kph in winter. The mainstream core is associated with the principal troughs of the Rossby long waves. This circulation governs the movement of surface weather systems. The reason for the restriction of the winds to a narrow core is not fully understood. Moreover, the structure can be complicated, especially in the northern hemisphere winter when the jet stream often has two branches (the subtropical and polar front jet streams).

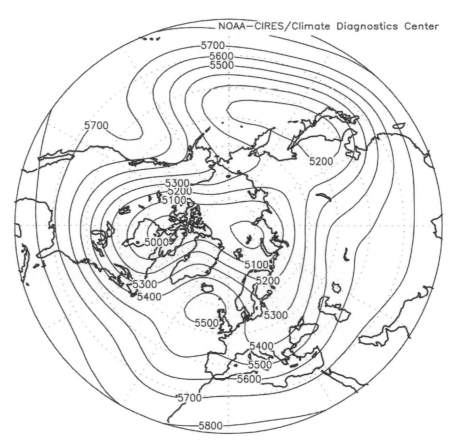

Fig. 3.7 The mean height of the 500-hPa pressure surface (in decametres) for January 1963, showing the pronounced wave pattern in mid-latitudes due to *blocking* off the west coast of the United States and close to the British Isles. (Data from NOAA Climate Diagnostic Center.)

Circulation patterns, which are radically different from those in Figs. 3.5 and 3.6, can last a month or two. They occur irregularly but are more pronounced in the winter when the circulation is strongest. Ridges and troughs can become accentuated, adopt different positions and even split up into cellular patterns. An extreme example of such a pattern occurred in the winter of 1962–63. Figure 3.7 shows the mean height of the 500-mb surface in January 1963. The striking features are a pronounced ridge off the west coast of the United States (40° N, 125° W) and a well-defined anti-cyclonic cell just to the south of Iceland (60° N, 15° W). This set the stage for the extreme weather. Cold arctic air was drawn down into the central United States and into Europe. Conversely, warm tropical air was drawn far north to Alaska and western Greenland. The net effect was that while an extreme

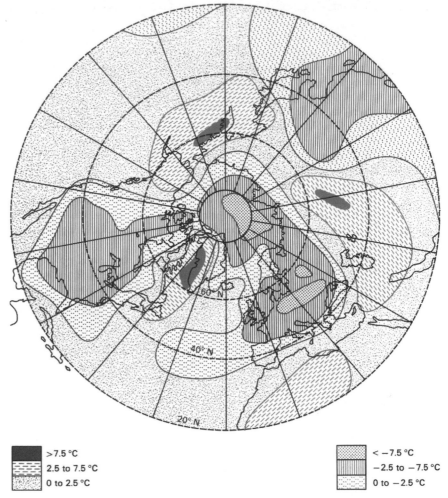

>7.5 °C
2.5 to 7.5 °C
0 to 2.5 °C

< −7.5 °C
−2.5 to −7.5 °C
0 to −2.5 °C

Fig. 3.8 When a well-established blocking pattern develops, it can have global implications. The winter of December 1962 to February 1963 is an extreme example of such a global circulation anomaly. This diagram shows that as a consequence of this pattern adjacent regions of the mid-latitudes of the northern hemisphere experienced compensating extremes with northern Europe, the United States, and Japan being exceptionally cold and Greenland, central Asia, and Alaska being very mild. (Burroughs, 1991, Fig. 12.4.)

negative anomaly of −10 °C for the month was observed in Poland, an equal positive departure occurred over western Greenland (see Fig. 3.8), and the mean hemispheric temperature was close to normal (see Section 2.1.4.).

Such an extreme pattern, which is usually termed '*blocking*', shows how important it is to understand the causes of anomalous atmospheric behaviour. Changes in the incidence of extreme seasons, like the winter of 1962–63, are a major factor in variations of the climate during recent centuries in Europe (see Section 8.9). So it is important to know what defines their occurrence as opposed to stronger, more symmetric circulation patterns that bring much milder winters to Europe and the eastern United States. So we need to address the basic questions of what causes the number, amplitude and position of the Rossby waves to change and then to remain stuck in a given pattern for weeks or months.

As noted earlier, in the northern hemisphere the distribution of land masses and the major mountain ranges plays a major role. This does not, however, explain why the number of waves around the globe may range from three to six or why they can vary from only small ripples on a strong circumpolar vortex to exaggerated meanderings with isolated cells. It follows from the description of the requirement to conserve vorticity in the high-level flow that an important factor is the speed of the upper atmosphere westerlies. There is some evidence that when the wind speeds assume a critical value this enhances the chances of strong standing waves building up downstream from the troughs at $70°$ W and $150°$ E, but this begs the question of what governs the changing speed of the winds from year to year and within seasons.

Any explanation of what causes different circulation patterns must also show why their incidence changes. In winter in the northern hemisphere just four or five categories of circulation regimes occur about three-quarters of the time. This fact may hold the key to understanding what controls the switches between regimes, as the limited number of states suggests there are tight constraints on which patterns are permissible in strong circulation systems. Any insight may then be extended to other times of the year and other parts of the world where circulation regimes exercise less influence on the climate. So the essential issues are why a given regime becomes established, what causes it to shift suddenly to a different form, and above all what defines the incidence of given regimes.

The obvious place to look for controlling influences is in the slowly varying components of the climate system. This means, first and foremost, the oceans, but could also involve the extent of snow and ice in high latitudes. But before doing so, the analysis must be set in the context of the physical processes outlined in Chapter 2. The atmosphere is driven principally by the solar energy absorbed at the surface at low latitudes, and the terrestrial energy radiated at higher latitudes (see Fig. 2.5). So in explaining how the meanderings of the jet stream in the upper atmosphere may lead to variations in the climate, we must not fall into the trap of assuming that small changes at high altitudes can exert a disproportionate impact at the surface. The fundamental point is that the weather is driven by the energy available at the bottom of the atmosphere. Although the upper atmosphere winds appear to be steering the progress of the surface weather, they are essentially the product of the amount and distribution of the solar energy absorbed into the atmospheric heat engine. But, the devil is in the detail: in explaining changes in climatic patterns there are all manner of subtle alterations of the radiation and water balance which may exert longer term influences, some of which may occur in the upper atmosphere.

There is one other aspect of upper atmosphere winds that needs to be mentioned here. This is the periodic reversal of the winds in the lower stratosphere at levels between around 20 and 30 km over equatorial regions. These winds go through a cycle every 27 months or so from being strongly easterly to nearly as strongly westerly (Fig. 3.9). The wind regime propagates downwards with the reversal starting at higher levels. This behaviour is known as the quasi-biennial oscillation (QBO) and appears to have weak

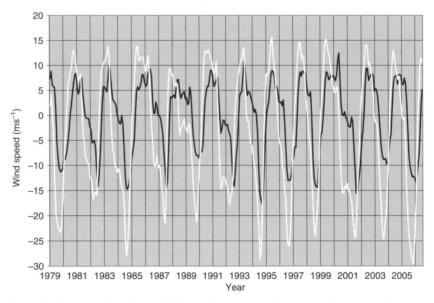

Fig. 3.9 The quasi-biennial oscillation of the winds in the stratosphere over the equator shows a pronounced periodic reversal. The scale of this oscillation is greatest at around 30 hPa (white line) and reduces at lower levels, as shown by the values at 50 hPa (black line) as the periodic feature migrates downwards over time, so that the peaks occur first at high levels. (Data from NOAA, available on websites: ftp://ftp.ncep.noaa.gov/pub/cpc/wd52dg/data/indices/qbo.u30.index, ftp://ftp.ncep.noaa.gov/pub/cpc/wd52dg/data/indices/qbo.u50.index.

echoes in many features of the weather at lower levels, including many temperature records, the behaviour of the El Niño southern oscillation (see Section 3.7), and the incidence of hurricane activity in the Atlantic. It is also the only clearly defined cycle in the weather that is longer than the dominant annual cycle.

What is particularly relevant about the QBO is that it is the product of internal variability of the atmosphere. The accepted physical explanation involves a combination of processes in the dissipation of upward-propagating easterly Kelvin waves (see Section 3.7) and westerly Rossby waves in the stratosphere. These waves originate in the troposphere and lose momentum in the stratosphere by a process of radiative damping. This involves rising air cooling and radiating less strongly than the air that is warmed in the descending part of the wave pattern. Westerly momentum is imparted by the decaying Kelvin waves in the shear zone beneath the downward-propagating westerly phase of the QBO. Rossby waves perform a similar function in respect of the easterly phase. The fascinating feature of this process is that these two types of wave combine to produce such a regular reversal of the upper atmosphere winds.

3.3   Radiation balance

As described in Chapter 2, the proportion of incoming solar radiation absorbed by the Earth depends on the absorption, reflection and scattering properties of the atmosphere and the surface. Satellite measurements have now provided a global picture of these processes and confirm the broad figures quoted in Section 2.1.4 and Table 2.1. Where there are no clouds the oceans are the darkest regions of the globe. They have albedo values ranging from 6 to 10% in the low latitudes and 15 to 20% near the poles. Ocean albedo increases at high latitude because at low sun angles water reflects sunlight more effectively (see Table 2.1). The brightest parts of the globe are the snow-covered Arctic and Antarctic that can reflect over 80% of the incident sunlight. The next brightest areas are the major deserts. The Sahara and the Saudi Arabian desert reflect as much as 40% of the incident solar radiation. The other major deserts (the Gobi and the Gibson) reflect about 25 to 30%. By comparison, the tropical rain forests of South America and Central Africa, as the darkest land surfaces, have albedos from 10 to 15%.

The pattern of outgoing long-wave radiation is more systematic. This reflects the fact that the temperature of the surface and the atmosphere decreases relatively uniformly from the equator to the poles. In effect these surfaces are close to being black bodies when it comes to emitting long–wave radiation and the emissivity of various surfaces does not vary appreciably. The average amount of energy radiated to space decreases from a maximum of $330\,\mathrm{W\,m}^{-2}$ in the tropics to about $150\,\mathrm{W\,m}^{-2}$ in the polar regions. This gives a general pattern of the net heating of the atmosphere at low latitudes and cooling at high latitudes when the combined figures for the amount of solar radiation absorbed at the surface and in the atmosphere are combined.

Before reviewing the impact of clouds on the energy balance both globally and regionally, it is necessary to consider the implications of the clear sky albedo measurements for long-term fluctuations in the weather. The most highly reflecting areas are also among the most variable climatic elements. For instance, increases in the extent of winter snow and ice cover in polar regions will have a significant effect on the amount of sunlight reflected into space. This on its own will produce a cooling effect so, the longer anomalous cover lasts, the greater the impact will last. More complicated is the effect of forests across the northern continents. As shown in Table 2.1, the albedo of snow-covered areas varies considerably. The average figure for forests is roughly half that for open country (around 35% as opposed to nearer 70%), so the removal of forests at high latitudes would increase the albedo in winter and spring and have an appreciable cooling effect.

A similar but somewhat more surprising effect relates to an expansion of the major deserts of the world. Here again the change will lead to an increase in albedo (see Table 2.1) and hence to more solar radiation being reflected into space. This effect can be seen in Figure 3.10 as the region of net cooling over the Sahara and the Middle East. So while deserts are regarded as hot places their expansion could lead to a general cooling, unless associated with some compensating changes in cloudiness. Therefore, if either natural climatic variability or human activities lead to an increase in the world's deserts, this is likely to have a cooling effect. Hence, if global warming leads to expanding deserts, any consequent cooling will be a negative feedback mechanism tending to damp down the warming trend.

The global distribution of clouds shows that they are most common over the mid-latitude storm tracks of both hemispheres (Fig. 3.11). They also have a less striking maximum over the tropics, especially the region around south-east Asia. On average, about 65 % of the Earth is covered by clouds at any given time. Clouds are almost always more reflective than the ocean surface and the land except where there is snow and so cloudy areas reflect more solar energy into space than do places that have clear skies. Overall their effect is approximately to double the albedo of the planet from what it would be in the absence of clouds to a value of about 30 % (see Section 2.1.4.).

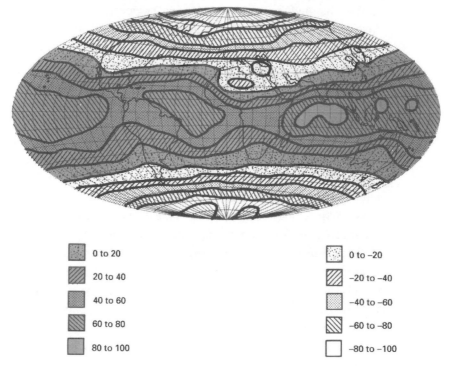

	0 to 20	
	20 to 40	
	40 to 60	
	60 to 80	
	80 to 100	

	0 to −20	
	−20 to −40	
	−40 to −60	
	−60 to −80	
	−80 to −100	

Fig. 3.10 Satellite measurements over the years show how the net balance between incoming solar radiation and outgoing terrestrial radiation varies with latitude and longitude. While these observations show considerable variations between different parts of the tropics and subtropics, the important feature is that energy is pumped into the climate system at low latitudes and escapes at high latitudes. (From Burroughs, 1991, Fig. 13.4.)

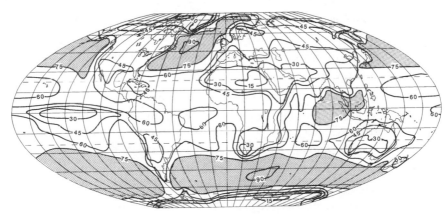

Fig. 3.11 Global distribution of annual mean cloud amount expressed as a per cent cover. Areas greater than 75% are shaded. (From Bryant, 1997, Fig. 2.10.)

The overall effect of clouds depends, however, on their net effect on both incoming and outgoing radiation. Under clouds less thermal energy is radiated to space than where the skies are clear. It is the net difference between these two effects that establishes whether the presence of clouds cools or heats the planet. The scale of the blanket-like warming effect depends on the thickness of the clouds and the temperature of their tops. High clouds radiate less than low clouds, and thick clouds are more efficient radiators than thin clouds. The average figures for clouds globally is to reduce the amount of solar radiation absorbed by the Earth by $48\,\mathrm{W\,m^{-2}}$ and to reduce the outgoing heat radiation to space by $31\,\mathrm{W\,m^{-2}}$. So satellite measurements confirm that clouds have a net cooling effect on the global climate.

While the global effect of clouds is clear, their role in regional climatology and in feedback mechanisms associated with climate change is much more difficult to discern. The blanketing effect of clouds reaches peak values over tropical regions and decreases towards the poles. This is principally because clouds rise to a greater height in the tropics, and the cold tops of deep clouds radiate far less energy than shallower, warmer clouds. Therefore, where there are extensive decks of high cirrus clouds, the amount of energy radiated to space, compared with clear skies in the same regions, is reduced by 50 to $100\,\mathrm{W\,m^{-2}}$. These thick high clouds occur in three main regions. The first is tropical Pacific and Indian Oceans around Indonesia and in the Pacific north of the equator where rising air forms a zone of towering cumulus clouds. The second is the monsoon region of Central Africa and the region of deep convective activity over the northern third of South America. The third is the mid-latitude storm tracks of the North Pacific and North Atlantic oceans.

The pattern of increased albedo due to clouds is different. The regions associated with the tropical monsoon and deep convective activity reflect large amounts of solar radiation, often exceeding $100\,\mathrm{W\,m^{-2}}$, as do the clouds associated with the mid-latitude storm tracks in both hemispheres and the extensive stratus decks over the colder oceans. The important difference is that these clouds at high latitudes have less impact on the outgoing thermal radiation as the underlying surface is colder and hence emits less energy whether or not there are clouds. So in the tropics the net effect of clouds is effectively balanced out, but over the mid- and high-latitude oceans polewards of $30°$ in both hemispheres, clouds have a cooling effect. This negative effect is particularly large over the North Pacific and North Atlantic where it can be between 50 and $100\,\mathrm{W\,m^{-2}}$.

The satellite records of global cloudiness are surrounded by scientific argument. Although weather satellites have been collecting regular pictures of clouds since the early 1960s, which might be expected to provide a continuous record of global cloudiness, this is not the case. Problems arise from the optical properties of different types of clouds and the changing sensitivity of satellite equipment during its lifetime. In recent years a major international effort [the International Satellite Cloud Climatology Project (ISCCP)] has, however, shown that there has been an overall change in the cloud amount in the tropical and subtropical regions. The observations show that in the period 1983 to 2005 cloud amount increased by about 2% during the first 3 years of ISCCP and then decreased by about 4% over the next 15 years and then recovered to close to the long-term average. Re-analysis data (see Section 4.3) suggests that these trends have been associated with changes in vertical motion associated with the Hadley Cell. In addition, there appear to have been feedback processes by which clouds control tropical SSTs.

The unravelling of these processes is complicated by the fact that ISCCP began immediately after one of the strong El Niño event in 1982–83 (see Section 3.7). In addition, the volcanic eruption of El Chichon in 1983 may have caused some changes in clouds (see Section 6.4). Other weaker El Niño events in 1986–87, the early 1990s and a major event in 1997–98, plus a major volcanic eruption (Mt. Pinatubo) in 1991 further complicated matters. Furthermore, there is still some debate over whether the trends shown are real, or if it is simply an artifact of the satellite analysis.

To make matters worse, it has yet to be determined what kinds of clouds are experiencing the changes, where these clouds are located, and whether or not the trends are more pronounced at certain times of the year. Furthermore, previous studies have not attempted to examine mechanisms that might be driving such changes either on the local or global scales. All of these issues should be addressed in order to characterise the nature of the cloud amount changes in the climate system. This will improve our ability to model them and to understand their impact on the climate system. So, we will need more like 50 years records before we can unscramble the relative contributions of natural and anthropogenic contributions to changing global cloudiness.

One entirely different way to measure the Earth's cloudiness is to observe sunlight reflected from the Earth onto the Moon. Known as *earthshine*, this phenomenon, which was first explained by Leonardo da Vinci, is the light reflected by the earth that illuminates the portions of the moon that are not sunlit. Described poetically as 'the new moon has the old moon in its arms', it can be seen as the ghostly image of the Moon, as if cradled in the arms of the new moon.

As the amount of sunlight reflected by the Earth is a measure of global cloudiness offers the opportunity for making independent observations of changes in the Earth's cloudiness. Comparing the amount of sunlight reflected by the illuminated part of the moon with the amount of earth-shine from the unlit portion provides an absolute check on changes in global cloudiness. So far, the limited measurements suggest that there have been changes in the Earth's cloud cover in the last decade or so that are similar to the ISCCP results.

3.4   The hydrological cycle

In placing so much emphasis on the radiative balance in understanding the elements of climatic fluctuations, we should not lose sight of how this is interlinked with the behaviour of the hydrological cycle. Water is the most abundant liquid on earth. The total amount with the earth–atmosphere system is estimated to be some $1384 \times 10^6 \, \text{km}^3$; of this amount 97.2% is contained in the oceans, 0.6% is ground water. 0.02% is contained within rivers and lakes, 2.1% is frozen in ice caps and glaciers (the *cryosphere*), and only 0.001% is in the atmosphere. Nevertheless, as noted in Chapter 2, water vapour is by far the most important variable constituent of the atmosphere, with a distribution that varies in both time and space. In addition, on time-scales of decades and longer, the proportion of water locked up in the cryosphere can change appreciably with significant consequences for the Earth's climate.

At any one time the mean water content of the atmosphere is sufficient to produce a uniform cover of some 25 mm of precipitation over the globe – equivalent to some 10 days of rainfall. This means that there is a continual recycling of water between the oceans, the land, and the atmosphere, which is known as the *hydrological cycle*. Moreover, because water has such a high latent heat of fusion and evaporation, the phase transitions between ice and liquid water, and between liquid water and water vapour involve large amounts of energy. So the process of evaporating large amounts of water into the atmosphere, and its subsequent precipitation as rain or snow is a major factor in the energy transport of the climate. The basic elements of this cycle are well understood and fully described in many meteorological texts. Here we are concerned with the less well-understood aspects of how various components of the cycle can vary and their implications for the climate.

Important climatic points about the potential variability of the hydro-logical cycle relate principally to how much water passes into the atmosphere, the nature and form of the clouds it forms, and how quickly it is precipitated out again. Over the oceans the most important factor is the temperature of the surface, with wind speed playing a secondary role. Over the land the level of soil moisture is a controlling factor, but the presence of living matter complicates the issue considerably (see Section 3.5). The amount of water vapour passing into the atmosphere is a combination of evaporation from the soil and transpiration from plants (this combination is defined as *evapotranspiration*), and depends on soil moisture, the air temperature, the temperature of the soil, which is related to the amount of sunlight absorbed by the soil, and wind speed. As noted earlier, the levels of water vapour in the atmosphere affect its radiative properties, while its condensation to form clouds is a critical factor in defining the climate.

3.5   The biosphere

The involvement of plants on land in the hydrological cycle is only part of the story. The totality of living matter, both on land and in the oceans (the *biosphere*), has the capacity to influence the climate in a variety of ways. The most important is in terms of the control it exerts over certain greenhouse gases. In particular, through photosynthesis the biosphere acts as the fundamental control over the level of CO_2 in the atmosphere. Similarly, the production of CH_4 by the anaerobic decay of vegetation maintains the natural level of this important greenhouse gas (see Section 2.1.3). The totality of the biosphere is not an easy concept to grasp. In climatic terms it is most obvious in how it alters the CO_2 content of the atmosphere throughout the year (see Fig. 7.3). During the northern hemisphere growing season the biosphere draws down CO_2 from atmosphere only to release much of it during the winter.

The potential of the biosphere to absorb additional CO_2 is an important factor in considering climate change. It is a temporary sink for a proportion of any additional CO_2 injected into the atmosphere through the *carbon cycle*. This sequestration is central to any discussion of the climatic implications of past variations in concentrations due to shifts in the overall productivity of the biosphere, or emissions of this gas into the atmosphere as a result of human activities. Because the productivity of the biosphere rises with increasing CO_2 levels, this slows down the build-up in the atmosphere. This negative feedback mechanism has exerted a major impact on past climates, and is a factor delaying some of the consequences of emissions of this greenhouse gas due to the combustion of fossil fuels.

The formation of particulates as a result of activity in the biosphere is another potentially important aspect of climate change. On land the emission of many volatile organic substances (e.g. terpenes from fir trees) leads to the formation of particulates and in hot sunny conditions in forested areas to natural photochemical smogs that may have an impact on the climate. In the oceans there is an analogous process involving the production of dimethylsulphide (DMS) by phytoplankton. This gas is a breakdown product of a salt, dimethylsuphoniopropionate that marine algae produce to maintain their osmotic balance with seawater. It escapes into the atmosphere, where much of it is converted into sulphate particulates that may be a major factor in the formation of clouds over the oceans. So it is possible that changes in SSTs will lead to varying production of algae and hence alter the amount of DMS released into the atmosphere. If this influenced cloud formation it could act as a negative feedback mechanism to stabilise the climate.

In a more direct way the biosphere can influence the climate by changing the albedo of the Earth's surface (see Table 2.1). This affects land surfaces and is of most concern where the removal of vegetation leads to the lasting formation of deserts (*desertification*). Because desert sand reflects so much more sunlight than savannah vegetation the consequences of desertification are to alter the radiation balance of the areas affected by this desiccation process. At the same time the removal of vegetation will lead to reduction of water vapour passing into the atmosphere by the process of evapotranspiration. This will have an additional impact of altering the radiative properties of the air and reducing the possibility of cloud formation.

3.6   Sustained abnormal weather patterns

The possibility of processes, which result in long-term changes in cloudiness, leads naturally back into the consequences of abnormal weather patterns. This in turn links into many aspects of the attempts to conclude whether periodic behaviour in the weather is tied into the underlying tendency of the climate to have a 'memory'. This reflects the fact that many components of the climate (e.g. snow cover, polar ice, SSTs and soil moisture) have a built-in inertia. This has the effect of damping out short-term fluctuations and making lower frequency variations more probable, so that the current weather is, in part, driven by the stored effects of past changes (see Section 5.2).

The scale and significance of sustained periods of extreme weather are reflected in, say, winter snow cover or the extent of polar pack ice. The even more important issue of SSTs will be considered separately in the Section 3.7. Here we will concentrate on the issue of whether a prolonged spell of extreme weather can affect the underlying components of the climate long enough to influence the weather in subsequent seasons, and conceivably lead to more prolonged changes.

Studies of severe winters show that extreme snow cover prolongs cold weather in the northern hemisphere. Analysis of fluctuations in the longer term of the extent of the annual average snow cover in the northern hemisphere suggests that these may have significant consequences. Because of its high albedo more extensive snow cover has a cooling effect. In principle, this should lead to a colder climate and so produce more snow. This could lead to a positive feedback mechanism that drives the climate into a much colder regime. Conversely, a series of mild winters, which result in well below average snow cover, could reinforce a warming trend.

To have a lasting climatic impact, changes in snow cover in the northern hemisphere must last an appreciable time. The area covered by snow varies by a factor of over ten between winter and summer (Fig. 3.12), reaching a maximum extent of around 45 million square kilometres in late winter and decline to less than 5 million square kilometres in late summer. Therefore, the impact of anomalies is greatest in the spring and summer when the amount of incident solar radiation at high latitudes is greatest. Abnormal snow cover is, however, a relatively transient phenomenon, with variations from year to year ranging up to $\pm10\%$ in winter, whereas in summer the extent can vary by as much as a factor of two (Fig. 3.12). Interannual variability is largest in autumn (in absolute terms) or summer (in relative terms). Monthly standard deviations range from 1.0 million km^2 in August and September to 2.7 million km^2 in October, and are generally just below 2 million km^2 in non-summer months. There is, however, little evidence that large anomalies are sustained for more than a few months (Fig. 3.13).

In the longer term, surface observations provide some indication trends in northern-hemisphere snow cover from around 1920 to the present. These show relatively little change in the first half of the century. It was not until the advent of satellite observations in the 1960s, however, that accurate measurements became possible. These show that the extent of snow cover declined abruptly in the late 1980s. Since then, it has, if anything, increased a little (see Fig. 3.13), while remaining well below the levels of the 1970s and early 1980s. The overall trend correlates closely with changes in the average temperature at higher latitudes of the northern hemisphere.

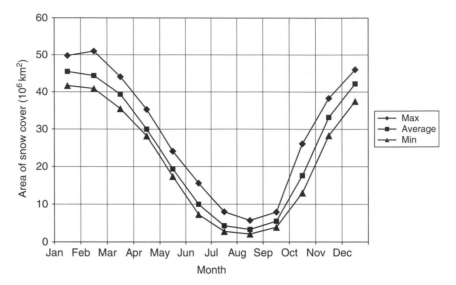

Fig. 3.12 The annual cycle of snow cover across the northern hemisphere shows how the mean areal extent (b) varies by over a factor of ten, while the variations from year to year, during the period 1973 to 2006, fluctuated between a maximum of monthly value of (a) and a minimum of (c). (Data from NOAA: ftp://ftp.cpc.ncep.noaa.gov/wd52dg/snow/snw_cvr_area/NH_AREA.)

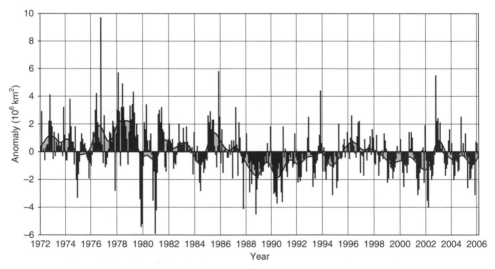

Fig. 3.13 The difference between the average monthly area of snow cover in northern hemisphere compared with the monthly average between January 1972 and Sept 2006 (vertical bars) and the smoothed average value (heavy line). These fluctuate dramatically from month to month and show a long-term decline in snow cover. (Data from NOAA: ftp://ftp.cpc.ncep.noaa.gov/wd52dg/snow/snw_cvr_area/NH_AREA.)

Fluctuations in snow cover at high latitudes exert greatest influence on the amount of sunlight reflected back into space around the spring equinox when the cover is still extensive and the sun is high in the sky by day. The average cover in April across the northern hemisphere is currently around 31 million km^2. Available data suggest that this figure has declined by about 2 million km^2 in the twentieth century (Fig. 3.14). These observations provide support for the view that reduced snow cover is amplifying the current global warming. Conversely, studies of longer-term climate change suggest that snow cover variations may be driven principally by longer-term factors. For example, if extent of snow cover in the summer half of the year responds markedly to changes in the amount of sunlight falling at high latitudes during the summer, due to orbital variations, it could act as a positive feed-back mechanism and have a major climatic impact (see Section 2.1.4).

The consequences of changes in the extent of polar pack ice are similar. In some sectors of the northern hemisphere, notably the North Atlantic, the changes in ice cover can affect weather patterns (Fig. 3.15). On average the area of Arctic pack ice ranges from a minimum of around 7 million km^2 in late summer to a maximum of some 15 million km^2 in early spring. The scale of these changes is, however, small compared with the variations in snow cover, which reach a maximum extent of around 45 million km^2 in late winter and decline to less than 5 million km^2 in late summer. In the southern hemisphere the reverse is true. Because the Antarctic snow cover is permanent, and winter snow in South America, Australia and New Zealand is small, the most important variations are associated with the extent of Antarctic pack ice. The annual cycle has an amplitude of some

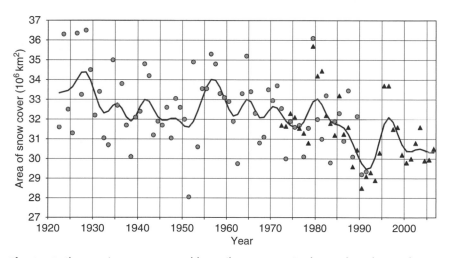

Fig. 3.14 Changes in average monthly April snow cover in the northern hemisphere.

15 million km^2 from a maximum extent of about 18 million km^2 and around 3 million Km^2 in late summer (see Fig. 3.15). From year to year the extent of the ice cover can fluctuate by several million km^2 (see Section 4.2).

Accurate measurements of the global extent of both snow and ice are only available for the satellite era from the early 1970s (see Figs. 3.13 and 4.7). The most striking feature of the records is the marked decline in the extent of sea ice in the Arctic, especially in the summer months. This trend is in line with the warming of high latitudes of the northern hemisphere in recent decades. Less easy to explain is the puzzling tendency for changes in sea ice cover to show different trends in the northern and southern hemisphere.

In summary, while snow and ice cover changes have a substantial short-term impact on the energy balance during the winter half of the year; the longer-term variability of the global climate appears to overwhelm the

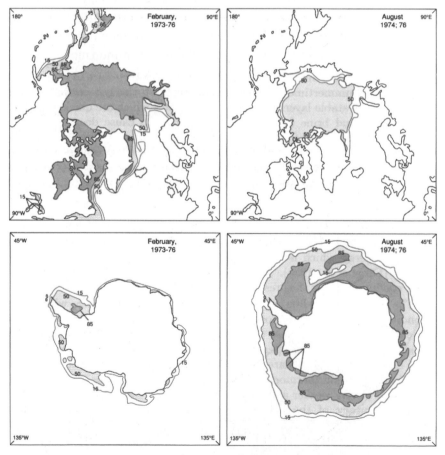

Fig. 3.15 Sea ice distributions in February and August in both hemispheres. The contours displayed are 15, 50 and 85% coverage. (From Trenberth, 1992, Fig. 4.12.)

temporary influence of these variations. In effect the climatic inertia of a metre or so of snow or pack ice is small compared with the temperature fluctuations of the top hundred metres of the oceans. There remain, however, many unanswered questions about how other components in the climatic system will respond to changes in snow and ice cover. For instance, the shift in the extent of Antarctic pack ice could produce parallel shifts in the storm tracks at lower latitudes and hence alter the cloud cover. Depending on the form of these additional responses, the net effect could be to either reinforce the changes in the pack ice extent or largely cancel them out. More important may be how these changes couple with shifts in the behaviour of the oceans.

3.7   Atmosphere–ocean interactions

How the atmosphere affects the oceans and vice versa is the next step in understanding climate change. Because of the much greater heat capacity of the oceans, how heat is taken up, stored and released by them is bound to have a bigger impact on longer-term climate change. Large-scale temperature anomalies can last much longer than the more fleeting change in snow cover and pack ice. For this reason, the processes, which control the surface temperature of the oceans, hold the key to many aspects of climate variability and climate change in timescales from a few years to centuries.

The changes in the oceans, cannot, however, be considered in isolation. They are linked with the effects that have been discussed earlier. Long-term fluctuations in cloudiness affect how much energy the oceans, especially in the tropics, absorb. Changes in the extent of pack ice may influence the rate at which cold dense water descends into the depths in polar regions. Sustained changes in precipitation and rates of evaporation at high latitudes may have similar consequences. Because these changes may take decades or centuries before influencing the temperature of upwelling of cold water at lower latitudes, they have the capacity to establish longer-term fluctuations. But most important of all is that changes in atmospheric conditions can lead to changes in the oceans' surface, which in turn can alter the weather patterns. These atmosphere–ocean feedback mechanisms have the potential to set up oscillatory behaviour and so produce periodicities or quasi-periodicities in the weather, and are central to much of the observed climate variability over the last 100 years or so.

The most celebrated of these feedback mechanisms occurs in the Pacific Ocean. 'When pressure is high in the Pacific Ocean, it tends to be low in the

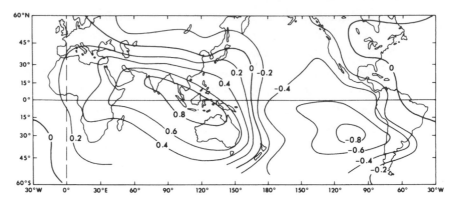

Fig. 3.16 The correlation of monthly mean surface pressure with that of Jakarta. The correlation is large and negative in the South Pacific and large and positive over India, Indonesia and Australia. This pattern defines the Southern Oscillation. (From Burroughs, 1994, Fig. 3.6.)

Indian Ocean from Africa to Australia'. This is how Sir Gilbert Walker described what he called the Southern Oscillation (SO) in the 1920s. He defined the SO in terms of differences in pressure observations at Santiago, Honolulu and Manila, and those at Jakarta, Darwin and Cairo. Subsequently, in the 1950s the Dutch meteorologist Berlage updated the index in terms of the overall tropical pressure field. Taking Jakarta as his reference station, he produced a map of the correlation of annual pressure anomalies (see Fig. 3.16) that showed that the value at Easter Island had a surprisingly large value of −0.8 (see Section 5.3). This analysis shows the SO is a barometric record of the exchange of atmospheric mass along the complete circumference of the globe in tropical latitudes. The change in atmospheric patterns is intimately linked with fluctuations in the surface temperature of the tropical Pacific known as the *El Niño*. For this reason, the overall behaviour is generally described as the El Niño Southern Oscillation (ENSO).

The name El Niño comes from the fact that a warm current flows southwards along the coasts of Ecuador and Peru in January, February and March; the current means an end to the local fishing season and its onset around Christmas means that it was traditionally associated with the Nativity (El Niño is Spanish for the Christ Child). In some years, the temperatures are exceptionally high and persist for longer, curtailing the subsequent normal cold upwelling seasons. Since the upwelling cold waters are rich in nutrients, their failure to appear is disastrous for both the local fishing industry and the seabird population. The term El Niño has come to be associated with these much more dramatic interannual events.

In normal circumstances an ENSO follows a rather well defined pattern (see Fig. 3.17a). In the ocean above-average surface temperatures off the coast of South America mark the onset in March to May. This area of abnormally warm water then spreads westwards across the Pacific. By late summer it covers a huge narrow tongue stretching from South America to New Guinea. By the end of the year the centre of the elongated region of warm water has receded to around 130° W on the equator and temperatures are returning towards normal along the coast of South America. Six months later the warm water has largely dissipated and in the eastern Pacific has fallen below the climatological normal.

In parallel with these changes in SST, large atmospheric shifts are in train. The surface pressure and wind and rainfall records reveal that, starting in the October and November before the onset of the El Niño, the pressure over Darwin, Australia, increases and the trade winds west of the dateline weaken. At the same time, the rainfall over Indonesia starts to decrease, but near the dateline it increases. In addition, the narrow band of rising air, cloudiness and high rainfall known as the Intertropical Convergence Zone (ITCZ), which girdles the globe, shifts position. Normally, it migrates seasonally between 10° N in August and September and 3° N in February and March. As a precursor to an ENSO it shifts further south in the eastern Pacific, to be close to or even south of the equator during the early months of El Niño years.

As the area of anomalous SST spreads westwards, a region of exceptionally high rainfall associated with the shift of the ITCZ accompanies it. During the mature phase of the event, most of the tropical Pacific is not only covered by unusually warm surface water but has also unusually weak trade winds associated with the southwards displacement of the ITCZ. Moreover, the heat transfer from the ocean means that the entire tropical troposphere in the region is exceptionally warm. This maintains the abnormal rainfall until the temperatures return to more normal values. With this return to normality the atmospheric patterns lapse back into a more standard form.

In parallel with these sea-surface and atmospheric changes, important developments occur beneath the surface. The tropical Pacific can be regarded as a thin layer roughly 100 m thick, of warm light water sitting on top on top of a much deeper layer of colder denser water. The interface between these two layers is known as the *thermocline*. High SSTs correlate with a deep thermocline and vice versa. As an ENSO warm event develops, the easterly trade winds that normally drive the currents in the equatorial Pacific become exceptionally weak. The sea level in the western Pacific falls

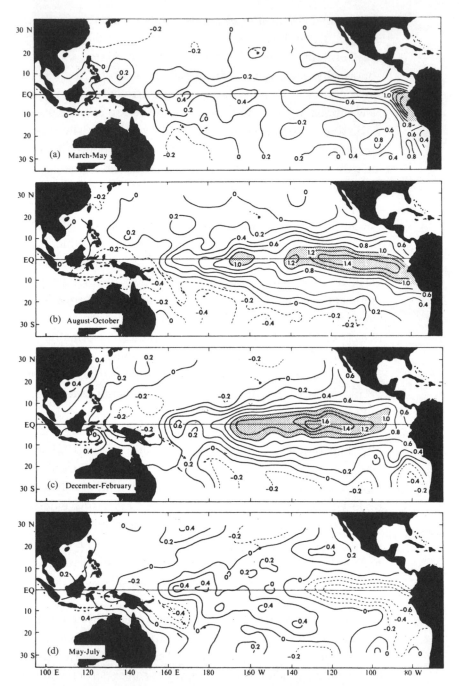

Fig. 3.17 Sea-surface temperature anomalies (°C) during a typical ENSO event obtained
by averaging the events between 1950 and 1973. The progression shows (a) March, April
and May after the onset of the event; (b) the following August, September and October;
(c) the following December, January and February; and (d) the declining phase of May,
June and July more than a year after the onset. (From Burroughs, 2003, Fig. 5.6.)

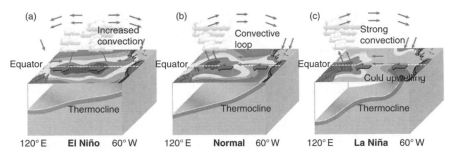

Fig. 3.18 Schematic illustration of the changes that take place between (a) El Niño conditions, (b) normal conditions and (c) La Niña conditions. The important features in (a) are the thermocline is less tilted than usual as the sea-surface temperature is above normal in the eastern Pacific and there is increased convection in the central and eastern Pacific, whereas in (c) the reverse is the case as the thermocline steepens from west to east and convection moves farther west. (From IPCC, 1995, Fig. 4.7.)

and the depth of the thermocline is reduced (Fig. 3.18a). Intense eastwards currents between the equator and 10° N carry warm waters away from the western Pacific. Along the western coast of the Americas there is an increase in sea level that propagates polewards in both hemispheres. This motion, which may be associated with cyclone pairs in the atmosphere north and south of the equator (see Section 6.2) that reinforces the early flow, propagates an eastward motion or wave. This called a 'Kelvin wave' (named after Lord Kelvin in recognition of his fundamental work in wave dynamics).

The changes in the ocean contain two important pieces of information. First, the observed movement in the oceans is a consequence of the alteration of the winds that normally drive the currents away from South America. This standard pattern produces lower sea levels in the east than in the west. It also means that cold water is drawn in from higher latitudes and also from greater depths. As the wind weakens so does the current. Sea levels rise and warm water spreads back to cover the cold water. This leads to a second counter-intuitive observation. The development of the SST anomaly appears to reflect a westward movement of warm water: that is not the case. The anomaly first appears off the coast of Peru, reflecting the fact that a small reduction in the overall movement westwards can lead to the cold Humboldt Current being capped by warmer waters. But the much more extensive region of abnormally warm water across most of the equatorial Pacific only develops after a sustained movement of warm water from the western Pacific. So, although the anomaly appears to move westwards, it is the counter-movement of water that is causing the observed effects.

This underlines the central fact about all aspects of the ENSO that only by considering the combined atmosphere–ocean interactions is it possible to understand the overall behaviour of the phenomenon.

When the ENSO swings away from El Niño conditions it may either spend some time in what constitute normal conditions (see Fig. 3.18b) or pass rapidly on into to *La Niña* conditions (see Fig. 3.18c). At these times the sea-surface temperatures of the central and eastern equatorial Pacific Ocean are cooler than normal. Although less well known than El Niño events, they are an essential part of the ENSO phenomenon. Just like El Niño, the La Niña events disrupt the normal patterns of tropical convection and constitute the opposite phase of the Southern Oscillation circulation. The core feature of these events is that cooler than normal ocean temperatures develop in the central and eastern equatorial Pacific Ocean. Atmospheric pressure over the cooler water rises but over Indonesia and northern Australia it falls below normal. Overall, the pressure difference across the Pacific increases and strengthens the easterly trade winds.

While the nature of the El Niño can be described in terms of changes in the tropical Pacific, its impact spreads far and wide. The effects are most noticeable elsewhere in the tropics, as can be seen in the patterns of rainfall (see Figs 3.19 and 3.20). The distribution of rainfall shifts all around the globe as the changes in pressure associated with the Southern Oscillation alter not only the position of the ITCZ in the Pacific but also less well-studied longitudinal atmospheric circulations. This leads to the region of heavy rainfall over Indonesia moving eastwards to the central Pacific. At the same time there is a smaller but significant move of the heaviest rainfall over the Amazon to the west of the Andes. More important is that a descending motion replaces the region of ascending air over Africa. This partially explains how the prolonged drought in sub-Saharan Africa since the late 1960s has been related to ENSO events in the Pacific.

The ENSO is also involved in another great mystery of tropical meteorology – the Indian monsoon. Ever since Edmund Halley proposed in a paper in the Philosophical Transactions of the Royal Society in 1686 the broad mechanism for the monsoons, there has been speculation as to what caused its fluctuation from year to year. During the period from the 1920s and the 1970s, evidence grew that the ENSO is a major factor in the summer monsoon across the Indian subcontinent. When the tropical Pacific is warmer than usual (El Niño years) rainfall is often below normal in India and vice versa. In more recent years the picture has become more complicated. The gathering of more comprehensive measurements of SSTs, it has become evident that large-scale variations in the Indian Ocean are also an important factor in defining monsoon rainfall.

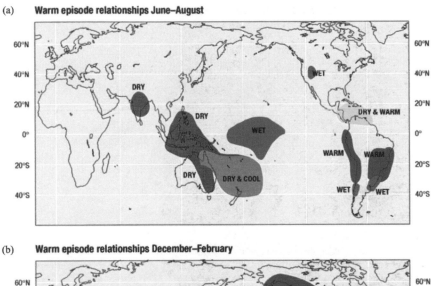

(a)   **Warm episode relationships June–August**

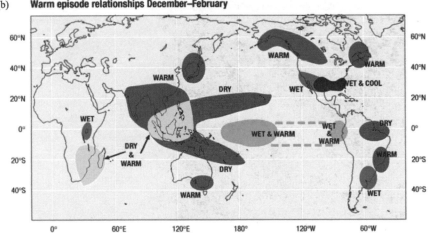

(b)   **Warm episode relationships December–February**

Fig. 3.19 Schematic diagrams of the areas that experience a consistent precipitation or temperature anomaly during an El Nino event in (a) the northern summer, and (b) the northern winter. (After NOAA CPC NCEP.)

The scale of these changes in the tropics plus the fact that in recent decades global temperature anomalies have been closely linked to ENSO events suggests there should be parallel disturbances at higher latitudes. These are not immediately obvious. While some features of year-to-year variability in the climate of mid-latitudes can be linked with events in the equatorial Pacific, they constitute only a small part of the story. Indeed, in the North Atlantic in particular, there is another less well-understood *oscillation*, which may be a further important aspect of atmosphere–ocean interactions. Since the eighteenth century it has been known that when winters are unusually warm in western Greenland, they are severe in

(a) **Cold episode relationships June–August**

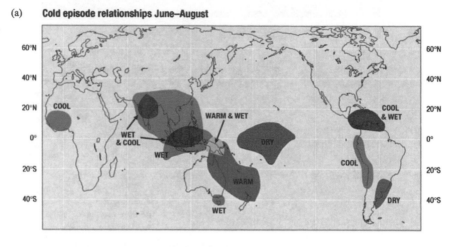

(b) **Cold episode relationships December–February**

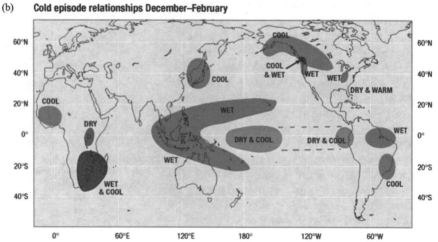

Fig. 3.20 Schematic diagrams of the areas that experience a consistent precipitation or temperature anomaly during a La Nina event in (a) the northern summer, and (b) the northern winter. (After NOAA CPC NCEP.)

northern Europe and vice versa. This see-saw behaviour was quantified by Sir Gilbert Walker in the 1920s in terms of pressure differences between Iceland and southern Europe and defined it was the *North Atlantic Oscillation (NAO)*.

The climatic influence of the NAO is greatest in the winter half of the year. It shifts between a deep depression near Iceland and high pressure around the Azores, which produces strong westerly winds (defined as the *positive* phase), and the reverse pattern with much weaker circulation (the *negative* phase). The positive phase pushes mild air across Europe and into Russia, while pulling cold air southwards over western Greenland. The strong westerly flow also tends to bring mild winters to much of North America. One significant climatic effect is the reduction of snow cover, not only during the winter, but also well into the spring. The negative phase often features a blocking anticyclone (see Section 3.2) over Iceland or Scandinavia, which pulls arctic air down into Europe, with mild air being funnelled up towards Greenland (see Fig. 3.8). This produces much more extensive continental snow cover, which reinforces the cold weather in Scandinavia and Eastern Europe, and often means that it extends well into spring as long as the abnormal snow remains in place (see Section 3.6).

Examination of re-analysis data of daily weather patterns since the 1950s (see Section 4.3) shows that the negative phase of the NAO features 67% more blocking days than when it is the positive phase. Equally important is that the duration of blocking episodes during the negative phase is nearly twice as long (11 days) than in the positive phase (6 days). On longer time-scale, since 1870, the NAO has fluctuated appreciably over periods from several years to a few decades (see Fig. 3.21). It assumed the positive strong westerly form between 1900 and 1915, in the 1920s and, most notably, from 1988 to 1995. Conversely, it took on a sluggish negative form in the 1940s and during the 1960s, bringing frequent severe winters to Europe but exceptionally mild weather in Greenland. So it appears to be a more persistent phenomenon than interannual fluctuations in snow and ice cover. This combination highlights the climatic significance of incidence of blocking in climatic fluctuations.

Changes in sea-ice cover in both the Labrador and Greenland Seas as well as over the Arctic appear to be well correlated with the NAO, and the relationship between the sea-level pressure and ice anomaly fields suggests that atmospheric circulation patterns force the sea ice variations. Feedbacks or other influences of winter ice anomalies on the atmosphere have been more difficult to detect, but the frequency of depressions appears to have increased and atmospheric pressure decreased where ice margins have retreated, although these changes differ from those directly associated with the NAO. If, however, it can be demonstrated that a period of one phase of the oscillation produces the right combination of patterns of SSTs and deep water production, eventually to switch it into the opposite phase, then this may provide insight into more dramatic changes in ocean circulation which can occur in the North Atlantic (see Section 3.8).

Recent research clearly shows that the North Atlantic responds to changes in the overlying atmosphere. The leading mode of SST variability over the region during winter consists of a tripole. This pattern can switch from year to year between a form where there is a cold subpolar region,

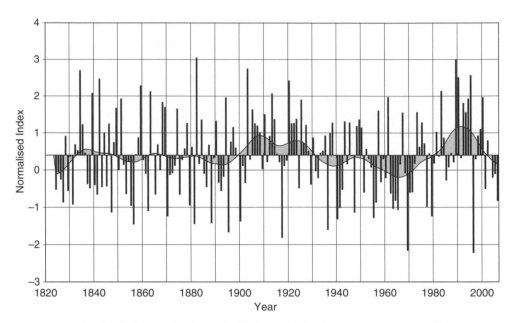

Fig. 3.21 The North Atlantic Oscillation in winter (December to February), based on the normalised pressure difference between Gibraltar and Reykjavik, provides a measure of the strength of westerly circulation in the mid-latitudes of the northern hemisphere. (Data from the Climatic Research Unit, University of East Anglia.)

warmth in the middle latitudes, and a cold region between the equator and 30° N and the reverse combination. The corresponding NAO index for these two patterns is respectively negative and positive. On a timescale of several months the tripole SST tripole appears to drive the NAO, but on the timescale of days to weeks the atmospheric consequences of a given NAO feeds back into the SST patterns.

Whatever the impact of the NAO on longer-term developments, it remains central to understanding recent climatic events, because of the influence it exerts on average temperatures in the northern hemisphere. Of all seasons, winters show the greatest variance, and so annual temperatures tend to be heavily influenced by whether the winter was very mild or very cold. When the NAO is in its strong westerly phase, its benign impact over much of northern Eurasia and North America outweighs the cooling around Greenland, and this shows up in the annual figures. So, a significant part of the global warming since the mid-1980s has been associated with the very mild winters in the northern hemisphere. Indeed, since 1935 the NAO on its own can explain nearly a third of the variance in winter temperatures for the latitudes 20° to 90° N.

While the ENSO and the NAO have attracted a great deal of attention in recent years, there is no reason why atmosphere–ocean interactions in other parts of the world should not produce similar longer-term climatic fluctuations. Indeed, the upsurge in interest in these interactions has led climatologists to look much more closely for evidence of such connections elsewhere. The product of this work has been a growing list of *oscillations* from the Arctic to the Antarctic. The real test of these new teleconnections will be whether they offer new insights into the workings of the climate and better still if they able to provide useful forecasts months and years ahead.

The first of these new oscillations, which is the subject of a concerted debate as to whether the NAO should be incorporated into a hemispheric *annular mode*, is known as the *Arctic Oscillation (AO)* that reflects the wider implications of circulation patterns around the northern hemisphere. This construct has theoretical attractions, as many of the physical mechanisms associated with annular modes are useful in explaining the existence of the NAO. Nevertheless, practical analysis suggests that the NAO is more physically relevant and robust for considering northern hemisphere variability than is the AO. So, here we will stick to the tried and tested NAO. This conservative approach also has the benefit of staying with a measure that is easily interpreted in terms of the nature of the circulation in the North Atlantic and also reflects the dominant part this aspect of the hemispheric circulation plays in the interannual fluctuations of the climate at high latitudes of the northern hemisphere.

In addition to the NAO, there is considerable evidence of 65–80 year oscillation in temperature records for the North Atlantic and its bounding continents. Although this basin-wide pattern of SST anomalies is probably related to the NAO, it has now become known as the *Atlantic Multidecadal Oscillation (AMO)*. This periodicity has also been identified in other records for the region, including sea-level measurements and has been linked with rainfall patterns over the United States.

In concentrating on the North Atlantic, we must not lose sight of the Pacific Ocean, which covers nearly a third of the Earth's surface. This huge area, combined with its roughly symmetrical form (both latitudinal and longitudinal) makes it the most obvious region for longer-term coupled interdecadal variations of the atmosphere and ocean. While shorter-term fluctuations in the 2- to 8-year range are dominated by the ENSO variation, interdecadal changes have the largest amplitude in the North Pacific rather than in the eastern tropical Pacific. In addition, there is a coherent pattern of surface temperature variability in the southern hemisphere with cold and warm anomalies in the region of New Zealand alternating in opposition to warm and cold anomalies in the south-eastern tropical Pacific.

In the North Pacific the principal fluctuation was originally identified by Sir Gilbert Walker in the 1930s, and defined as the *North Pacific Oscillation (NPO)*. It has now become known as the *Pacific Decadal Oscillation (PDO)*. A measure of this phenomena, developed by Nate Mantua and colleagues at the University of Washington, Seattle, is based on SST anomalies in the North Pacific: *warm events* are defined as when the temperature of the northeast Pacific and much of the tropical Pacific is above normal while much of the eastern Pacific north of $20°\,N$ is below normal, and atmospheric pressure is below normal over the Aleutian Islands. During *cool events* the reverse conditions apply. Typical PDO events persist for 20 to 30 years. Their climatic impact is most evident in the North Pacific/North American sector, while secondary signatures exist in the tropics, in contrast to the dominantly tropical nature of ENSO. During the twentieth century cool events prevailed from 1890–1924 and again from 1947–1976, while warm events dominated from 1925–1946 and from 1977 until around 1998. Since then there appears to have been a move back towards cool PDO conditions.

The low-frequency variability over the North Pacific in wintertime is closely linked to the intensity of the semi-permanent low-pressure system near the Aleutian Islands (the *Aleutian Low*). During warm events the Aleutian low is deeper than normal and the westerly winds across the central North Pacific strengthen. This leads to above-normal surface air temperatures over much of northwestern North America and below-normal precipitation, while over the south-eastern United States the reverse is the case. During cool events the opposite patterns occur more often over North America. This pattern also has hemispheric implications. When the warm PDO conditions are in phase with a positive NAO they reinforce the strength of the hemispheric circulation and hence the global warming, as happened during the 1980s and 1990s.

In the South Pacific there is a coherent pattern of surface temperature variability in the southern hemisphere with cold (warm) anomalies in the region of New Zealand and warm (cold) anomalies in the southeastern tropical Pacific. Known as the *Interdecadal Pacific Oscillation (IPO)*, these changes exhibit an approximately 20-year periodicity with the possibility of an underlying longer 50–60-year variation. Recent work suggests that these longer period variations can also modulate ENSO fluctuations.

More generally, variability of the extratropical southern hemisphere circulation is linked to ENSO. In mid-latitudes these circulation changes affect the frequency of winter storms and rain to central Chile and east of the Andes. Farther south, SST anomalies tend to move eastward in the general flow of the Antarctic Circumpolar Current suggesting a coupling in the ocean–atmosphere system of the region. This has become known as the *Antarctic Circumpolar Wave*. Four huge pools of alternating above- and below-normal temperature water appear to circulate every eight years or so, and appear to be linked with changes in rainfall patterns over the southern continents.

There is also an annular mode in the pressure patterns at high latitudes of the southern hemisphere. Known as the *Antarctic Oscillation (AAO)*, it can be regarded as mirroring the Arctic Oscillation, and its better-known components, the NAO and PDO, in the northern hemisphere. Indeed, as long ago as 1928, Sir Gilbert Walker noted that 'Just as in the North Atlantic there is a pressure opposition between the Azores and Iceland, . . . there is an opposition between the high pressure belt across Chile and the Argentine on the one hand, and the low pressure area of the Weddell Sea and Bellingshausen Sea on the other.' It was not until the 1980s that sufficient data became available to establish the reality of this oscillation and its influence on the seas around Antarctica and its links with ENSO.

Long-term fluctuations in the behaviour of the tropical Atlantic and Indian Oceans exert a major influence on weather patterns in Africa, India and South America. In some instances these effects can cancel out the influence of ENSO events. In particular, a pattern of variations in the Indian Ocean seem to play an important part in the variety of responses to ENSO phases that occur in southern Africa and the monsoon in India. Unlike the Pacific Ocean, the surface wind flow over the tropical Indian Ocean does not have a prevailing easterly component. Strong upwelling off the Horn of Africa during the summer monsoon provides seasonal cooling of SSTs and reinforces the surface pressure patterns that drive the monsoon winds of India. Recent analysis of historical ocean and atmospheric data suggests that every few years or so there is an east-to-west oscillation of warm waters similar to the El Niño and La Niña events of the Pacific. This behaviour may explain why the strength of the monsoon over India sometimes responds unexpectedly to El Niño and La Niña events in the Pacific.

3.8   The Great Ocean Conveyor

So far the consideration of the part played by the oceans has concentrated largely on their surface properties, horizontal motion and the interaction with the atmosphere. This has effectively restricted the analysis to the mixed layer of the oceans generated by the action of the winds. From top to bottom of this mixed layer there is little temperature difference. As noted in Section 3.7, beneath it is the *thermocline* – a narrow zone over which there is a rapid drop in temperature. The thickness of the mixed layer, and hence the depth of the thermocline, depends on the wind speed, thermal mixing where the surface waters are heated by the sun or altered by the passage of warmer or colder air, and by the advection of warmer or colder water or the upwelling of cold water. But this involves only small part of the oceans. Beneath the seething surface layer is a much more gradual but equally important set of motions.

An essential part of the process driving the abyssal waters of the oceans is *thermohaline circulation (THC)*. Combined with the action of the surface winds and possibly the effects of the dissipation of tidal energy in the deep ocean (see Section 6.6), THC is an important component of the global patterns of ocean currents, often termed the *Great Ocean Conveyor (GOC)* (Fig. 3.22). An alternative definition of the surface transport of ocean waters to high latitudes and their return at depth is *meridional overturning circulation (MOC)*.

The process of THC results from changes in seawater density arising from variations in temperature and salinity (see Box 3.1). Where the water becomes denser than the deeper layers it can sink to great depths. The temperature depends on where the surface waters come from and how much heat the oceans either pick up or release to the atmosphere in both sensible heat and evaporative loss. The salinity of a given body of water depends on the balance between losses through evaporation as opposed to gains from either rainfall, or freshwater run-off from rivers and melting of the ice sheets of Antarctica and Greenland plus the pack-ice of the polar oceans. In practice there are few regions where sinking waters have a major impact. *Deep waters*, defined as water that sinks to middle levels of the major oceans, are formed only around the northern fringes of the Atlantic Ocean. *Bottom waters*, which constitute a colder denser layer below the deep waters, are formed only in limited regions near the coast of Antarctica in the Weddell and Ross Seas.

The basic observations can be made about these figures. First, where freshwater enters the oceans (e.g. rivers, melting ice and rainfall) it can float on the top of seawater preventing deep circulation. At the same time any surface warming due to either solar heating of advection of warmer water from lower latitudes will tend to form a stable surface layer. The stability of this surface layer, and hence the depth of the thermocline will be controlled by the amount of surface mixing due to winds. But deep mixing depends on the formation of high salinity water either by the process of

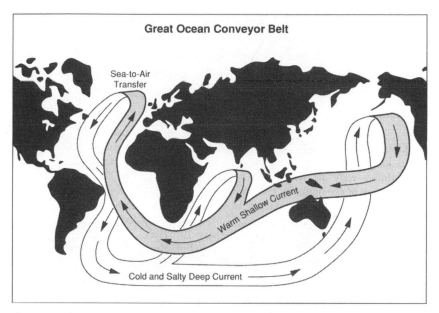

Fig. 3.22 The Great Ocean Conveyor Belt – a schematic diagram depicting global thermohaline circulation. (From Trenberth, 1992, Fig. 17.12.)

freezing as salt is shed into the surrounding cold water or in areas of high evaporation. These processes (e.g. the creation of North Atlantic Deep Water, and Antarctic Bottom Water, and the flow in and out of the Mediterranean, whose dense salty water spills out into the deep Atlantic and is replaced by less dense Atlantic surface water) are the principal mechanisms driving larger scale thermohaline circulation of the oceans.

For the rest, the surface motions of the oceans reflect the effects of the winds and by the differing salinity and temperature of various bodies of water that mix over time. This is a difficult process to represent in computer models (see Section 10.1) because much of this mixing takes place in the form of relatively small eddies. So the density of seawater must be calculated as its properties change. This involves the calculation of *isopycnic surfaces* (defined by points of equal density). For instance seawater with a salinity of 32.5‰ and a temperature of 0 °C has the same density as that with values of 35‰ and 14.5 °C, or 37.5‰ and 22 °C (see Table B3.1).

The basic GOC model, which was developed by Wallace Broecker at the Lamont Doherty Laboratory, has had an immense impact on thinking about the nature of climate change. There is, however, a debate as to whether this is an oversimplified representation. In particular, there is a chicken-and-egg question about whether THC is in the lead, or whether the wind fields remain the dominant factor in driving the ocean circulation, especially in

Box 3.1 Thermohaline circulation (THC)

The density (ρ) of seawater is not a simple function of temperature and salinity. So the easiest way to visualise how changes in these parameters control thermohaline circulation is to consider the actual values of the density set out in Table 3.1. These show that while freshwater has a maximum density at 4 °C, at normal levels of salinity (32.5 to 37.5‰) the density increases with declining temperature right down to the freezing point around −2 °C. The changes in density with salinity are simpler as, the saltier the water, the denser it becomes.

Table B3.1 Density of water of different salinities at different temperatures (Density units are sigma values,[*] at atmospheric pressure)

Temp. (°C)	Salinity (‰)							
	0	10	20	30	32.5	35	37.5	40
30	−4.3	3.1	10.6	18.0	19.9	21.7	23.6	25.5
20	−1.6	5.8	13.3	21.0	23.0	24.7	26.6	28.5
10	−0.3	7.6	15.2	23.1	25.2	26.9	28.8	30.8
5	0	8.0	15.8	23.7	25.7	27.6	29.6	31.6
0	−0.1	8.0	16.1	24.1	26.1	28.1	30.2	32.2
−2	−0.3	7.9	16.0	24.2	26.1	28.2	30.3	32.3

[*] The sigma value (σ) is way of presenting the density (ρ) of sea water relative to that of distilled water at 4 °C, where $\rho = 1.000$:

$$\sigma = 1000 \, (\rho - 1).$$

Thus the densities given in the above table range from 0.9957 (at 30 °C and zero salinity) to 1.0323 (at −2 °C and 40‰ salinity).

the North Atlantic. How winds and the THC combine to maintain the Gulf Stream ties in neatly with the surface ocean currents, which transport so much energy polewards (see Section 2.1.4.).

Carl Wunsch, at Massachusetts Institute of Technology (MIT), argues that the Gulf Stream, as part of a current system, is forced by the torque exerted on the ocean by the wind field. Heating and cooling affect its temperature and salinity, but not its basic existence or structure. As long as the sun heats the Earth and the Earth spins, there will be a Gulf Stream and other ocean currents around the world. The primary mechanism of heat transport in the ocean is the wind-forcing of currents that tend to push warm water toward the poles, cold water toward the equator. He maintains that without the effect of the winds, a sinking motion at high latitudes could not drive ocean circulation; with no wind, heating and cooling could produce a weak flow, but not the observed circulation.

If the sinking motion at high latitudes were switched off, by covering that part of the ocean by sea ice for instance, there would still be a Gulf Stream to the south, and maybe an even more powerful one as the wind field would probably then become stronger. If the sinking were stopped by adding fresh water, the Gulf Stream would simply respond to changes in the wind system. If the storm track moved further south, the amount of heat transported by the system would shift, and the net result would be that the northern North Atlantic was left out in the cold.

The fact that the ocean current system can exist in different states is a fundamental factor in climate change. It is of particular relevance to the North Atlantic where the current circulation carries the majority of heat to the Arctic (see Fig. 2.9). This warm water gives up its heat to the arctic air through evaporation. Then its low temperature and high salinity enables to sink and form deep water that flows all the way to Antarctica. Here it is warmer and less dense than the locally frigid surface waters and so rises to become part of a strong vertical circulation process. Descending cold water from around Antarctica flows northwards into the Pacific and Indian Oceans where there is no descending cold water.

One of the unresolved questions about the THC in the North Atlantic is the part it plays in rapid climate change. Evidence of changes during the last ice age (see Section 8.4) shows that in certain circumstances the circulation of the ocean underwent rapid and substantial shifts. Measurements of the strength of the flows of water in the North Atlantic in modern times are few and far between. As a consequence we have only fragmentary snapshots of changes in this important climatic variable, and so little evidence concerning its stability. What is disconcerting is that both deep-water flow measurements and satellite studies of the Gulf Stream suggest that the THC in the North Atlantic has slowed down appreciably in recent decades. At the same time the subtropical gyre appears to have strengthened. Whether these are part of a natural fluctuation of, say, a periodic nature (e.g. the AMO), or part of a more radical circulation shift, is not yet clear.

Computer models used to predict the impact of human activities of the global climate (see Chapter 10) show that one of the consequences of global warming could be a weakening of the THC and a reduction of ocean heat transport into high latitudes of the northern hemisphere. The reassuring aspect of these models is that, even where the reduction of heat transport is

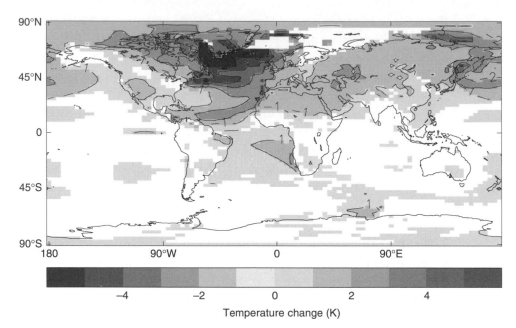

Fig. 3.23 A computer model simulation of the change in temperature that might occur if the thermohaline circulation in the northern North Atlantic were to shut down. (With permission from the UKMO.)

substantial, there is still warming over Europe. What these models do not show is any propensity for the climate, in its current state, to undergo sudden shifts of the type observed during the last ice age.

In contrast, a number of experimental models, which have been designed to explore abrupt regime shifts in the climate, do show a marked propensity for the climate to undergo more dramatic changes. In its most extreme form such a change could consist of effectively 'switching off' the Gulf Stream. Such a shift could, in principle, occur within a matter of years and would then have dramatic consequences for the climate of the northern hemisphere over the subsequent decades (Fig. 3.23). Because such shifts are essentially chaotic, they are unpredictable. So, even if we conclude that global warming increases the possibility of a sudden shift in the climate, we have no way of telling whether it is an immediate prospect or a more distant threat.

3.9   Summary

The essential feature of the various elements of the Earth's climate is how they are all connected to one another. So, it is not realistic to isolate one element and consider what changing it might do to the climate. Although this may provide some interesting insights, failure to estimate how its knock-on effects will affect the rest of the system is a recipe for failure. Only by considering the Earth's climate as a whole is it possible to form a sensible appreciation of how it has changed in the past and how it may alter in the future.

The network of links in climate is the key to the balanced view. In building up this picture we have considered how each element affects:

(a) the amount of solar energy that is absorbed, noting, in particular the importance of clouds, snow and ice and land-surface effects, including the role of the hydrological cycle and the biosphere;

(b) how energy is transported across the face of the Earth by both the atmosphere and the oceans underlining the importance of the atmosphere to assume long-lasting circulation patterns and the oceans to alter how they move and store energy; and

(c) how the amount of energy re-radiated to space is controlled by various elements, again placing particular emphasis on the role of clouds.

The timescale of the response to all these processes to change is also central to how they contribute to fluctuations in the climate. For the most part, the atmosphere reacts more rapidly to various stimuli, but is capable of getting stuck in long-term states that are capable of producing significantly different climatic conditions over large parts of the globe. But, in every aspect of atmospheric variability, the role of clouds is crucial. By comparison, oceans effectively respond more slowly but are capable of producing much greater changes. The speed of the response of land surface conditions, and snow cover, tends to fall in between. These differing time constants mean that the various climatic elements will not react in a uniform way to change. Instead there will be lags and leads between them, which are capable of producing erratic and sometimes contradictory responses throughout the system.

How the various elements of the climate can combine to produce change, has to be put in context with what is known about past changes in the climate. This involves reviewing the evidence, consequences and measurement of variability and change to establish what can reasonably be expected to happen. This is a journey through vast stretches of time, viewing an extraordinary range of clues, which provide some indication of the complexities which must be addressed. None of these, however, alters the fact that what we are really concerned about is how the fluctuating behaviour of the atmosphere and the oceans combine to produce such a variety of climatic regimes.

1. Explain the radiative arguments for the statement that low clouds cool the Earth's climate whereas high clouds warm it.
2. With the aid of a globe explain why the jet stream has a wave-like pattern rather than being circular as it circles the Earth.
3. It can be argued that reductions in the extent of sea ice will lead to a strong positive feedback as the exposed sea-water will be able to absorb more solar energy and so lead to further warming. What other processes might lead to a negative feedback that results in much of the sea ice effect being cancelled out?
4. Using the albedo figures in Table 2.1 make an estimate of the difference between the amount of energy absorbed by a unit area of the Sahara desert and the same area covered by savannah vegetation. Is this difference significant, and, if so, is it a realistic assessment of the climatic impact of the expansion of desert in this part of the world?

FURTHER READING

A complete reference list is available at the end of the book but the following is a selection of the best books or articles to follow up particular topics within this chapter. Full details of each reference are to be found in the Bibliography.

Barry and Chorley (2003): An excellent textbook for discovering the basics of how the global climate functions, but best read in conjunction with other texts when it comes to climatic change.

Bigg (2003): Particularly valuable in providing analysis of the growing understanding of how the atmosphere and oceans combine to govern many aspects of the climate.

Bryant (1997): A textbook which takes a slightly different approach to the question of the climate change in exploring some of the difficult questions of whether we adequately understand the causes of change and can model their consequences. This inquisitive approach, combined with a clear and direct style, makes it a valuable source for getting to grips with a number of difficult issues.

Diaz & Markgraf (eds) (2000): A comprehensive set of papers about various aspects of El Niño and the Southern Oscillation, which provides a useful guide to the many features of this phenomenon and its global implications.

Open University Oceanography Series (2001): The ocean circulation volume, in particular, provides a clear and concise introduction to the basic aspects of ocean dynamics.

4 The measurement of climate change

The one duty we have to history is to rewrite it.   **Oscar Wilde.**

To be certain about how the climate has changed in the past we would need reliable measurements of the relevant parameters (e.g. temperature, rainfall, cloud cover, extent of winter snow and sea-ice, etc.) for representative points around the globe, for regular intervals of time going back as far as we wanted. In practice we have none of these things. Even with modern observations systems there are gaps in our knowledge of how the global climate is changing. Going back in time the problems mount. The whole process of improving the measurement of climate change is designed to fill in gaps in our knowledge of the past and establish a better foundation for the theories of why changes have occurred. The objectives of this chapter are:

(a) to explore briefly the most essential features of various methods of measuring past climatic conditions;
(b) to establish their strengths and weaknesses; and
(c) to investigate whether the various measurements can be combined to provide a more coherent picture of climate variability and climate change.

This involves going over some well-worn ground concerning the measurement of meteorological parameters. To achieve our objectives we must identify those features of measurement that limit our ability to draw conclusions about how the climate has behaved in the past and how it is currently changing.

4.1   *In situ* instrumental observations

In principle, modern instruments are capable of making most of the observations needed to study climate change, but to do so they must measure the same thing under the same conditions, wherever they are. Temperature measurement provides good examples of how difficult this basic requirement can be. While a properly calibrated thermometer can provide an accurate observation, ensuring that measurements are always made under the same conditions is less easy. For this reason, surface observations are required to be made at a specified height of the *shade air temperature*, preferably over grass. This specification is designed to ensure that what is measured is the climatologically important parameter – the temperature of the air – and not the capacity of the thermometer to absorb sunlight or the potential of the ground to heat up and so influence what is observed. So thermometers should be mounted in a well-ventilated, louvered, white shelter, which prevents either direct sunlight or terrestrial radiation from the ground reaching the instruments. The standard design for such an installation is known as a *Stevenson shelter*.

Even where there are well-maintained temperature records, other requirements can cause problems. As regular human observations in remote locations are expensive, there are strong economic pressures to use automatic systems that can be interrogated electronically. Although, electrical measuring devices (*thermistors*) are capable of accurate and reliable measurements, the switch to such instruments can produce significant shifts in observed temperatures. Unless there is careful calibration of the change-over, there is a risk of these instrumental effects going undetected and prejudicing the quality of the observations.

Although modern measurements are made under standardised conditions, many earlier observations were made in less uniform arrangements. So it requires careful detective work to construct reliable time series that enable earlier measurements to be compared with current figures. In addition, while the siting of modern instruments can be checked, changes in land use and construction of buildings over longer periods can exert important influences on local temperatures. Therefore, what were initially rural sites can be modified by changes in agricultural practice, removal of tree cover and local construction.

These effects are more important with many early observations that were made close to towns. As these grew they developed an *urban heat island*, which has produced a marked rise in observed temperatures within the city and the suburbs that could be interpreted by the unwary as climatic warming. Although this warming is real, the fact that urban areas cover so little of the Earth's land surface means the results of urban sites can be misleading. So observations have to be checked to understand what processes have been at work. Similarly, the movement of an observation site from a city centre to the local airport could lead to a significant shift that must also be carefully calibrated. More insidious is local spreading urbanisation and changes in land use in the vicinity of the measurement site, which can alter local climatic conditions in subtle ways. The extent to which these effects are affecting measurements is the subject of an ongoing debate.

The other problem with standard land-surface temperature measurements is that, although they have been made in a few places for up to 300 years, they do not provide an adequate coverage of the Earth's surface. Until the late nineteenth century only parts of the northern hemisphere had any semblance of a network of observations. The remaining gaps have been largely filled since, but it requires considerable interpolation to build up the record of global land temperatures going back to the middle of the nineteenth century.

Using temperature measurements at sea have a comparable but different set of challenges. There are a large number of measurements of both air temperatures and surface-water temperatures available from the middle of the nineteenth century. Air temperatures made during the daytime are now regarded as unreliable because sunlight absorbed by the decking produced exaggerated figures but, because the difference between day and night temperatures is much smaller at sea (see Section 2.1.4), night-time observations provide a valuable source of climatic information.

Measurements of sea-surface temperatures (SSTs) provide a different set of challenges. In the nineteenth century many ships kept records of the temperature of water collected in buckets over the side of the ship. To make use of these observations corrections need to be made for whether the samples were collected in wooden or uninsulated canvas buckets, as the latter cooled more quickly. A bigger puzzle was a sudden jump in temperature of nearly 0.5 °C between 1940 and 1941. It has been established that this change was due to a sudden undocumented shift from using canvas buckets to measuring the temperature at engine intake. This change took place because during wartime using lights to make measurements over the side at night was too dangerous. Applying these corrections to some 80 million observations led to the conclusion that observations before 1941 had to be corrected by an amount that rises from +0.13 °C in 1856, when wooden buckets were widely used, to +0.42 °C in 1940, when canvas buckets were standard. Given that the overall warming of global SSTs since the late-nineteenth century amounts to about 0.5 °C (Fig. 4.1), these corrections are a major adjustment and show how essential it is to standardise instrumental observations.

The other limitation with SST measurements is the gaps in their geographical coverage. Away from the main shipping lanes, there are many areas with few data, especially in the early part of the record. Coverage was affected by two World Wars and changing trade routes (e.g. the opening of

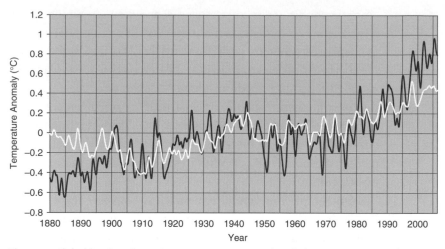

Fig. 4.1 Global land-surface air temperature anomalies (°C) 1880 to 2006, relative to 1961 to 1990 (black line) and sea-surface temperature anomalies (white line). (Data from NOAA: http://www.ncdc.noaa.gov/oa/climate/research/ghcn/ghcngrid.html.)

the Suez and Panama canals in 1869 and 1914, respectively). Even in recent decades there is virtually no data for the Southern Ocean south of 45° S. Given that the oceans cover 71% of the Earth's surface, this is a major deficiency. While additional records may be discovered, there is no prospect of many of these gaps being filled in, so all that can be done is to extrapolate the available data on the basis of known temperature patterns to produce an agreed series for global temperatures.

The global record extending back to 1860 combines measurements on both land and sea. The principal uncertainty in the land surface temperature observations is the effect of urbanisation. Overall, out of the observed warming of around 0.5 °C during the twentieth century, less than 0.05 °C can be attributed to urban influences. The incorporation of marine night-time temperatures and SSTs into the global temperature records introduces additional uncertainties. The early data has limitations, but from the beginning of the twentieth century, the estimated errors for the decadal SST anomalies are less than 0.1 °C for those areas with adequate data (about 60% of world's oceans). The warming observed in these observations at sea is less than that detected over the land (Fig. 4.1). The overall trend is broadly similar. The principal difference is since the 1970s when the oceans have not warmed as rapidly as the land. This slower response probably reflects the thermal inertia of the oceans. The agreed curve for global warming is the combination of these two sets of measurements. This analysis also shows broadly the same trend for both hemispheres, and confirms that the warming is a global phenomenon. The total uncertainty in measurements is reckoned to be $\pm 0.2\,°C$ with a consensus of a rise in global temperature of $0.65\,°C$ since the beginning of the twentieth century.

Using instrumental observations before the middle of the nineteenth century poses additional challenges. Lengthier temperature records are available for Europe and eastern North America. The longest of these is the series that has been produced for rural sites in central England. This series is a classic example of the care that is needed to detect climate change. It was the product of many years of diligent scholarship by the late Professor Gordon Manley and involved a number of interlinked efforts. The first task was to search out and bring together all the records that had been accumulated by a bewilderingly diverse array of amateur observers before the days of official meteorology.

The gathering together of the records was only the start of the analytical problems. First, there was the question of how the measurements were made. Back in the early nineteenth century the combination of reasonable standard observations plus sufficient numbers of overlapping records enabled useful checks to be made of the reliability of the observations and adjustments made for, say, measurements at different times of the day, but earlier records posed greater problems. Before 1760, some of the best-kept records depended on having thermometers exposed in well-ventilated, north-facing, fireless rooms. A further complication was the change from the Julian to the Gregorian calendar in 1752. Prior to this date it was not possible to compare monthly means with those of England today, neither could they be compared with those of contemporary Western Europe unless there were daily observations, as by 1752 the difference in the calendars amounted to 11 days. This series is considered in more detail in Section 8.9.

A similar set of challenges has to be confronted in analysing other instrumental records. Rainfall measurements on land have been made for at least as long as temperature observations. Here again there are problems with the reliability of early observations, many of which probably understate rainfall amounts because of the siting of gauges. Interference by wind eddies near the gauge is capable of seriously distorting the amount of rain caught. These problems are compounded by growth of vegetation and/or building development around the site. Furthermore, early measurements of snowfall did not catch much of the precipitation. There are also issues with the change of design of instruments. All records in the former Soviet Union must be modified for changes in gauge design in the 1960s.

The other challenge is that rainfall is spatially much more variable than temperature and so in measuring long-term trends requires observations from a large number of sites. Where there are few measuring sites the chances are that drawing conclusions about trends will be bedevilled by local fluctuations. For instance, in the case of the England and Wales monthly rainfall series, which goes back to 1772, the standard error in the annual figures for the early years is more than halved by increasing the number of sites from five to over 20. So only where there are copious reliable records of rainfall is it possible to draw conclusions about long-term changes.

Against this background it is hardly surprising that attempts to measure global trends in precipitation amounts are bedevilled by uncertainty, and so far less progress has been made in identifying how precipitation has changed with the rising temperatures this century. Prior to the 1970s measurements are restricted to land areas but this is changing with the advent of satellite measurements (see Section 4.2). Indeed there is no accepted value for the average annual global precipitation. Various estimates since 1960 have varied from 784 to 1041 mm per year with little progress towards agreement. Over land areas some estimates have been made for changes in precipitation for different parts of the globe. But it is a measure of the scale of the problem that there is no agreement as to whether the continents are getting wetter or drier (see Section 8.10).

At the regional level, there are many lengthy precipitation records available. For a few places in Western Europe these extend back to the early eighteenth century and for eastern North America from the early nineteenth century onwards, but only limited conclusions can be drawn from these. There is no clear trend to wetter or drier conditions. In England and Wales, where reliable records extend back to the mid-eighteenth century, there is, however, some evidence of a shift towards wetter winters and drier summers. In the United States the 1830s and the 1850s to 1880s tended to be wet and there has been a marked increase in precipitation during the twentieth century (Fig. 4.2).

As for other forms of instrumental observations, the principal shortcoming is that they have only been maintained extensively for a relatively limited period. So, while local effects can be measured (e.g. changes in sunshine amounts in urban areas provide a good indication of how increases or decreases in pollution levels have affected the local climate), the absence of comprehensive regional and global networks mean little can be said about how other meteorological parameters have varied in the past, in spite of the existence of a considerable quantity of instrumental data.

There are, however, new systems that are bringing an entirely different perspective on the climate. One such system is the Argo global array of 3000

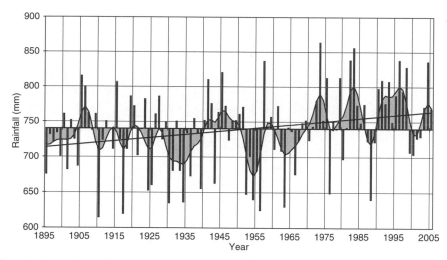

Fig. 4.2 Changes in annual precipitation in the USA since 1895, together with a 9-year binomial smoothing of the data, showing a marked rise over the period. (Data from NOAA.)

free-drifting profiling floats that measures the temperature and salinity of the upper 2000 m of the ocean. Distributed roughly every 300 km, the system can, for the first time, continuously monitor the temperature, salinity and velocity of the upper ocean on a genuinely global basis. Designed to provide a quantitative description of the changing state of the upper ocean and the patterns of ocean-climate variability from months to decades, including heat and freshwater storage and transport, it is already producing new insights into global ocean temperature patterns (see Section 11.4).

4.2 Satellite measurements

Since the beginning of the 1960s, weather satellites have offered the potential for global observations of the Earth's weather. Early satellites provided only rudimentary measurements of cloud cover and temperatures of the surface and cloud tops. This was expanded in the 1970s with the development of infrared radiometers that could analyse the amount of terrestrial radiation upwelling from the surface and the atmosphere at different wavelengths (see Fig. 2.3) to make measurements of temperatures at different levels in the atmosphere. In addition, microwave radiometers were developed which worked on the same principles, and had the huge advantage of seeing through clouds. This meant that not only could they measure

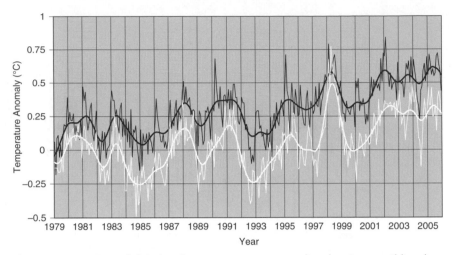

Fig. 4.3 Comparison of global surface-temperature anomalies showing monthly values (thin black line) and values smoothed with a 25-term binomial filter (thick black line) and global temperature anomalies as measured by satellite radiometers (white lines). (Data from: http://www.ncdc.noaa.gov/oa/climate/research/ghcn/ghcngrid.html, and http://vortex.nsstc.uah.edu/public/msu/.)

temperatures in cloudy conditions, but also make accurate observations of the extent of snow cover and pack-ice in polar regions which are frequently cloud-covered.

Although satellites offer the prospect of global coverage of climate change, when it comes to temperature measurements, they have two basic limitations. First, there is the obvious fact that they have only been making observations for a short time and so can only really contribute to the future monitoring of the climate. Secondly, radiometers measure upwelling radiation from a relatively thick slice of the atmosphere and so their results are not directly comparable with surface results and hence require careful calibration. Nonetheless, satellite measurements of the lower troposphere have already made a significant contribution to the analysis of global warming in recent years (Fig. 4.3). They are capable of making accurate global measurements of atmospheric temperature. But they produce different results for warming trends since 1979 as compared with surface observations. The explanation for the difference lies, in part, in which levels of the atmosphere are being observed, and in the regional nature of the warming since the 1970s. Even after careful examination by the climatological community and a number of corrections, this discrepancy does, however, remain an unresolved issue in establishing the scale and causes of current global warming.

The most significant contribution of satellites is in monitoring the change in climatic parameters that were not observed before. Microwave radiometers now regularly measure the extent of snow cover in the northern hemisphere (see Figs 3.13 and 3.14) and sea ice in polar regions (Fig. 4.4). In the case of snow there are large short-term fluctuations and some evidence that snow cover in the northern hemisphere has declined in line with global warming. This is hardly surprising given the rise in temperature in the winter half of the year across the northern continents since the 1970s. What is not clear is whether the decline has acted as a positive feedback to amplify the warming or is merely a symptom of the stronger westerly circulation in the late 1980s and early 1990s (see Section 3.6). The continued decline in snow cover in April (see Fig. 3.14) is, however, seen as particularly significant.

In the case of sea ice, the measurements have transformed the monitoring of polar regions. The figures for the period November 1978 to December 2005 show that the extent of ice in the northern hemisphere has decreased by a rate of 2.9% a decade, whereas around Antarctica it has increased at a rate 1.3% a decade. The short duration of these records exposes the frustrations of drawing inferences about longer-term fluctuations. Estimates of earlier shifts in the extent of sea ice depend on sporadic observations from occasional scientific expeditions and the interpretation of other data such as the records of the whaling industry. In the Antarctic these records suggest that between late the 1950s and the early 1970s there was a decline of some 25% in the extent of the ice (a reduction in area of 5.65 million km^2), which far exceeds the increase in area that has occurred since around 1976 (see Fig. 4.7b below). While such observations provide valuable clues about longer-term trends, it is hard to be certain that they have captured an accurate picture of what was going on at the time. They do, however, show how essential it is to build up more extensive observations using satellites before jumping to conclusions about the significance of recent trends.

Since January 1998, microwave radiometers, onboard the Tropical Rainfall Monitoring Mission (TRMM) satellite, have been measuring rainfall between 35 °N and 35 °S. Although these observations will require careful calibration before any inferences can be drawn about trends in rainfall, this technology offers the possibility of monitoring precipitation on a global basis. Nevertheless detailed results of monthly global rainfall measurements are now being published and, over time, they will provide a much better indication of how regional and global rainfall patterns vary in the longer term.

Hailed as the ideal way for the measurement of cloud-cover, satellites have been slow to produce evidence of long-term changes. Surface measurements

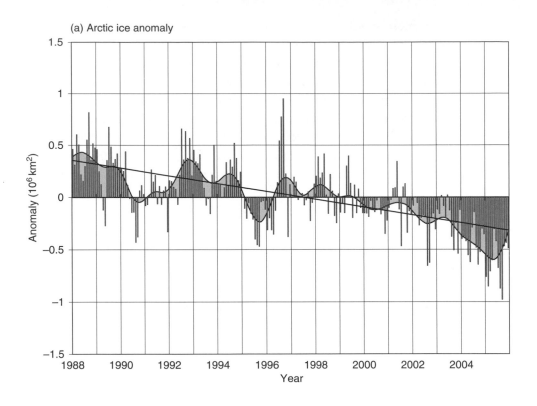

(a) Arctic ice anomaly

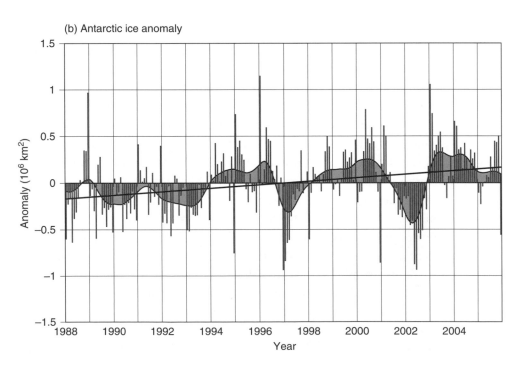

(b) Antarctic ice anomaly

over land suggest an increase of a few percent during the last 100 years. This increase is closely related to the decrease in diurnal temperature range that has been observed over the same period (see Section 8.10). Since the first weather satellite was launched in 1960, the images obtained have provided large amounts of information about clouds, but early attempts to produce analyses of changing cloud cover produced equivocal results. In part, this is a problem of drift as, during their lifetime, instruments became less sensitive to the amount of light reflected from clouds. In addition any analysis has to discriminate accurately between cloud types and their altitude and many observations are incapable of providing this information. Coordinated efforts since 1983 have, however, made considerable progress (see Section 3.3).

Where satellites have made a major contribution is in measuring the vegetation on land. These measurements have been of particular value in two areas. The first was in investigating the causes of desertification, notable in sub-Saharan Africa (the *Sahel*). The claimed human contribution to this environmental disaster became a major international issue in the 1970s (see Section 7.3). Satellite data collected since the 1970s suggests, however, that the central role of natural climate change had been underestimated. Many of the observed shifts in the desert were in fact largely due to annual fluctuations in rainfall. Satellite observations provided evidence of the re-establishment of vegetation in wetter years.

These results were supported by more detailed studies at local level, which showed an astonishing capacity for what looks like complete desert to spring to life when heavy rain falls and long-dormant seeds germinate. This undermines the implicit assumption of many environmentalists in the 1970s that not only were the changes the result of human activities, but also that they were irreversible. This latter point is central to much of the debate of climatic change. There is a risk of underestimating the capacity of many forms of life to adapt to sudden changes in the climate. We should not forget that they evolved through periods of far greater climatic variability than have been experienced in recorded history. This means that they contain genetic defences that enable some of their species to survive through a wide variety of extremes.

Fig. 4.4

Sea ice extent anomalies relative to 1988–2005 for (a) the northern hemisphere and (b) the southern hemisphere. Smooth lines generated using a 25-term binomial filter applied to the monthly anomalies to filter out periodicities of 12 months or less. (Data from NOAA NSIDC: http://nsidc.org/dat.)

Satellites have also been used to build up a comprehensive record of the vegetative cover of the Earth. Using the difference in reflected sunlight in the visible and infrared wavelength regions it is possible to obtain detailed observations of the amount and type of vegetation. Known as the *normalised difference vegetation index (NDVI)*, it is used to build maps of the extent of vegetation during the year and from year to year. Two measurements are of particular climatic interest. First, estimates have been made of the total primary production by land-based vegetation. Although there are substantial fluctuations, notably as a result of ENSO in the tropical Pacific (see Section 3.7), these measurements suggest a 6% increase in total productivity during the period 1982 to 1999. The largest increases have been in the tropics. Amazon rain forests accounted for 42% of the global increase. This increase in productivity appears to be a consequence of decreased cloud cover and the resulting increase in solar radiation.

The other interesting result concerning vegetation is the change in timing of the *green wave* of spring that moves northwards across the continents of the northern hemisphere. During the 1980s and 1990s the date of the start of spring advanced average by about 8 days in North America and 6 days in Eurasia. The decline in plant growth in autumn was delayed by 4 days in North America and 11 days in Eurasia. Therefore, the growing season is 18 days longer on average in Eurasia, compared with two decades ago, while in North America it has become 12 days longer (see Section 8.10).

4.3   Re-analysis work

A major climatological development in recent years has been the re-analysis of the data obtained in the production of weather forecasts since around 1950. This work has been conducted by the National Centers for Environmental Prediction (NCEP) and National Center for Atmospheric Research (NCAR) in the US and the European Centre for Medium Range Weather Forecasts (ECMWF). It has produced improved global analyses of atmospheric fields, which provide new climatic insights.

The process of re-analysis involves the recovery of land surface, ship, radiosonde, balloon, aircraft, satellite and other data. These data are then quality controlled and assimilated with a system that handles the data in a standard way over the re-analysis period. This eliminates the impact shifts over time associated with changes in the operational data assimilation system. The re-analysis is, however, still affected by changes in the observing systems (e.g. the early data have far fewer upper-air data observations, whilst the introduction of satellite temperature sounding into forecasting techniques was a major shift in the measurement system). So, the early re-analysis is less reliable than for the later 40 years.

The great benefit of this re-analysis work is that it exploits all the data collected over the years. This has led to the production of more reliable measurements of the behaviour of the global climate over the annual cycle and from year to year. These measurements are of particular value in defining the conditions around the globe at any given time of the year and giving improved measurements of the transport of energy and water vapour around the globe. When it comes to interpreting the trends observed in various climatic parameters much greater care is needed as the changes in the measurement techniques may have been a significant factor in subtle changes in the observed climate.

4.4   Historical records

Where instrumental observations do not exist, information on the climate can sometimes be extracted from historical records, including personal diaries, of both the weather and weather-related activities, but to make a significant contribution to quantitative measurements they must meet sterner tests. Nonetheless, in Europe and China detailed reports of many annual events can provide valuable additional insights. In particular, this is possible where historical records overlap instrumental observations and so can be calibrated against these numbers. Most frequently, these records relate to agricultural activity. For example, the price of wheat is a good indicator of the abundance of the harvest, as bumper years led to low prices and dearth produced high prices, but these figures relate only to the summer half of the year and are a complex product of temperature and rainfall. This means that, while the extreme years stand out, many of the fluctuations cannot be attributed to one or other variable. A better measure is the date at which grapes were picked to make wine – the *wine-harvest date*. This is a good measure of the temperature of the growing season (April to September).

The best-known example is of wine-harvest dates that has been built up from records kept in various parts of northern France and adjacent wine-growing regions by the French historian Emmanuel Le Roy Ladurie and co-workers. The latest set of these cover the period 1370 to 2003 for the Burgundy region, where officially decreed dates have been carefully registered in parish and municipal archives since at least the early thirteenth century. The extent to which they provide a direct measure of the

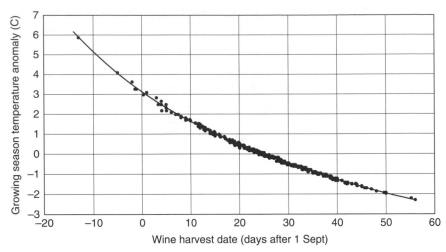

Fig. 4.5 The correlation between the date of the wine harvest in Burgundy between 1370 and 2003 compared with the temperature anomaly during the growing season (April to September) derived from a set of proxy records and temperature observations. (Data published to accompany Chuine *et al.*, 2004.)

temperature in the growing season is shown in Figure 4.5 by comparing annual harvest dates with temperature figures for April to September derived from a set of proxy (see Section 4.5) and instrumental records. Although the relationship is not linear, the correlation is remarkably high, and so they do provide an excellent guide to temperature in the summer half of each year (see also Section 8.9).

The limitation of many other historical records is that often they only note exceptional events. For example, the incidence of Frost Fairs on the Thames implies colder winters in London, but does not give a measure of how much colder they were in the seventeenth century. With a thorough examination of all the extremes recorded it is sometimes possible, however, to build up a more complete picture of notable changes. A good example of how this approach can produce a lot of information is the work of Christian Pfister on changes of temperature and rainfall in Switzerland since the early sixteenth century (Fig. 4.6). But in many cases the records are few and far between and then great care has to be exercised in not attaching too much importance to fluctuations in rare events. Only where there are sufficient observations to construct statistically significant series can these records be used to draw conclusions on climate change. Otherwise, their principal value is to be used in conjunction with proxy data to build up a more complete picture of change.

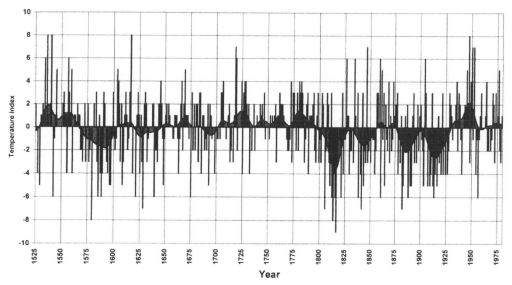

Fig. 4.6 The summer temperature index for Switzerland, together with smoothed data showing longer-term fluctuations. (Data from Pfister's paper in Bradley & Jones, 1995.) (From Burroughs, 1997, Fig. 2.9.)

4.5   Proxy measurements

Where there is neither instrumental observations nor historical records, we have to rely on indirect measurements of climate change. For the vast span of the Earth's history these are the only ways to find out how the climate has changed. Fortunately, there are a wide variety of techniques for approaching this task. Before examining these in more detail it is illuminating to establish the strengths and weaknesses of using indirect (*proxy*) data to infer climatic conditions in the past. The products of this analysis are often termed *palaeoclimatic data*.

The first point to emphasise is that most proxy data is rarely ever a direct measure of a single meteorological parameter. For instance, the width of tree-rings is a function of both temperature and rainfall over not only the growing season, but also on ground-water levels, which may relate to rainfall in earlier seasons. Only where the trees are growing near their climatic limit can most of the growth be attributed to a single parameter (e.g. summer temperature). Similarly, ice-core data, apart from the amount of snow accumulating in annual layers being a direct measure of precipitation, other properties are a measure of processes taking place far away from where the snow fell. As for many other records (e.g. analysis of the pollen content and the thickness of annual layers in lake sediments or the foramininfera deposited in ocean sediments), the climatic inferences that can be drawn from changes depend on subtle links between what is being measured and how it was influenced by climate change at the time it was laid down.

The second limitation is that some sources of information may be blurred by natural processes. For instance the resolution of events in deep ocean sediments is affected by three things; the slow rate of deposition; the fact this deposition rate may have varied with time; and the possibility that burrowing creatures may have mixed up the layers as they formed. All of these effects mean only changes on timescales longer than around a thousand years can be resolved by these measurements. If the consequence of any churning is to transport microfossils into different layers it can make the interpretation of mass extinctions (see Section 9.3) much more difficult. The transport of fossils downwards into older sediments, which predate some assumed extinction horizon (e.g. the Cretaceous–Tertiary boundary) is often termed *backwards smearing*, whereas the lifting of other fossils above their horizon is termed *forward smearing*. These processes make it much harder to tell whether relatively rapid changes in population levels were the product of a catastrophic event (see Section 6.9), or took place more gradually.

A third problem is that some records have major interruptions. Dating the organic debris in the terminal moraines of glaciers provides an insight into the timing of their major advances, but there is no way of knowing whether intermediate, less substantial advances occurred as all evidence of these are scrubbed out by later surges. Similarly, the early over-simplified model of the four major ice ages in the Pleistocene (see Section 8.4) was a result of the fact that the geological evidence of more rapid fluctuations had been erased in those parts of the world where the original work on these events was carried out.

Finally, in many cases, proxy measurements are only available in limited locations, which makes it difficult to draw generalised conclusions about change. Tree-ring or ice-core data can only be found in some particular parts of the world. Many fossil assemblages are the product of exceptional circumstances, which may well mean that the populations preserved are not representative of the conditions prevailing more widely at the time. Only in the case of ocean sediments do we have sufficient data to provide something approaching a truly global picture of certain types of change.

4.5.1   Tree rings

The response of trees to climatic fluctuations is recorded in the annual growth rings they produce. The most direct measure is in terms of the

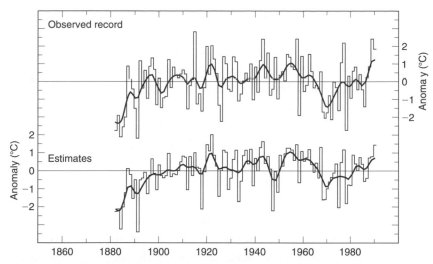

Fig. 4.7 Actual and estimated July/August temperatures for northern Fennoscandia, showing how the calibration of tree-ring widths against instrumental observations can be checked by reconstructing the temperature record on the basis of the established relationship. The estimates are based on the regression equation calibrated over the period 1876–1975. The smoothed curve shows 10-year filtered values. (With kind permission from K. Briffa, Climatic Research Unit, University of East Anglia.)

thickness of the rings, which not only provides insights into the climate, but also provides the basis of dating ancient trees (see Section 4.6). The challenge facing palaeoclimatologists is to establish what combination of temperature, rainfall and soil-moisture levels led to changes in ring thickness or changes in the wood structure within rings. This is achieved by comparing the observed ring widths in a locality for the period since standard meteorological measurements started in the area. Statistical analysis of the correlation between various meteorological parameters (e.g. summer temperatures, or annual rainfall) then establish how much of the variance can be attributed to different climatic factors. This calculation is then used to reconstruct the variation of the given parameter to check how effective the use of the tree-ring data is in representing the past (Fig. 4.7). On this basis, the established statistical link can be applied to earlier tree-ring data to establish how much the climate changed from year to year before instrumental observations were available.

Where trees are near their climatic limits of either rainfall or adequate summer warmth then it is possible to attribute most of the variability in ring width to one or other meteorological parameter. In many parts of the world ring-width fluctuations are, however, due to a combination of factors which make trees grow faster or slower than normal in any year. So it is only possible to draw more general conclusions about how a combination of weather conditions altered tree growth. Moreover, in tropical regions, where there is no pronounced annual cycle in growth, it is not possible to obtain information about climate change.

More detailed information can be obtained from analysis of the structure of the wood within individual rings. Such parameters as maximum late-wood density, minimum early-wood density and width of early- and late-wood growth can provide insight into weather fluctuations within individual seasons. For instance, stunted early growth could indicate late cold springs while lack of late growth could reflect cool, short summers. Also the isotope ratios of the wood (see Box 4.1) should provide additional information about the environmental conditions at the time of growth. In practice, in all these areas only relatively little work has been done and, in a number of cases, has produced equivocal results. So, although this type of investigation has considerable potential, more research is needed before any of these techniques can become a larger part of the standard armoury of climatologists. One area that has, however, developed well is the detection of evidence of frost damage in late-spring growth, which is evidence of an exceptional cold spell.

One other aspect of tree-ring studies requires careful thought. This is the fact that all analyses have to take account of the fact that as trees mature their ring widths naturally decrease. In the case of conifers this change tends to be exponential, but in broad-leaved species (e.g. oaks) it is a more complicated process. Standard practice is to calibrate growth curves for given species and then analyse the annual fluctuations about the normal. Where this process involves overlapping of series of tree samples, the age of which is not known, this standardisation process tends to iron out longer-term fluctuations and hence may produce misleading information about climatic variability. Often known as the 'segment length curse', this process can lead to tree-ring series having a very stable appearance over the longer term (Fig. 4.7), and seriously underestimating the climatic variance on time-scales of 50 years and longer. The only way to avoid this pitfall is to restrict the analysis to wood of known age, but in the case of ancient timbers this rules out many samples which have no indication of age of the tree from which it was hewn and so greatly reduces the scope of climatic studies.

4.5.2   Ice cores

Snow deposited on the ice sheets of Antarctica and Greenland, and in glaciers in mountain ranges around the world, contains valuable

Box 4.1 Information from isotope ratios

Many elements exist in a number of stable isotopic forms, each having a different atomic weight. This weight difference between isotopes (e.g. hydrogen [H] and deuterium [D], carbon-12 [^{12}C] and carbon-13 [^{13}C] and oxygen-16 [^{16}O] and oxygen-18 [^{18}O]) alters the physical properties of molecules containing different isotopes. For instance, water vapour molecules containing deuterium (HD^{16}O) will evaporate a little less readily than normal water (H$_2{}^{16}$O), as will water containing ^{18}O (H$_2{}^{18}$O). Similarly, the rates at which different isotopes of hydrogen are taken up in the formation of wood are likely to be dependent on the temperature. This means that measuring the ratio of certain isotopes in, say, ice cores, ocean sediments and tree rings can provide information on the physical conditions at the time these materials were formed.

In the case of isotopic fractionation in natural water, three principal molecular species (H$_2{}^{16}$O, HD^{16}O and H$_2{}^{18}$O) matter most. In order to describe completely the processes involved, all three species must be considered: so both the difference of deuterium (δD) and of oxygen-18 (δ^{18}O) in any sample from normal is measured. Both may be expressed as deviations per mil (‰) from the SMOW (Standard Mean Ocean Water), where

$$\delta D = 1000 \left\{ \frac{HD^{16}O/H_2{}^{16}O_{sample}}{HD^{16}O/H_2{}^{16}O_{SMOW}} \right\} - 1$$

and

$$\delta^{18}O = 1000 \left\{ \frac{H_2{}^{18}O/H_2{}^{16}O_{sample}}{H_2{}^{18}O/H_2{}^{16}O_{SMOW}} \right\} - 1.$$

Expressing these deviations in terms of ‰ produces manageable units as H$_2{}^{16}$O makes up 99.76% of SMOW, while H$_2{}^{18}$O constitutes 0.2% and HDO makes up 0.03%. Most of the water vapour precipitated on the ice sheets of Antarctica and Greenland originates from the oceans in mid-latitudes. The process of evaporation depletes the vapour of about 10‰ of H$_2{}^{18}$O. Then as the vapour is precipitated as rain or snow it is further depleted of H$_2{}^{18}$O because the isotope condenses out preferentially. Although some additional water vapour will be added from the oceans at higher latitudes, the remaining vapour is increasingly depleted as it rises over the ice sheets, and the value of δ^{18}O in the snow is a measure of the temperature at which it precipitated. For snow falling on the Greenland ice sheet δ^{18}O typically ranges from -23 to -38‰ and for

Box 4.1 (cont.)

Antarctica the range is from around -18 to -60‰. The process for depletion of HDO is the same and the value of δD is given by the expression

$\delta D = 8\delta^{18}O + 10$‰.

Because the value of $\delta^{18}O$ or δD in ice cores is related to the route by which the vapour followed, it is not a precise measure of the temperature where the snow fell. So, while changes to these parameters provide a good indication of how temperature has changed over time, they are not an absolute measurement and so must be treated with care. This challenge of identifying which physical factor was responsible for the observed change in isotope ratios assumes greater significance with other measurements. The changes in the $^{16}O/^{18}O$ ratio in the shells of benthic foraminifera reflect how amount of these isotopes locked up in the ice caps fluctuated between glacial and interglacial periods. Because the ice will be depleted in $H_2^{18}O$, the levels of this molecular species in the oceans will increase as the ice volume increases. This is a measure of ice volume that can be translated into an indication of global temperatures. But when it comes to measuring the H/D, $^{12}C/^{13}C$, or $^{18}O/^{16}O$ ratios in tree rings and other organic material, the temperature relationship is more complicated. Although there appear to be identifiable temperature relationships between the observed isotope ratios and ambient temperatures, there is considerable scientific argument as to what precisely is being measured.

There is one important additional feature of the $^{12}C/^{13}C$ ratio. This relates to how these two isotopes are fractionated in inorganic and organic reactions. In general, carbonates are enriched in ^{13}C, while organic matter is enriched in ^{12}C. The difference of the $^{12}C/^{13}C$ ratio of a specimen and that of a standard, expressed in parts per thousand, is defined as $\delta^{13}C$. This ratio gives an idea of the magnitude of the biomass present on the Earth, as recorded in sediment of a given age, and hence, a measure of climatic conditions at the time.

information about the climate at the time it fell. Where there is no appreciable melting in summer, the accumulation of snow, which is compressed to form ice, contains a continuous record of various aspects of climate variability and climate change. This includes evidence of changes in temperature, the amount of snow that fell each year, the amount of dust transported from lower latitudes, fall-out from major volcanoes and the composition of air bubbles trapped in the ice. The best results are, however, restricted to Antarctica and Greenland, with more limited results from glaciers and ice caps elsewhere around the world. Hence many features of regional climate change are not covered.

The most direct measure of the climate is obtained from the isotopic composition of the water molecules in the ice, and, in particular, the ratio of the stable isotopes of oxygen (^{16}O and ^{18}O) (see Box 4.2.). By measuring this ratio it is possible to observe both the annual cycle in the temperature and longer-term changes in temperature (Fig. 4.8). By counting in three ice core records, which were linked by volcanic reference horizons, have been used for the annual layer counting of the most recent 8000 years of the Holocene, while measurements of chemical impurities are used for the dating of the early part of the Holocene. The maximum counting error is 0.5% or less for the past 6900 years, increasing to 2.0% in some of the older sections of the timescale.

In Antarctica where the rate of snow accumulation is much slower, it is not possible to measure annual layers for more than a few hundred years, and most dating depends on the behaviour of the ice sheet. This does not cause major problems for dating back to around 100 kya for the best sites near the centre of the ice sheet in Greenland, while in east Antarctica a record of changes over more than 800 kyr has been built up (Fig. 4.9). As the core approaches bedrock, however, the risk of the ice being distorted by the topography of the bedrock in its flow outwards from the centre of the ice sheet can cause major discontinuities in the layers and make the interpretation of change more difficult, if not impossible.

The dust content of the ice provides a measure of atmospheric circulation. Stronger winds in mid-latitudes increase the amount of continental dust stirred up and transported to high latitudes. The changes in the amount of dust fluctuate sharply with shifts in temperature, showing that the variations in isotope ratios are also reflected in switches in atmospheric circulation patterns.

Measurements of the acidity of different layers of the ice core provide clear evidence of major volcanoes. As the most climatically important volcanoes are those that inject large quantities of sulphur compounds into the stratosphere to form long-lasting sulphuric acid aerosols (see Section 6.4), the acidity of precipitation at high latitudes is a good measure of the climatic impact of these eruptions (Fig. 4.10).

The analysis of the trace constituents of the atmosphere in the air bubbles trapped in the ice provides a direct sample of the amount of each gas

Box 4.2 Extracting ocean temperatures from sediment records

In principle, the amount of oxygen-18 ($\delta^{18}O$) in foraminifera in ocean sediments provides a measure of water temperature at the time of formation. But, as noted in Box 4.1, when there are major fluctuations in the amount of ice locked up in the great ice sheets of the world, using $\delta^{18}O$ levels to infer ocean temperatures is complicated by the fact that much of the $\delta^{18}O$-change is defined by the amount of ice locked up in the ice sheets. Apart from the types of foraminifera found in the sediments and their isotope ratios, other measurements include the ratios of various elements in their skeletons (e.g. Sr/Ca, Cd/Ca and Mg/Ca), which provide information about ocean temperatures. Then the levels of certain types of organic chemicals, known as *alkenones*, formed from the decay of dead algae that lived in the surface waters, are an independent measure of SSTs.

Mg/Ca ratios of benthic foraminiferal tests have become an increasingly valuable proxy to reconstruct past temperatures. In conjunction with foraminiferal $\delta^{18}O$ values it can be used to assess changes in the $\delta^{18}O$ of seawater through time. The basis for this method of thermometry lies in the laboratory experiments which show that the partition coefficient of Mg_2^+ ions into inorganic calcite strongly correlate with temperature, and in empirical and culture studies of temperature-dependence of Mg in marine biogenic calcite.

An additional source of independent information about SSTs can be obtained from the levels of certain types of organic chemicals, known as *alkenones*, formed from the decay of dead algae that lived in the surface waters. Alkenones are long-chained ketones synthesised by a limited number of algal species; their unsaturation ratio (determined by gas chromatography) provides good estimates of the growth temperature. Because these organisms must live in the photic zone, their growth temperature is closely tied to SST. The alkenone method sidesteps ambiguities inherent in using foraminiferal $\delta^{18}O$ and/or faunal assemblages as paleotemperature proxies, as both can be influenced by variables other than near-surface temperature.

present in the atmosphere when the snow fell. This gives a history of the past fluctuations of trace gases such as carbon dioxide (CO_2) and methane (CH_4) (Fig. 4.11), and is an indication of how these changes may have contributed to past climate change (see Sections 2.1.3. and chapter 7).

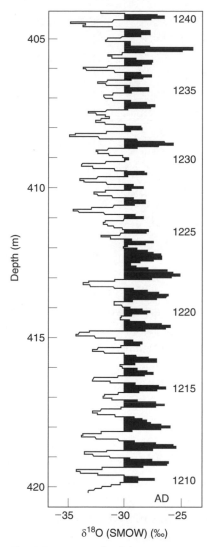

Fig. 4.8 An example of the measurement of the $^{18}O/^{16}O$ isotope ratio in an ice core from the Greenland ice sheet, showing the annual layers. (Source: Wigley, Ingram & Farmer, 1981, Fig 5.5.)

4.5.3   Ocean sediments

The sediments that collect on the bed of the deep oceans provide one of the most valuable sources of information on longer-term climate change. Because this ooze is formed largely of the bodies of the fossil shells of both pelagic and benthic foraminiferal species, it provides a measure of the conditions when these tiny creatures were living in either the surface or deep waters of the oceans. So taking a core down through this sediment, which has collected over

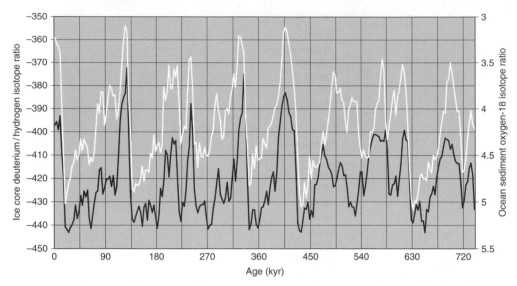

Fig. 4.9 A comparison of ice-core and ocean-sediment records over the last 736 kyr (an increase in isotope ratio means an increase in temperature. The black line shows the change in the deuterium/hydrogen ratio measured in the EPICA ice core drilled at Dome C in Antarctica (data from EPICA Community Members (2004), supplementary information, www.nature.com/nature) and the white line show the proportion of oxygen-18 (δ^{18}O) in the foraminifera sampled in 57 ocean-sediment records (from Lisiecki & Raymo, 2005).

hundreds of thousands of years, provides a record of the species living above a spot in the past. The fact that these sediments occur across the major ocean basins means that it is possible to build up a picture of global changes drawing on a substantial ocean-drilling programme over the last three decades.

Two principal factors are used in the analysis of cores. First, the types of species living in the surface at any given time and their relative abundance is a guide to surface temperatures – whether the various species only survive in warm or cold water is a direct measure of the conditions when they were alive. So by mapping the populations of different creatures at different times using cores taken from around the world, it is possible to construct a picture of how the temperature of the oceans' surface waters varied over time. The second factor is the ratio of oxygen isotopes in the calcium carbonate in the skeletons of the foraminifera living in the deep water. Changes in this ratio are a longer-term consequence of the fractionation process described in Box 4.2. As the ice sheets in the northern hemisphere grew, the amount of ^{16}O locked in the ice was proportionately greater than the amount of ^{18}O. So the ratio of these oxygen isotopes in

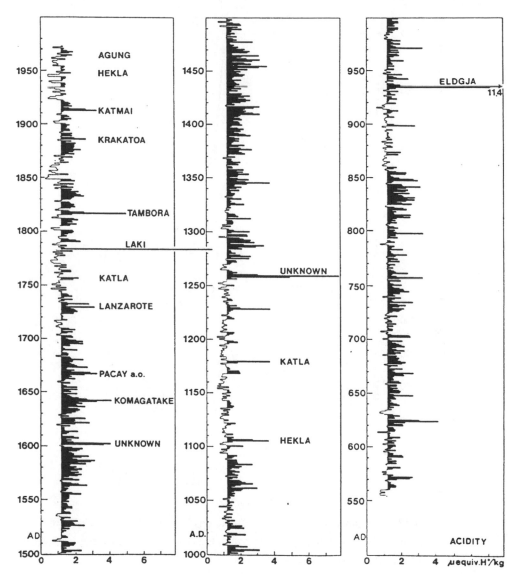

Fig. 4.10 Measurements of the acidity of a Greenland ice core showing the peaks that occur after a major volcanic eruption. (From Hammer *et al.*, 1980. With permission from Macmillan Magazines Ltd.)

the oceans changed as the amount of ice rose and fell, and these variations were reflected in the shells of the foraminifera which formed using oxygen in the oceans, whether they were living close to the surface or deep down. This means that changes in the $^{16}O/^{18}O$ ratio is a measure of the size of the ice sheets at high latitudes, and hence indirectly of fluctuations in global temperatures.

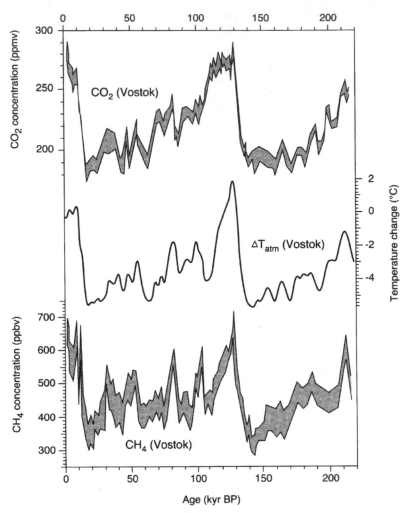

Fig. 4.11 A comparison between the CO_2 and CH_4 concentrations in an ice core obtained at Vostok, Antarctica, and the estimated changes in temperature over the last 220 kyr. (From IPCC, 1994, Fig. 1.6.)

The dating of these sediments is a more complicated issue. Carbon dating can handle the last 40 kyr or so, and in the longer term magnetic reversals can act as markers. But for the climatically interesting period covering recent ice ages dating depends initially on assumptions about sedimentation rates. Climatologists are, however, fortunate in this context as the data from a large number of cores shows beyond peradventure that changes on timescales from 10 to 100 kyr are dominated by the Earth's orbital parameters (see Section 6.7). This means that while the assumption of constant sedimentation rates is a reasonable first assumption for many cores, once the orbital link has been established, it is possible to refine the timescale to fit the orbital parameters more closely. This 'tune up' process has led to an agreed dating for the major changes in ocean isotopic ratios. The latest example of this process was published by Lorraine Lisiecki and Maureen Raymo, at Boston University, in 2005. It has been constructed from 57 globally distributed cores and provides a measure of the major fluctuations in ice-sheet volumes over the last 5.3 Myr (see also Section 8.2).

Comparison of ice-core and ocean-sediment records (see Fig. 4.9) provides an indication how well this calibration process works. The correlation between the changes in the isotope ration of snow that fell on the Antarctic ice sheet over the last 738 kyr and an ocean-sediment isotope ratio record created by the of average of 57 benthic $\delta^{18}O$ records from globally distributed sites is striking. This clearly shows that these different measurement techniques reinforce one another and together provide a coherent picture of global climate change for over 700 kyr.

4.5.4   Pollen records

Where sediments have formed in shallow lakes or bogs, the presence of pollen grains can provide similar information on land, as foraminifera provide in the oceans. If these sediments have been laid down in a regular manner, the abundance of pollen from different species of trees and shrubs provide details of climate change. Because different species have distinct climatic ranges, it is possible to interpret their relative abundance in terms of shifts in the local climate. Given that this process can occur almost anywhere on the continents, it means that pollen records have the potential to fill many of the geographical gaps that ice cores and ocean-sediment records cannot cover.

Pollen records are of particular historical importance in studying climatic change because of the early work charting the emergence of northerly latitudes from the last glaciation (see Section 8.5). Most pollens (from flowering trees and plants) and spores (principally ferns and mosses) are tiny. Few exceed 100 μm (0.1 mm) in diameter and the majority are around 30 μm. It is the outer portion of the cell (*exine*) that is preserved by means of a waxy coat of the material *sporopollenin*. The size and shape of this outer wall, along with the number and distribution of apertures in it, are specific to different species and can be readily identified. A typical pollen diagram (Fig. 4.12) will plot the proportion of the principal species found in a stratigraphic sequence and can be used to draw inference about how the climate has changed over the period covered by the record.

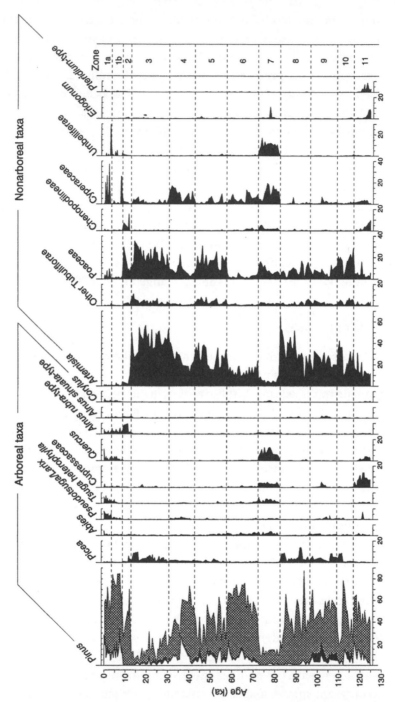

Fig. 4.12 An example of the information that can be extracted from pollen stratigraphy, using a core take from the bed of Carp Lake in the Cascade Range of north-western USA. (From Whitlock and Bartelein, 1997. With permission from Macmillan Magazines Ltd.)

In spite of their potential, most climate-change studies using pollen records have focused on alterations in regional vegetation since the end of the last ice age. This work has established that these shifts are controlled by broad global patterns of climate change. For longer-term changes there was until recently greater uncertainty as to whether other factors involving the migration and competition of species may have played a more significant role. There are two principal explanations for this uncertainty. First, almost all the longer pollen records related to northern Europe and hence there were doubts about whether the results were representative of global changes. Secondly, many of the cores contained high frequency variations during the last glaciation. These raised doubts about the dating of the strata in the cores and were not readily explained until more recent ice core and ocean sediment data became available (see Fig. 8.14). Recent pollen results obtained from cores drilled in Carp Lake in the Cascade Range in north-west USA (see Fig. 4.12) have confirmed that pollen records can be accurately dated back to the last interglacial, some 125 kya, and provide detailed information about how the climate has changed. Furthermore, the close correlation between these results and those from cores from lakes in Europe, and variations observed in ice cores and ocean sediments, means that future research into pollen records will play an increasing part in identifying global patterns of climate change during the last ice age.

4.5.5   Boreholes

Holes drilled deep into the ground offer a completely different, and potentially valuable, approach to estimating past climates. By making measurements down boreholes, it is possible to relate profiles of rock temperatures with depth to the history of temperature change at the surface. With certain assumptions this change can be converted to the changes in air temperature. In principle, by making measurements to depths of a kilometre or more, this can provide insights into local temperature variation over time intervals of a few hundred years to over a millennium.

Because borehole are drilled in so many parts of the world, temperature–depth profiles from some 1000 boreholes have been used for climate research. The geographic coverage is most dense in North America and Europe, but substantial data have also been obtained from Australia, Asia, Africa, and South America, and new borehole measurements are continually being obtained.

The calculation of temperature records shows substantial sensitivity to the assumptions made to convert the core temperature profiles to atmospheric temperature changes. The usual initial assumption employed, when setting up the mathematical procedures to convert borehole temperature profiles to temperature changes at the surface, is that surface temperature has remained constant. Then, using certain simplifying assumptions about the geothermal properties of the earth in the area of the core to calculate what the temperature profile should be under isothermal conditions, the deviations from this expected profile can be translated into past changes in the surface temperature.

A number of additional factors do, however, complicate the interpretation of these measurements. First, the temporal resolution decreases as one goes back in time. Secondly, long-term variations in winter snow cover and in soil moisture change the sub-surface thermal properties of the ground. In addition, non-climatic factors such as land-usage changes and natural land cover variations can affect the response to surface temperature. For instance, land-usage changes may explain why some borehole estimates of warming in North America are 1–2 °C greater than instrumental records of local air temperatures show to be the case.

4.5.6   Speleothems

Wherever a natural process lays down long-term deposits of material at an approximately constant rate, there is the prospect that some climatic information will be recorded. The deposition of calcium carbonate encrustations by running water in caves (*speleothems*), better known as stalagmites and stalactites, record changes in the isotopic ratios in precipitation. So, where such encrustations build up over a long time, they have the potential to provide useful climatic information. An increasing number of high quality speleothems have been obtained from caves around the world. These are producing detailed information about climatic conditions throughout the last ice age and into the Holocene.

The particular value of speleothems is the stable isotopic content of the calcite. This provides not only a record of variations in temperature and other climatic parameters but also an absolute timescale using uranium-series dating, which can be compared with other proxy measurements. In addition, their isotope ratios (e.g. $^{16}O/^{18}O$) provide a measure of temperature and it is even possible to identify types of pollen trapped in the calcite. Their limitation is that often the amount of running water creating them may vary appreciably over time, and in extreme cases cease altogether. This is particularly true of lengthy records. For instance, during the ice ages where permafrost formed above the cave the record could be interrupted for thousands of years.

Recently, an increasing number of speleothems have been studied to provide insight in climate change over the last millennium or two. These offer the possibility of obtaining an independent measurement of climate change, and in some cases have clear annual rings in their formation. In particular, they are more likely to record centennial and longer-term fluctuations. This is valuable in forming an accurate picture of climate change over timescales from a few years to many millennia.

4.5.7   Corals

Another burgeoning area of proxy records comes from the tropics. Here there are sufficient differences in the temperature and sunshine levels at various times of the year for certain forms of coral (e.g. the star coral *Pavona clavus*) to produce seasonal growth rings. The coral grows faster in warm sunny conditions than when is cooler and more overcast. The resulting bands of alternating density can be measure using X-ray techniques and used to draw conclusions about changes in seasonal water temperature. In addition, the changes in the $^{16}O/^{18}O$ ratio in the coral are another measure of the temperature at the time of formation. Using these techniques measurements of temperature fluctuations in the Galápagos Islands have been obtained back to AD 1600, which provide an independent record of ENSO events since then.

Corals can also record variations of freshwater run-off because they may extract and co-deposit river-borne organic compounds in their annual growth rings. The amount of organic material from the freshwater can be measured by recording the fluorescence in the yellow–green part of the visible spectrum when the coral is illuminated with ultraviolet (UV) light. Studies inside the Great Barrier Reef off the coast of Queensland, Australia, have shown that the large slow-growing (5–25 mm per year) corals of the genus *Porites*, which may live for up to a 1000 years and grow to 10 m across, provide an accurate record of the amount of run-off from local rivers. This is a measure of rainfall in tropical Australia, which provides further evidence of ENSO events. This technique has also been used to examine ancient exposed coral reefs associated with past high sea levels.

4.6   Dating

Clearly, dating proxy records is no easy task. Only where it is possible to build up on unbroken sequence of measurements to the present is it possible to create an absolute chronology. For the rest, there are two principal limitations. First, where there are gaps in the record (e.g. lack of overlapping tree-ring series), some way has to be found to link the measurements with a reliable timescale. Secondly, where there is some physical means of measuring the age of the sample (e.g. carbon dating: see Box 4.3) there are questions of both the accuracy of the technique and the problems of contamination.

The best examples of creating an absolute timescale are achieved with tree rings and ice cores. In the case of ice cores, this is a matter of counting the layers in any given core, but with tree rings it is possible to build up chronologies using different samples that have grown in a given location. Because the annual growth of trees is directly related to fluctuations in the weather, and these are never the same over a long run of years, the variation of tree-ring widths is a unique record of how any tree has responded to local climatic fluctuations in its lifetime. Moreover, because anomalous weather patterns extend over considerable areas, it is possible to use trees from up to several hundred kilometres apart to build up a chronology. So by sampling living trees in a region a series can be extended back several hundred years, or with certain trees much longer (e.g. the Bristlecone pine – *Pinus longaeva* – in the White Mountains of California, which can live several thousand years Fig. 4.13). Records can then be extended back using timber from trees that grew in the locality and were felled long ago and either preserved in buildings and artefacts, or fossilised. By overlapping these samples with each other, and in the case of relatively recently felled timber with living samples, it has been possible to produce chronologies extending back between 8000 and 9000 years in the case of European oaks, and 8000 years in the case of Bristlecone pines. What is more, by using a large number of trees, it is possible to overcome problems of missing or very narrow rings in single samples that experienced particular stress in some years.

Beyond the range of tree rings, ^{14}C-ages have been estimated using a variety of records including annually laminated lacustrine deposits (varves), where the unprecedented length of the sequence from Lake Suigetsu in Japan has the potential to extend the calibration back to around 45 kya (see Box 4.4). Another prolific source of information is from the Cariaco Basin, off the coast of Venezuela. Here rapidly deposited (~3 mm per decade) organic-rich sediments contain visible annual laminations. What is more they are devoid of preserved benthic faunas. This means the sediments were laid down in anoxic conditions, so there were no burrowing creatures scrambling the evidence. These deposits are composed of light (plankton-rich)

Box 4.3 Radioactive isotope dating

In principle the age of any sample containing a radioactive isotope can be calculated on the assumption that at some point in time the amount of the isotope was fixed. The most direct approach is to estimate the amount of the isotope present now (Q) as a proportion of the amount originally present (Q_o), as the relationship between the two quantities is given by the expression:

$$Q = Q_o \exp(-t/\tau),$$

where Q is the quantity present at time t

Q_o is the quantity present at time $t = 0$, and

τ is the average life expectancy of the isotope and is the time taken for the amount of the isotope to fall to $1/e$ of its original quantity, where e is the natural logarithm (2.718).

The lifetime of the isotope (τ) is related to the half-life ($t_{1/2}$) by the expression

$$t_{\frac{1}{2}} = 0 \cdot 693\tau.$$

In the case of ^{14}C the value of $t_{1/2}$ is 5730 years and so τ is 8267 years, and so the age of a sample is given by the expression

$$t = -8267 \ln(Q/Q_o),$$

where ln is the natural logarithm.

So, providing we know what Q_o was when the sample was formed, measurement of Q will give a direct measurement of its age. In the case of radiocarbon dating this analysis is limited to values of t less than about 40 000 years, because beyond this the amount of ^{14}C even in a large sample becomes very difficult to measure. Other isotopic combinations can, however, be used to date much older samples. For example, the use of the decay of potassium-40 to argon-40, which is often used to date the solidification of molten lava following volcanic activity, has a half-life of 1250 Myr. This much slower decay process can be used to date samples extending back over much greater lengths of time but, because of the sensitivity of mass spectrometry to detecting ^{40}A, it is also possible to date eruptions as recent as some 30 kya.

Fig. 4.13 Bristlecone pine (*Pinus longaeva*), native to mountainous areas in the southwestern USA. These trees grow at altitudes above about 3000 m. Because of their extreme longevity and the fact that they live near the limit of their climatic tolerance, the are uniquely important in dendroclimatological studies.

and dark (mineral-rich) layers. This provides insights into what was going on in the surface waters as that result from strong seasonal fluctuations in trade-wind-induced upwelling and regional precipitation. The mineral sediments come from the surrounding watersheds and provide an accurate picture of regional hydrologic conditions. Cores from the Cariaco Basin have been used to provide detailed analysis of interannual climate change over the last 15 kyr, and of decadal fluctuations over the last 500 kyr.

In the longer term the problems of dating become much greater. These challenges were discussed in terms of ocean sediments earlier (see Section 4.5.3.) One valuable tool in addressing this issue is to measure the residual magnetic field in the sediments. Because the magnetic polarity of the Earth's magnetic field reverses from time to time, this can leave an indelible record in rocks laid down at the time. On the scale of geological time reversals are common, occurring irregularly, but on average about every 700 000 years. Indeed, the last reversal occurred 700 000 years ago, and this marker, together with earlier reversals during the last 5 million years, which have been dated with considerable accuracy, are central to the accurate dating of ocean sediments.

4.7   Isotope age dating

Dating rocks using their radioactive content is a simple concept. Certain widely distributed elements (e.g. carbon, potassium, thorium and uranium) occur as radioactive isotopes as well as stable isotopes. Radioactive isotopes decay (uranium isotopes decay to lead, potassium -40 (^{40}K) to argon -40 (^{40}A) and carbon -14 (^{14}C) decays to nitrogen -14 (^{14}N)). The rate of decay is characteristic for each isotope, which cannot be changed by any known force, and is defined as the *half-life* – the time taken for half the number of atoms originally present to decay to daughter atoms. So providing we can assume that only the parent radioactive isotope was present when the sediments were laid down, and neither the parent or the daughter elements have been lost or gained since the mineral was formed, it is possible to calculate its age (see Box 4.3). If the daughter element leaks away the date will be too young; if the parent leaks away the date will be too old.

Carbon-14 (^{14}C) is different. It has not been present since the Earth formed. Indeed it is created continuously in the upper atmosphere by collisions between cosmic rays and the nuclei of nitrogen. The ^{14}C created diffuses through the atmosphere as carbon dioxide (CO_2), is dissolved in the oceans, converted by plants into organic matter, and ingested by animals. It is capable of decaying to ^{14}N as soon as it is formed, but in living organisms an equilibrium is maintained with the levels in the atmosphere or the oceans. Once the organisms form lasting tissue or when they die, however, the radioactive decay of ^{14}C continues without replenishment. So, providing the amount of ^{14}C in oceans and atmosphere has remained constant it is possible to use the residual radioactivity in samples of dead tissue, wood or shell to determine their age (see Box 4.4).

The importance of carbon dating for climate studies is in examining organic samples, including some tree ring series and peat deposits, which are not part of a continuous record. These can be dated directly using radiocarbon measurements. The accuracy of dating depends on both the technology of both measuring and calibrating carbon-dating systems, and ensuring that the samples are not contaminated by extraneous sources of carbon of a different age.

One other form of isotope dating, which is growing in climatic importance, uses the ratio of ^{230}Th to ^{234}U in the series of decay products from uranium to lead. This method depends on finding materials which form a *closed system* in which the amount of uranium incorporated at some given

Box 4.4 Accuracy of Carbon Dating

Because the half-life of ^{14}C is 5 730 years (Box 4.2), the carbon-14 atoms in any sample of material will decay at an immutable rate of 1% every 83 years. So, at the most basic level, the dating of a sample depends on measuring either the rate at which ^{14}C atoms are decaying, or the proportion of ^{14}C atoms remaining in the sample. If the rate of decay is measured, then to obtain an accuracy of $\pm 0.5\%$ in the number of atoms present would require the observation of 40 000 decaying carbon atoms. This would, in principle, allow the sample to be dated to an accuracy of ± 40 years and would require a sample of several grams to be measured for about a day. If, however, the proportion of carbon-14 to carbon-12 atoms in a sample is measured directly in a mass spectrometer the same accuracy can be achieved in a few hours using a sample of a few milligrams.

In practice, the absolute measurement of the proportion of ^{14}C present in the sample is only one of a number of sources of error in carbon dating. Most significant among these is the fact that the amount of ^{14}C created in the atmosphere has varied as a consequence of changes in the Sun's magnetic field (see Section 8.5). These fluctuations in ^{14}C production have been estimated on the basis of measurements of tree rings. These measurements have produced a calibration curve which compares the actual age of the tree rings with that inferred by measuring the proportion of ^{14}C remaining in a ring of a given age.

These curves show two important effects. First, over much of the calibration period, the production of ^{14}C was higher than in the recent past and so the carbon-dating process would, in the absence of the correction obtained from the calibration curve, underestimate the age of the sample. Second, the production of ^{14}C clearly fluctuated on time-scales of a century or more. These fluctuations (often termed *de Vries wiggles* after the scientist who first studied these variations in the calibration curve) have to be compensated for in any dating exercise. What is particularly interesting is that the most marked wiggles appear to coincide with periods of sudden climatic change, which has raised the question of a solar influence on the climate.

Beyond the range of tree rings ^{14}C-ages have been estimated using the annually laminated lacustrine deposits (varves) from Lake Suigetsu in Japan, which have the potential to extend the calibration back to around 45 kya. Thus far international collaboration have produced an agreed calibration (INTCAL98) that can be used to convert radiocarbon measurements into calendar years before the present back to 24 kya (see Fig. B4.1). Allowing for the wiggles in the ^{14}C-curve (Fig. B4.1) the two start to diverge before 2 kya. By 10 kya the ^{14}C-age underestimates the real age by 1100 years, and by 15 kya the gap has risen to 2.5 kyr, it widens more gradually to reach nearly 4 kyr at 24 kya. For dates that fall in the range 24 to 45 kya, results obtained from the Lake Suigetsu work provide a measure of how to convert radiocarbon measurements into calendar years. These figures bring the ^{14}C-age and the real age almost back together around 34 kya and then part a little, so that between 36 and 45 kya a standard correction of 2 kyr needs to be added to any radiocarbon date.

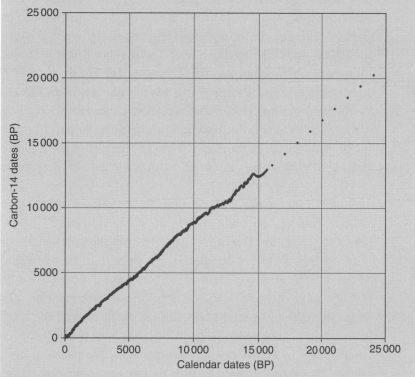

Fig. B4.1 The comparison of radiocarbon dates and calendar dates using the INTCAL98 figures back to 24 kya, showing how radiocarbon dating underestimates the antiquity of organic material with increasing age.

Other sources of error are more difficult to identify. Because carbon dating depends on the sample dying or being cut down at some given time we assume we are looking only at the remnants of the ^{14}C that was in the sample at this instant. Any addition of carbon from other sources (e.g. other organic material and, in particular, any recent material which

> **Box 4.4** (cont.)
>
> could be affected by such human activities as the combustion of fossil fuels and nuclear testing) will introduce errors. Subtler is where plants have been growing in hard water with particularly high concentrations of calcium carbonate that is depleted in ^{14}C.

time is known, and there is no external source of thorium. Living coral absorbs about 3 parts per million of uranium from sea-water, and this proportion has not changed appreciably over the timescale for which corals are used in climatic studies. There is negligible incorporation of thorium into the coral. So the ratio of ^{230}Th/ ^{234}U is a measure of when the coral was formed. This calculation is, however, complicated by ^{234}U having a half-life of 245 kyr and ^{230}Th having one of 75.4 kyr, but this is manageable providing there was no thorium present when the coral was formed. In these circumstances, the thorium will eventually reach an equilibrium level with the uranium. The rate at which the concentration of the daughter product (^{230} Th) approaches equilibrium with the parent (^{234}U) is defined by the half-life of the thorium (75.4 kyr). This technique is particularly well-suited to dating coral stands left by high sea levels during interglacials (see Section 4.4) in the last 250 kyr or so.

Finally, there is one other climatically interesting isotope dating method, which relies on the ratio of the two stable isotopes ^{86}Sr and ^{87}Sr of the element strontium. The isotopes derive from the weathering of continental and oceanic rock and are found everywhere in ocean water in minute quantities. Solutions leached out of continental rocks have a high ^{87}Sr/^{86}Sr ratio, whereas oceanic islands and volcanoes have a low one. The contributions from the two sources have altered over time and the ^{87}Sr/^{86}Sr ratio of ocean water has varied with them. So it has been possible to measure, using rocks dated by other means, how this ratio has changed over time. This agreed timescale can then be used to date calcareous fossils because strontium is chemically similar to calcium and is stored in these shells. The timescale based on the ^{87}Sr/^{86}Sr ratio is used to date fossils laid down over the last 65 Myr.

4.8   Summary

Exploiting a wide range of accurate measurements is the only way to build up a coherent picture of climate change. Even so there are substantial gaps in all aspects of our knowledge, and the further we go back in time the greater these become. New technologies (e.g. satellite radiometers) can improve the monitoring of current conditions or the precision of exploiting sources of palaeodata (e.g. accelerator mass spectrometers for improved carbon dating). New sources of past climatic information will be found to fill some of the gaps. This will involve extending the available proxy records of tree rings, ice cores, pollen and ocean sediments and by developing new sources of data (e.g. speleothems, corals and boreholes). Progress in understanding the causes of, and predicting the future course of climate change depends, in part, on them squeezing the most information out of this data. This takes us into the demanding world of statistics.

QUESTIONS

1. Identify the most significant causes of the urban heat island. What checks must be conducted to ensure that temperature readings taken in the vicinity of cities are not distorted by urban development? Are the same checks needed for rainfall measurements, and, if so, what corrections have to be applied?
2. Compare and contrast the strengths and weaknesses of various forms of proxy measurement of past climates. In the light of this analysis, identify the most important improvements needed to enable the different forms of measurement to be used together to provide a better picture of climate change.

FURTHER READING

A complete reference list is available at the end of the book but the following is a selection of the best books or articles to follow up particular topics within this chapter. Full details of each reference are to be found in the Bibliography.

Alley (2000): A vivid description by one of the foremost researchers in the field of ice-core studies of the challenges of extracting information from the world's major ice sheets and the invaluable information this can provide about past climate change and the potential implications this has for predicting future climatic developments.

Baillie (1995): A fluent and highly readable account of how tree-ring data is analysed and the contribution it can make to various historical studies including climate change.

Fritts *et al.* (1976): A standard text by a leading authority on the extraction of climatic information from tree rings which provides a comprehensive and informative review of dendroclimatology.

Karl *et al.* (1995): An illuminating discussion of the many problems that confront climatologists in making effective use of available data and ensuring that various forms of bias and inaccuracy do not creep into the analysis.

Pecker & Runcorn (1990): A set of papers which explore the links between the Earth's climate and variability of the Sun, and provide useful insights into the implications for carbon dating of changes in solar activity.

5 Statistics, significance and cycles

All is flux, nothing stays still.   **Heraclitus, c540–c480** BC

Squeezing useful information out of the available data is the essence of unravelling the causes of climate variability and climate change. The challenge is to exploit statistical techniques to tease out significant cycles, shifts or trends in the climate from the sea of noise that washes over all aspects of this subject. This requires a disciplined approach to avoid falling into the trap of attributing too much to what is nothing more than noise, because the one thing that is certain is that every aspect of the climate fluctuates on every timescale. So what really matters is to define specific meaning to the terms introduced in Chapter 1 (Figs 1.1 and 1.2) in order to substantiate any conclusions reached about past changes and to provide a benchmark for making predictions about the future.

This process of analysis is not just a matter of identifying significant changes in the climate (e.g. trends, cycles or sudden shifts) but also involves interpreting the properties of the associated variability. The latter can occur on every timescale within the period covered by the observations. The only way to conduct this analysis is to consider the statistical techniques designed to get the best out of noisy data. This involves a wide range of complicated mathematical techniques. These will not be discussed here. Instead we will concentrate on three practical matters. First, there is the problem of dealing with extreme events that are a normal part of climatic variability. Deciding whether the changing incidence of these rare events is a sign of a lasting shift in the climate is a fundamental issue. This leads into the second issue of identifying significant trends in the data as quickly as possible. This is particularly relevant to interpreting current events and deciding whether the observed behaviour are simply an expression of the natural variability of the climate or evidence of a more significant change. The third area is the techniques used to analyse data to identify the presence of cycles.

5.1   Time series, sampling and harmonic analysis

The starting point is the fundamental properties of the data under examination. Many of the examples already considered in earlier chapters provide an indication of the type of data we want. Ideally, it should consist of an accurate measure of whatever parameter (e.g. precipitation, pressure, temperature, wind speed etc.) at equal intervals of time, as often as needed for as long as was required. By 'accurate' we mean that errors in the measurement technique are small compared to the changes in the parameter we wish to study. Such a *time series* would then allow us to draw detailed conclusions about the behaviour of the parameter over the period of the observation (see Box 5.1). But, as we have seen, for all sorts of reasons the record is limited by many practical considerations. On the other hand, where the record is complete, there may be far more data than is needed to consider longer-term fluctuations in the climate. In these cases there are good reasons for working with the minimum amount of data required to meet certain criteria to obtain unambiguous information about real changes in the climate.

The simplest of these criteria relate to the sampling interval and the length of the series. Because of the dominant nature of the annual cycle, many series consider annual average figures when considering longer-term changes. The same applies to analysis of seasonal trends. Figures for, say, winter temperatures (i.e. the average of daily figures for December, January and February) will be quoted on an annual basis, independently of spring, summer and autumn values. Analysis of such series can tell us nothing about periodic variations shorter than 2 years in length. At the other extreme, if the length of the record is, say, 200 years, it is impossible to say anything about periodicities longer than this timescale, and difficult to draw reliable conclusions about any periodicity longer than 50 to 100 years.

These limiting conditions arise out of a fundamental mathematical property of time series. The French mathematician, Jean-Baptiste Joseph Fourier demonstrated that any time series consisting of $2N$ equally spaced points can be expressed as the sum of N harmonics of differing amplitudes (see Fig. 5.1). So in the case of a set of 200 successive annual observations this means the series could be expressed as the sum of 100 harmonics, with the first harmonic having a period of 200 years, the second 100 years, the third 66.7 years, and so on to the hundredth harmonic with a period of 2 years exactly. The process of calculating the amplitude of these harmonics is known as *Fourier analysis* and the resulting set of harmonics is known as

Box 5.1 Meteorological time series and variance

Any regular measurement of a meteorological variable (e.g. pressure, rainfall or temperature) can be expressed as a time series. This series can be defined as:

$$X(t) = X_0, X_1, X_2, \ldots, X_N,$$

where X_0, X_1, X_2, etc. are successive observations of the given meteorological parameter at equally spaced intervals at times O, Δt, $2\Delta t$, etc. The entire series consists of $N+1$ observations and covers a period P ($P = N\Delta t$). Given that we are usually concerned with how $X(t)$ varies from the normal, it is standard practice to define the series in terms of variations about the mean value $\overline{X}$, where

$$\overline{X} = \frac{\sum\limits_{n=0}^{n=N} X_n}{(N+1)}.$$

So the new series $x(t)$ can be defined as

$$x(t) = (X_0 - \overline{X}), (X_1 - \overline{X}), (X_2 - \overline{X}), \ldots, (X_N - \overline{X})$$

$$x(t) = x_0, x_1, x_2, \ldots, x_n,$$

where x_0, x_1, x_2, $\ldots$, x_N are the deviations of each successive observations about the mean $\overline{X}$, and can be either positive or negative.

The variance of any time series is defined as

$$\sigma^2 = \frac{\sum\limits_{n=0}^{n=N} x_n^2}{(N+1)}.$$

This definition of the variance includes all fluctuations about the mean $\overline{X}$, which could cover not only random variations, but also regular changes (e.g. the annual cycle) and long-term trends. Depending on the nature of the statistical studies being conducted, it may be more illuminating to remove these identifiable sources of change from the series, before considering the nature and scale of the residual variance. This process frequently involves the removal of the annual cycle and, in the case of spectral analysis (see Fig. 5.2) may also include removing any identifiable linear trend, especially if there is some doubt about its climatic reality (Box 5.2).

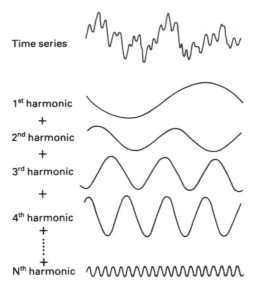

Time series

1st harmonic
+
2nd harmonic
+
3rd harmonic
+
4th harmonic
+
⋮
+
Nth harmonic

Fig. 5.1 A time series may be represented by the combination of a set of sine waves (harmonics) of differing amplitudes and phases. (From Burroughs, 2003, Fig. 2.1.)

the *Fourier transform* of the time series. Because the amount of variance in any time series, which is the sum of the squares of the difference of each point in the series from the average value (see Box 5.1), is the standard measure variability, it is usual to consider the square of the amplitude of each harmonic, and the values of these form the *power spectrum* of the variance in the time series.

Computer programs, which can calculate the Fourier transform of lengthy time series, are readily available for many personal computers (PCs). So it is easy enough to explore the harmonic components of time series, and the important discipline is to interpret accurately the meaning of the computed spectra. The easiest way to understand this discipline is to consider some examples of time series and their complementary power spectra. Starting with the most trivial example, the monthly temperature record for a mid-latitude site in the northern hemisphere would be dominated by the annual cycle (Fig. 5.2a), where virtually all the variance is in this cycle and the residual scattering of the other lesser components, reflect all the other fluctuations from month to month and year to year.

A slightly less trivial example of a Fourier transform can be found in sunspot numbers. Given that so much of the search for cycles in the weather has been associated with finding links with solar activity, it is a good example to consider. As Section 6.5 describes, sunspot numbers show pronounced cyclic behaviour with the major fluctuations having an

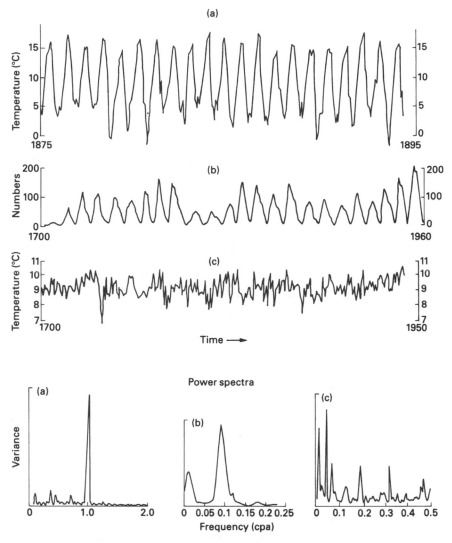

Fig. 5.2 Time series and their power spectra: (a) the monthly temperature record for central England between January 1875 and December 1895, (b) the number of sunspots during the period 1700 and 1960, and (c) the annual temperature for Central England during the period 1700 to 1950, showing how with the increasing irregularity in the time series the power spectrum becomes more complicated. (From Burroughs, 1994, Fig. 2.7.)

approximately 11-year period. In addition, successive 11-year peaks show a periodic variation in intensity that reflects a periodicity of around 90 years. So the power spectrum obtained by calculating the Fourier transform of a lengthy series of sunspot numbers shows two pronounced peaks (Fig. 5.2b). Because the cyclic variations are not precisely 11 and 90 years, the power spectrum shows relatively broad peaks that reflect the varying period from cycle to cycle. But the important feature is that the power spectrum confirms what is evident from inspecting the record of sunspot numbers – almost all the variance (see Section 5.3) in the last 200 years or so can be attributed to the 11-year and 90-year periodicities; about 65% of the variance is attributable to harmonics in the 9- to 12-year range, while some 20% is due to the 90-year feature. As in the case of the annual temperature cycle, the link between the time series and the power spectrum is relatively easy to see.

This direct link becomes much less obvious in the case of a typical meteorological record where we are interested in identifying periodicities in the range 2 to 100 years. Here there is no obvious cyclic behaviour in the variances of the year-to-year figures from the long-term mean. So the power spectrum (Fig. 5.2c) will contain a number of features of varying magnitude. In the example given, the most obvious features at 76, 23, 14, 5.2, 3.1 and 2.1 years account for only about a third of the variance. Deciding which of these features is both statistically significant and of physical significance requires careful analysis, and will be considered in more detail in the next section. For the moment the important fact is that by calculating the Fourier transform of a time series, it is possible to produce the power spectrum of all the harmonics that uniquely define the observed series. Conversely, if we knew only the power spectrum it would be possible to recreate the time series by the reverse calculation. This complementary nature of the time series and its power spectrum is not only an expression of the mathematical link between the two; if the observed power spectrum were a measure of the physical behaviour of the weather and climate in the future as well as in the past, then it could also be used to forecast future events. Successful forecasting is the true test of reality of the supposed cyclic behaviour in weather and climate statistics.

There are a variety of more sophisticated computational techniques which can squeeze some additional information out of time series, but they have to be viewed with caution in the case of examining climatic change. Because, for the most part the evidence of cycles is, at best, slight, there is a danger of attaching too much importance to what may be nothing more than noise (Section 5.2). Because these techniques can present the spectra so effectively it is easy to fall into a trap. So when confronted with what look like convincing examples of cyclic behaviour there are two basic guidelines to follow. First, the spectrum should be shown in its entirety so it is possible to establish how much of the total variance is found in the principal features (see Section 5.3). Second, where spectral features are identified as being highly significant, their importance is greatly enhanced by there being an established (*a priori*) physical proposal as to why such a periodicity should occur.

5.2   Noise

The examples presented in Fig. 5.2 provide evidence of the fact that the variability in weather and climate statistics consists of a complicated mixture of fluctuations. Leaving aside the recognisable regular variations (e.g. daily and annual cycles) and other identifiable examples of climate change (see Figs. 1.1 and 1.2), there are quasi-cyclic and apparently random fluctuations. The latter are an important aspect of analysing the nature of climate change. They are a consequence of the chaotic nature of the global weather system and are often termed *noise*. The meaning of this term does, however, warrant some closer examination.

If the fluctuations in weather and climate were truly random on every timescale the power spectrum of the fluctuations would have an equal probability of any frequency components being present. While any particular time series might exhibit a different mixture of spectral components, on average the power spectrum of such fluctuations would have equal amounts of variance in any unit interval of frequency. This distribution is often termed *white noise* – this expression is derived somewhat loosely from optical spectroscopy where *white light* contains all the visible frequencies although not in equal amounts.

The natural variability of the climate is not this simple. Many of the components of the system (see Chapter 3) have different frequency characteristics. Slowly varying factors such as snow cover, polar ice, SSTs and soil moisture build in inertia and mean that the weather has a 'memory' and so is more likely to exhibit greater low-frequency fluctuations than higher frequency ones. Again in the terminology of optical spectroscopy, such noise is defined as being 'red', denoting that its distribution is weighted towards lower frequencies. In effect, because the weather has a better recollection of recent events, the short-term variations are damped out more than the longer-term ones, as with the passage of time the connections become more tenuous. The theoretical distribution of red noise depends on the assumptions made about how any connections between successive events decay over time. These draw on available statistics to make an estimate of the typical time constant of the climate's memory. This construction may then be used to assess the significance of what appear to be real features in the calculated spectrum of any time series.

In practice, this means that lower frequency/longer period cycles have to contain a greater proportion of the observed variance to achieve the same significance as higher frequency/shorter period feature.

Fluctuations in instrumental records frequently exhibit red noise. In the case of proxy data the situation is more complicated. Because the link between the observed variable (e.g. tree-ring width) and the meteorological parameters (e.g. rainfall) is subject to uncertainty, there are additional random errors. While these errors can be reduced by the careful calibration of more recent proxy data using modern meteorological records, the problem cannot be eliminated. As a consequence, there is greater randomness in the inferred meteorological variability. This will produce white noise in any computed power spectrum. At the same time the underlying weather will have contained red noise. So spectral analysis of proxy data will contain both white and red noise; this combination is often referred to as 'pink' noise. This means that in any consideration of the significance of spectral features obtained from the analysis of proxy data an estimate of the pinkness of the background noise must be calculated.

5.3   Measures of variability and significance

In any meteorological series the properties of the identified components of the variability can be defined in terms of the standard deviation (σ) (Box 5.1). This definition has three important applications. First, if the fluctuations are considered to be randomly distributed about the mean then observations can be made about the probability of a single extreme event, or run of extreme events occurring by chance. The probability of any observation being a given amount from the mean is defined by a symmetrical *bell-shaped* curve (also known as the *normal* curve or mathematically as a *Gaussian* curve). There is a 32% chance that any observation will be one standard deviation (σ) from the mean, and approximately a 5% chance it will occur 2σ from the mean (Fig. 5.3).

The assumption that residual variance is randomly distributed only applies in limited circumstances. Two particular criteria are important. First, the series must be stationary. This means that there is no significant long-term trend or other form of significant change in the climate during the period covered by the series. Secondly, the fluctuations are evenly distributed about the mean. This is an approximation that can be made for temperature and pressure statistics. In the case of temperature extremes, however, it often proves to be an oversimplification. But in the case of

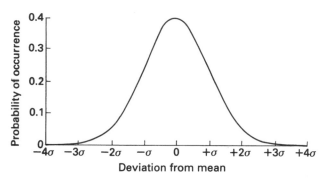

Fig. 5.3 The distribution of random fluctuations in a measured variable can be represented as 'normal' curve, which shows that the probability of any particular value being observed is related to the deviation from the mean. This distribution is usually expressed in terms of the standard deviation (σ), and shows that 68% of observations will be within one standard deviation ($\pm\,\sigma$) of the mean, 95% will be within two standard deviations ($\pm\,2\sigma$) and well over 99% will be within three standard deviations ($\pm\,3\sigma$). (From Burroughs, 2003, Fig. A.1.)

rainfall and wind speed where the distribution of almost all the figures is markedly skewed toward low values (Fig. 5.4) because there are, in most records, far more observations of, say, zero or low rainfall than there are of very wet periods. This type of curve, which called *non-Gaussian*, has a long tail at high values, which is a measure of the fact that there is always the possibility of extremely high but exceedingly rare observations. The key to using the best statistical techniques for such series is to establish the statistical nature of the basic distribution of the data. Only when it is clear what form the distribution about the mean takes, is it possible to decide what is the appropriate form of statistical analysis to apply to the series.

When examining instrumental records much of the variance is dominated by the natural variability of the climate. This makes the detection of any trend, periodicity or sudden shift hard to detect. In addition, extreme events exert a substantial influence on short-term trends, and their spacing has a special fascination for statisticians.

The best way to consider these issues is to consider specific examples. In a simple case (e.g. annual figures for average temperature), where the values are approximately randomly distributed about the mean, the significance of any linear trend in the figures can be assessed in terms of the size of the rise from the beginning to the end series as compared with the standard deviation about this trend. Standard curve-fitting programs using least-squares analysis to estimate trend lines are a standard feature of PC statistical packages and provide an estimate of the significance of any trend.

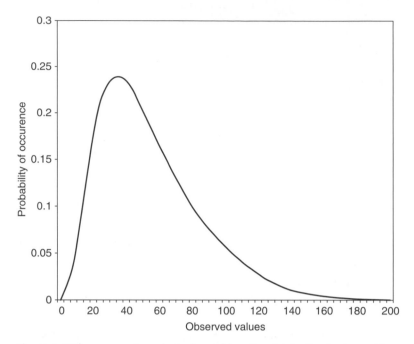

Fig. 5.4 With some meteorological variables, the distribution is skewed towards lower values and the scatter cannot be expressed in terms of the deviation from the most probable value.

They can also provide higher-order polynomial fits, but without some physical explanation as to why the time series should follow such more complicated patterns, this analysis is nothing more than a statistical refinement, which is unlikely to offer new insights.

Even the case of an apparently simple time series, which shows a fairly clear trend and a reasonably normal distribution about the mean, may disguise more complicated behaviour. For instance the central England temperature (CET) series of annual figures (Fig. 5.5a) shows a distinct upward trend (0.71 °C over the period). If we use the standard analysis of a least-square linear trend line, the correlation coefficient (see Box 5.2), we get has a value of $r = 0.32$ ($r^2 = 0.102$). While highly significant in purely statistical terms, this rise accounts for only 10% of the variance in the series. When, however, the monthly statistics are examined a more fragmented picture emerges. First, most of the temperature rise is in the winter half of the year (see Fig. 8.21), and, if the analysis is conducted on winter (December to February) temperatures, although the absolute rise in temperature (1.14 °C) is greater (Fig. 5.5b) the significance of the trend is rather less because the value of r is 0.24 ($r^2 = 0.06$), because of the greater variability

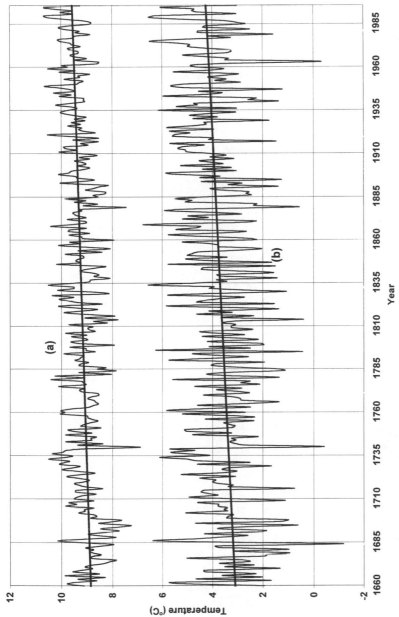

Fig. 5.5 The linear trends for the temperature of central England over the period 1660 to 1996 for (a) the annual data, and (b) the winter months (December to February) show a marked warming. In both cases this warming is significant, but although the temperature rise is greater in winter, this trend is less significant because the variance from year to year is correspondingly greater.

Box 5.2 Interpretation of correlation coefficients

A fundamental aspect of statistical analysis is to establish whether there is a significant link between two variables. Here this could be how a meteorological variable changes with time, or how some weather-sensitive parameter (e.g. cereal prices, tree-ring widths or wine-harvest dates) varies as a function of some meteorological variable. In the case of a meteorological time series $X(t)$ (see Box 5.1), the detection of a trend is possible by estimating whether there is a function which will fit the observed data better than the mean value $\overline{X}$. This curve could take many forms, but here we will consider only the linear trend line, which is calculated by the method of least squares to obtain the minimum value of

$$S_m{}^2 = \sum_{i=0}^{i=N}(X_i - X_i')^2,$$

where

$$X'(t) = a + bX(t)$$

and is the calculated trend line value of $X_i'(t)$ for the i-th observation as opposed to the observed value X_i. This calculation is a complicated process, which is described in standard statistical texts (see Further reading), and performed automatically by many statistical programs on standard PCs. The important feature of this calculation is that the minimum value of S_m is less than the same calculation using the mean value of $\overline{X}$:

$$S_0{}^2 = \sum_{i=0}^{i=N}(X_i - \overline{X})^2$$

where

$$\overline{X} = \frac{\sum_{n=0}^{n=N} X_n}{(N+1)}.$$

To the extent $S_m{}^2$ is smaller than $S_0{}^2$ is a measure how much better a fit $X_1(t)$ is for the time series $X(t)$, as opposed to $\overline{X}$, and this is usually expressed as the *coefficient of correlation*, which is defined as:

$$r = \pm\sqrt{1 - \frac{S_m{}^2}{S_0{}^2}}$$

If the fit is good the ratio of the sums of squares will be close to zero and r will be close to ± 1. If the fit is poor the ratio of the sums of the squares will be close to unity and r will be close to zero. The same type of analysis can be conducted to estimate the goodness of a fit between $(N+1)$ meteorological observations and some proxy measurements (e.g. growing season temperature and wine harvest date – Section 4.4) for the same period.

The interpretation of a correlation coefficient (r) requires care. As a general rule, the value r can be defined as a significant at a given level by a simple numerical relationship. For instance, a correlation is significant at the 95% level (i.e. there is only a 5% chance it is the product of chance) if:

$$r \leq -1.96\big/\sqrt{N} \ \text{ or } \ r \geq 1.96\big/\sqrt{N}$$

Because in many examples of studying climatic change the value of N is large (i.e. 100 or much greater) this means that values of r of 0.2 or considerably less can indicate, say, a significant linear trend in the rise or fall of some meteorological time series. But does this have a meaning in interpreting what is really going on?

The way to interpret r is in terms of how the least-squares fit of a linear relationship to a set of data reduces the variance compared with the value obtained with the original assumption of there being no identifiable relationship. This is given by the ratio of S_m^2 to S_0^2, the value of which is $(1 - r^2)$. In the case of a lengthy meteorological series where the value of r^2 is 0.1, a value of $r = \pm 0.31$ is highly significant (see Section 5.2). But in terms of the variance all it is saying is that the linear trend is explaining only 10% of the variance in the series, and the remaining 90% is due to other causes. For a time series of instrumental observations, the identification of a significant trend is illuminating, even though it represents so little of the variance. For proxy data, however, where there is the additional question of whether some of the long-term variance has been filtered out by the process of data extraction (see Section 4.4.1.), it may be even more difficult to attribute significance to any trend in time series.

from year to year. So, although most of the warming is concentrated in this part of the year, and the seasonal trend remains statistically significant, it now accounts for only 6% of variance. This means the greater fluctuations in winter temperature from year to year make it more difficult to establish the causes of variance and emphasise the importance of understanding underlying processes at work when interpreting statistics.

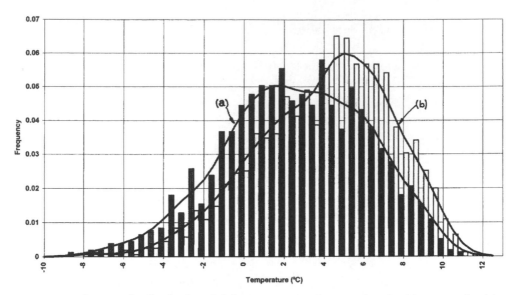

Fig. 5.6 The distribution of daily temperatures for central England in January for (a) the period from 1772 to 1821, and (b) from 1946 to 1995, showing a large shift in the median temperature between the two periods but relatively little change in the extreme values. (From Burroughs, 1997, Fig. 5.14.)

An additional insight into the nature of the variance of winter temperatures in England may be found in looking at the distribution of daily figures in the CET record, which have been produced by the UK Meteorological Office for the period 1772 to the present. In Figure 5.6 the distribution of values for January for the 50 years 1772 to 1821 is compared with those for 1946 to 1995. For both periods it is distinctly skewed towards higher temperatures with a long tail of low temperatures. But, although the latter period was nearly 1.4 °C warmer than the earlier one, the range of extremes is virtually unchanged. The real differences are in the central regions where, on the smoothed curves, the median has shifted 3.5 °C reflecting that there are substantially more cold days in the first period and more mild days in the latter. Analysis of December temperatures shows the same behaviour, but February has shown no appreciable warming and the distribution during the two periods is effectively identical.

A possible explanation of the shift in distribution may be found in terms of the changing incidence of winter weather regimes. If, for the British Isles, these represent, say, five broad patterns which result in different types of air masses dominating the weather in winter, these can be defined in terms of their mean temperature. The coldest of these is continental polar (cP) air that sweeps out of the icy vastness of northern Russia and Siberia when high pressure becomes established in the vicinity of Scandinavia. Next is maritime Arctic (mA), which comes straight down from the ice-covered Arctic ocean. Then there is less cold maritime polar (mP) air that sweeps down from the northwest when high pressure builds over the North Atlantic. Milder maritime polar westerly (mPw) air sweeps across the Atlantic from North America warmed along the way. The warmest air is defined as maritime tropical (mT), which comes up from the tropical Atlantic and on rare occasions appears in the form of intrusions of continental tropical air (cT) from the Sahara.

Using these definitions, it is then possible to attribute a standard temperature distribution to each type of air mass and consider how differing combinations of the prevalence of these regimes alters the overall distribution of temperature in central England. The results of this exercise are shown in Fig. 5.6 to the changing incidence as shown in Fig. 5.7a and b. Although this attribution is designed to reproduce the observed distribution, and is not linked to actual observations of how the incidence of particular weather regimes have changed, it illustrates an important physical point. This is that the oddities in the statistics become more understandable when explained in terms of a simple climatological model. In effect the substantial change in distribution without a comparable shift in the most extreme values can be interpreted in terms of a decline in easterly and northerly weather patterns, which bring cold Arctic air to the British Isles, and an increase in westerly and southerly patterns bringing milder air, but no significant change in the properties of the air masses involved. So this analysis suggests that the changes affecting winter temperatures in the British Isles over the last 200 years or so are a matter of a shift in weather patterns rather than a significant warming or cooling of the northern hemisphere. This observation is also consistent with the comments made on the importance of circulation regimes in Sections 3.2, 3.6 and 4.3.

The same care has to be exercised when interpreting the power spectrum of a time series. Assuming that one or two features, such as the solar cycles in Figure 5.2b, do not dominate the computed spectrum and the variance is spread over many frequencies, great care must be taken in how much significance is attached to the most prominent of the features. The easiest way to approach the analysis of spectral variance is to consider typical examples. Figure 5.8 shows two examples of the spectral distribution for time series. In the first (Fig. 5.8a) the spectrum is effectively 'white' and the expected variance is shown as the horizontal line. This is the level for which

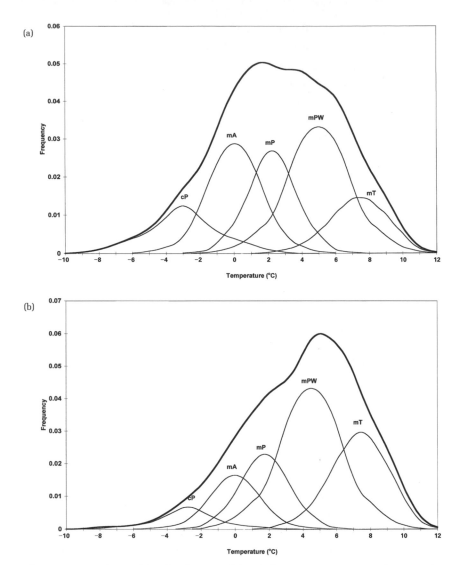

Fig. 5.7 An analysis of the distribution of the daily temperature curves for central England in January (see Fig. 5.6) for (a) the period from 1772 to 1821, and (b) from 1946 to 1995 showing how they might be made up of different incidences of certain types of air masses assuming that the temperature of these air masses is normally distributed (see Fig. 5.3).

there is an even chance of any particular spectral component occurring. In practice, there is a considerable scatter about the expected value reflecting the random nature of the time series analysed. The mean value of the variance in the spectrum is, however, the same as the horizontal line.

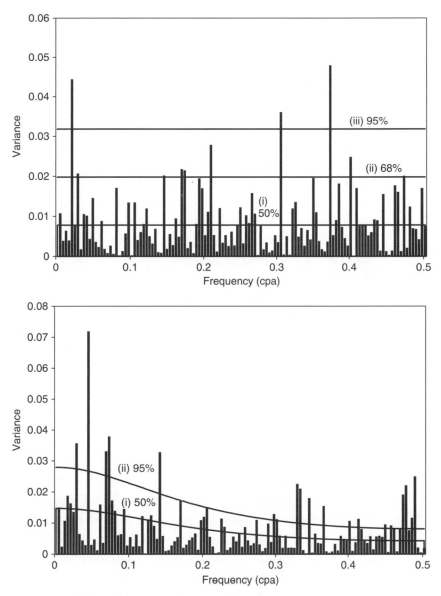

Fig. 5.8 Examples of typical power spectra exhibiting (a) 'white' noise, and (b) 'red' noise. The solid bars are the spectral components and the horizontal heavy lines are the levels that would be expected on the basis of a purely random process.

In the case of a 'red' spectrum (Fig. 5.8b) the expected variance increases at low frequencies. The shape of this curve (see Section 5.2) reflects the fact that the series measure a variable that has a significant 'memory'. Again the actual spectrum shows considerable fluctuations about the mean variance with the most substantial features occurring at the lower frequencies, but the average variance is the same as the horizontal mean curve.

It is possible to attach statistical significance to the features in the computed spectra in Figure 5.8. This can be done by calculating the probability of any spectral feature being more than a certain multiple of the expected variance. In many published papers these levels are shown by curves indicating the levels which would be reached by chance in 1 in 10 or 1 in 20 occasions, and often are labelled 90 or 95% to indicate their level of significance (i.e. with higher and higher levels there is less and less likelihood that they are the product of chance). But the important point is that for every 100 spectral components there is an even chance that one of them will exceed the 99% significance level. In Figure 5.8a, the features at 0.02 and 0.375 cpa, and in Figure 5.8b, the feature at 0.045 cpa would most certainly fall into this category, but are almost certainly the product of chance.

Against this background, when assessing whether any particular peak is truly 'significant' it is as well to ask two questions. First, as noted in Section 5.1, has the frequency in question been predicted, in advance, as being the product of some specific physical process? This criterion is often referred to as the *a priori* requirement. Second, and directly related to identifying a physical cause, does the frequency occur in other independent time series of the same, or other meteorological variables? If the answer to both these questions is no then the 'significance' of the observed feature may be nothing more than a consequence of the random nature of the time series under investigation.

The CET series of annual figures considered earlier in this section provides a good test for these principles. The Fourier transform of this series for the period 1700 to 1950 is shown in Fig. 5.2c. As noted in Section 5.1, the six features, which are significant at the 95% confidence level, explain about a third of the variance (in this particular calculation the trend was removed before the transform was computed, as leaving it in leads to additional 'reddening' of the spectrum). Of these features, the 76- and 23-year peaks may conceivably be linked to solar activity (see Section 6.5), while the 2.1-year peak is possibly an echo of the Quasi Biennial Oscillation (QBO) (see Section 3.2). The other features do not have *a priori* reason for occurring and so may be nothing more than random fluctuations.

This final examination of the CET annual figures provides a good measure of what can be extracted from a noisy time series. After identifying the trend and the six most significant features, of which only three might just be part of a periodic behaviour of the climate, we are left with nearly two-thirds of the variance apparently being a natural expression of the temperature patterns in central England.

5.4   Smoothing

The dominant influence of the random variations in many meteorological series means there is little purpose in going into great detail to analyse the spectral components. All that really matters is to get some sense of the underlying longer-term variations by smoothing out the shorter-term fluctuations. This approach helps pick out quasi-cyclic behaviour and possible significant shifts in the climate.

The simplest and most frequently used method of smoothing out a time series so that longer term fluctuations can be identified is to form a running mean of data. In its most basic form this method consists of forming the average of a given number of successive points in the time series to produce a new series. Known as the 'unweighted' running mean, this approach is widely used and easy to apply [many statistical packages associated with spread-sheets (e.g. Excel) on personal computers provide such smoothing as a standard feature]. This approach does, however, have a number of limitations, which need to be considered alongside the other methods of smoothing and filtering.

To appreciate the impact of any smoothing operation on a time series we must consider how it affects the various harmonic components of the series. As we have already seen, any series can be represented by the sum of a set of harmonics. The easiest way to explain this is to take an example. If we are taking a 10-year unweighted running mean, the first obvious feature is that it will completely flatten out a 10-year periodicity of constant amplitude. This is because it will always be forming the average of one whole cycle wherever it starts from. Similarly it will remove all the higher harmonics that are an exact number of cycles in the 10-year averaging period (i.e. 5 years, 3.33 years, 2.5 years, etc.). It may also be apparent that its effect will be approximately to halve the amplitude of a 20-year periodicity, as the 10-year running mean will take the average of half this cycle as it moves along the series.

So far, so good, but when we come to look at what it does to some of the shorter periodicities, the problems start. Take, for instance, a cycle that has a periodicity of 6.33 years (i.e. it completes 1.5 cycles each 10 years). The 10-year running mean will thus form an average that contains the net effect of the additional half cycle as it moves through the series. Not only will this

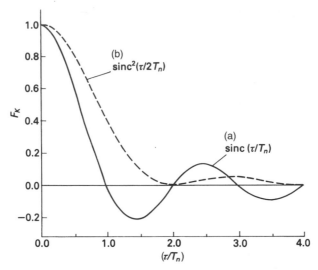

Fig. 5.9 The filtering function for (a) an unweighted running mean, and (b) a triangularly weighted running mean. The filtering function (F_k) is the ratio of the amplitude of the harmonics in the running mean to the amplitude of the corresponding components in the original time series. This ratio is shown as a function of the interval covered by the running mean (τ) divided by the period of the n-th harmonic (T_n) in the time series. (From Burroughs, 2003, Fig. A.4.)

cycle be present in the smoothed series but it will also be inverted with respect to its original phase. It can be shown by mathematical analysis (see Further Reading), that in the worst case, 22% of this harmonic passes through the smoothing process and turns up as a spurious signal completely out of phase with the original harmonic in the unsmoothed series (Fig. 5.9). This type of distortion, together with the presence of higher-frequency features in different amounts, makes the use of the unweighted running mean both inefficient and potentially misleading.

To see how more efficient smoothing can be achieved, it is illuminating to consider the characteristics of an unweighted running mean in another way. The reason that high-frequency fluctuations get through is because of the way in which the smoothing deals with the data. Take, for instance, a time series of average winter temperatures that can fluctuate dramatically from year to year. These fluctuations may be random or contain some significant periodicities. The unweighted running mean is like a 'box-car' running though the series. Every data point within its span is given equal weight. So an extreme winter will enter the running mean with a sudden jump and exit in the same way. This means its effect on the smoothed series will show up in a sharp way, even though the running mean is designed to

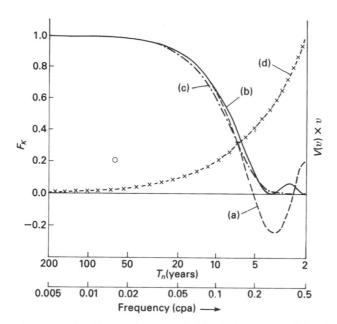

Fig. 5.10 The filtering function of: (a) a five-year unweighted running mean; (b) a seven-year triangularly weighted running mean; and (c) an eleven-year binomially weighted running mean. The frequency scale is logarithmic, so to show the impact these running means have on 'white' noise at (d), the curve of constant variance per frequency interval ($V(\nu)$) is multiplied by the frequency (ν) so that equal areas under the curve represent equal variance. (From Burroughs, 1994, Fig. A.5.)

remove all such sudden change. Given that we are only interested in the extremes to the extent that they are evidence of either longer-term periodicities or a sustained shift in the climate, it would be better if each data point came into the running mean gently, built up to a maximum in the middle and faded out again. Providing this approach solves the problems of the unweighted running mean and does not introduce other distortions, it should be a better way of examining time series.

A variety of weighted running means have been explored (see Further Reading). The simplest approach would be a triangular weighting with its peak in the middle of the running mean (see curve (b) in Fig. 5.10). It turns out, however, that it is possible to design more efficient running means to act as relatively sharp 'low-pass' filters that remove virtually all the harmonics above a certain 'cut-off' frequency. The remaining harmonics are present in the series without any phase distortion, but close to the cut-off frequency their amplitude is substantially reduced. The choice of the mathematical form of the smoothing operation is a balance between achieving a sharp cut-off and minimising both the computational effort and the number of terms needed to produce the required smoothing effect. The latter is important because in general the sharper the cut-off the larger the number of terms that have to be used. This means that the ends of the series are effectively wasted in achieving an efficient smoothing, and if there are only a limited number of observations in the series this can be a high price to pay. As a general observation, using a binomial or Gaussian weighting is a good compromise.

If the main interest is the frequency distribution, it is possible to adopt a more selective procedure. The straightforward operation of smoothing time series using either a weighted or an unweighted running mean is only a specific example of the more general technique of filtering. Instead of simply working with a 'low-pass filter' which leaves the low-frequency harmonics in the series unaltered and easier to see, there is no reason why this practice should not be extended to suppress both high- and low-frequency components and let only a limited range of frequencies through. The advantage of this process is that, unlike harmonic and spectral analysis (see Section 5.1), it permits the examination of the persistence of periodic features throughout the duration of the series. By comparison the power spectrum is only about the mean amplitude of apparently significant oscillations while any variations in their amplitude or frequency over time are transformed into other components of the spectrum. So if these changes are appreciable, spectral analysis makes it harder to identify the real nature of the fluctuations.

This distinction is important. Wherever the question of cycles is addressed in this book, it will become apparent that convincing evidence of periodic behaviour can come and go with tantalising regularity. After several periods a cycle can suddenly disappear, only to reappear at some unspecified interval later, or shift phase and amplitude, or disappear for good. So mathematical techniques, which expose different aspects of this frustrating behaviour, can help to pin down the physical reality of causes of any supposed cyclic behaviour.

Ideally, a filter should pass all frequencies within a narrow band without any change in amplitude and completely suppress all other frequencies (Fig. 5.11). In practice, this is impossible to achieve and compromises have to be made in choosing a filter which provides the best combination of removing unwanted frequencies and leaving largely unaltered the frequencies of interest. The underlying approach to narrow-band filtering is to construct a filter that is effectively an oscillation of the required frequency. The amplitude of this oscillation increases from a small value up to a maximum and then reduces again. The number of points used in the filter defines the bandwidth of the filter and hence the number of oscillations

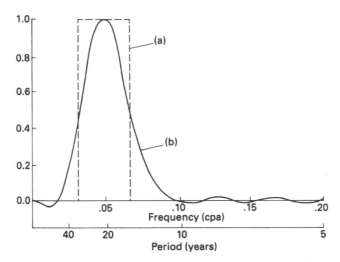

Fig. 5.11 A comparison between (a) an ideal statistical filter which removes all unwanted periodicities and leaves unaltered those periodicities which are of interest, and (b) what can be achieved in practice. In (a) the filter transmits periods between 15 and 30 years whereas (b) only transmits 20.6 years unaltered and reduces the amplitude of all other periodicities to a greater or lesser extent depending on how close they are to 20.6 years. (From Burroughs, 2003, Fig. 2.6.)

included in the computation – the greater the number of points used in the filter, the narrower its bandwidth. But, as with all smoothing and filtering operations, there is a pay-off between the narrow bandwidth of the filter and both the computational effort and the available data. In particular, if a sharp filter involves using a significant number of the available data points to compute a single point in the smoothed series, it limits the scope of the analysis. Moreover, if the data contain a considerable amount of noise, too precise a focus on a narrow frequency range may serve little purpose as it will only produce a beautifully smoothed picture of the level of noise in a narrow frequency range. So, as with so many aspects of the search for order in meteorological series, there is a compromise to be struck in dealing with the limitations of the data.

5.5   Wavelet analysis

In recent years, a refinement of spectral analysis has become widely used in climate-change studies. Known as *wavelet analysis* this technique examines how the power spectrum of a time series varies over the time of the record. It involves transforming a one-dimensional time series (or frequency spectrum) into a two-dimensional time–frequency image. Often this highlights the fact that evidence of cycles and quasi-cycles flits in and out of many climate data series. This unwelcome fact can be lost when examining the Fourier transform of a complete time series, which smears out these variations, and in the worst case effectively throws the baby out with the bathwater.

The simplest approach to studying the changing frequency response of some aspect of the climate system is to calculate a form of running Fourier transform of the available time series. This can be done by using a certain window size and sliding it along in time, computing the transform at each time using only the data within the window. This would give us information about the frequency spectrum, but results lead to an inconsistent treatment of different frequencies. For a given window width of N sampling intervals in the time series, the width of window would be too small to resolve different low-frequency oscillations, while at high frequencies although the resolution would be fine it would be better to have a narrower window to examine the shorter-term time variations of these oscillations.

What wavelet analysis does is to combine both a weighted window and a defined number of oscillations of a given frequency within this window. The convolution of this wave *packet*, often defined as a *Morlet* (Fig. 5.12) for the

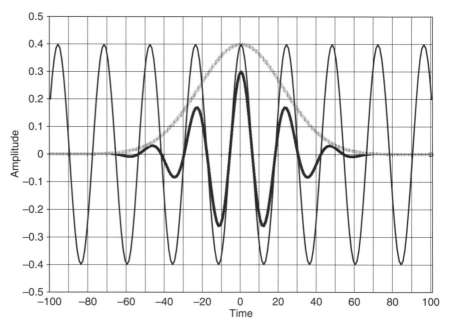

Fig. 5.12 A diagram showing how a 'Morlet' function (thick black line) is formed by the combination of a simple harmonic (thin black line) and a Gaussian function (fuzzy grey line) restricting the influence of the morlet to about six periods of the harmonic.

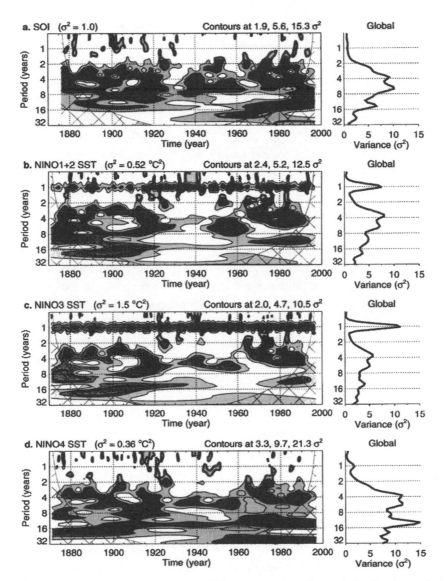

Fig. 5.13 Wavelet power spectrum of the Southern Oscillation index for the period 1876 to 1996. The contour levels are in units of variance and are chosen so that 50%, 25% and 5% of the wavelet power is above each level respectively. The thick contour shows the 10% significance level. The cross-hatched area indicates the *cone of influence* where the variance is reduced by extending the series beyond the observed range using zeros as 'padding'. (From Allan, 2000, Fig. 1.4a in Diaz & Markgraf, 2000, Cambridge University Press)

defined frequency provides a measure the variance of any periodic features within the series. This corresponding to the frequency range defined by the transform of the window, and how the amplitude variance changes with time. As such it is a version of the band-pass filter described in Section 5.4. The choice of the weighting of the window must be chosen to avoid the pitfalls of using a simple 'box-car' form that were identified in Section 5.4. The most obvious choice for studying weather cycles is to use a Gaussian envelope, as the Fourier transform of a Gaussian is another Gaussian.

The difficulty of using this approach to time series of limited duration is that examination of the longer periodicities is constrained by the length of the wavelet. For example, in the case of the various analyses of the CET record (see Section 3.1) even the examination of the 23-year periodicity requires a wavelet containing some 120 terms. This means that only the central part of the record can be examined without making some assumption about the behaviour of the CET record beyond the existing data. For this reason, the presentation of wavelet analyses adopts the convention of showing a cross-hatched area in the bottom corners of the diagrams to indicate a 'cone of influence' where the edge effects become important.

There is, however, one great additional advantage of wavelet analysis. This is that it is free of the assumption of stationarity (see Section 5.3). This is a particular benefit when examining proxy record, such as tree rings, which have been built up from overlapping records from different trees, and have suffered smoothing of longer-term variability (the segment length curse: see Section 4.5.1).

5.6   Multidimensional analysis

Although much of the evidence of climate change is derived from time series of a single meteorological parameter for a given site, where greater quantities of data are available it is possible to build up a spatial picture of change. This multidimensional analysis is central to unravelling the causes of climate change. For example, there is little doubt that there has been a significant warming of the global climate during the last 100 years or so (see Section 8.10). What is more difficult to establish, however, is how much of this warming is due to the natural variability of the climate and how much is due to human activities. In Chapters 10 and 11 we will discuss the various attempts to model climatic change and explain the form of rise in global temperature during the twentieth century. But, in statistical terms, we do not have an adequate knowledge of the natural variability of the climate on the timescale of decades to centuries. Since it may take a very long time to establish this background variance, a short cut is to examine whether natural variability will exhibit different spatial patterns to the impact of human activities. This requires three-dimensional statistical analysis of patterns of change, as compared with the patterns that are predicted using computer models. This analysis is often termed *fingerprinting* (see Section 11.4).

The handling of the variation over time of two- or three-dimensional patterns of meteorological variables takes us into the realms of matrix algebra. Because this involves large amounts of data, with too many factors varying, the essence of the statistical techniques is to reduce the number of variables without discarding essential information. The most widely used techniques is known as *Principal Component Analysis* (PCA). This involves computer programs which manipulate a matrix of the data to estimate how much of the variance can be attributed to each one of a set of *orthogonal* principal components. This is done by the linear transformation of the data to identify the first principal component, which represents the greatest proportion of the variance in the data. The second stage of the analysis is to calculate the next component, which is uncorrelated with the first principal component, and estimate the variance. The value of this approach is that in many cases it homes in on the important features of change as a large part of the total variance can be attributed to a few principal components.

Perhaps the easiest way to consider how this process works is to describe a general example. Often PCA is applied to the variation over time of the geographical distribution of climate changes. This requires the analysis of, say, pressure, rainfall or temperature data for, say, a region such as northwest Europe or the USA from around 100 stations for the last 100 years or so. If these were studied station by station the overall patterns of change are particularly difficult to identify as there are as many variables as stations. With PCA what can happen is a remarkable distillation of the data. The first principal component is often a simple measure of whether conditions across the region, as a whole, were above or below normal (i.e. wetter or drier, or warmer or colder than normal) explains a considerable proportion of the variance. The second component tends to identify how much variance is due to one-half of the region being above normal while the other half is below normal. This second principal component often reflects well-known features of the circulation patterns of the region. The third component may then be the variance attributable to fluctuations, which are perpendicular to the changes measured in the second principal component, and this accounts for rather less of the variance. Subsequent components will locate more complicated patterns, but often over 80% of the variance has been identified with the first three components and the analysis of the higher principal components yields diminishing returns. So, in effect, most of the analysis of all the data from so many stations can be squeezed down into three variables, with relatively little loss of information.

The reason for presenting PCA in this way is that it is widely used and inevitably the product of computer analysis. So what has been done to the data is not evident to either those exploiting statistical programs or to readers of scientific papers. It is all too easy to bung large quantities of data into the computer and press the button. Out comes a string of figures defining the coefficients of a large number of principal components (*eigenvectors*) and estimates of the amount of variable attributable to each principal component (*eigenvalues*). What can be overlooked is what precisely these figures are representing. Without a physical understanding of what the data analysis is doing for you, or what lies behind the presentations of principal components in papers, there is a risk of being misled. So, unless you are prepared to really get to grips with PCA and understand the mathematics of the analysis, (see Further Reading), simple guidelines to its use are twofold. First, you restrict your analysis to those first few components that can be readily defined in terms of physical features in the weather. The second is complementary to the first, in that the analysis is unlikely to be particularly illuminating unless most of the variance can be accounted for in these few components.

5.7   Summary

The brief review of statistics in this chapter is designed to emphasise two essential features of dealing with data that may contain information about climate change. The first is that there are a wide variety of statistical techniques that can be applied to the large quantities of data available. But, once you get beyond relatively simple approaches to smoothing, filtering and spectral analysis, more advanced techniques should only be used with great care, because, without a thorough understanding of what can be achieved, there are real dangers of getting out of one's depth. Part of the problem is the ready availability of computing power and statistical programs which can handle large quantities of data quickly, efficiently, but without any insight as to what is the meaning of any apparent significant variations.

The identification of physical reasons for climate change is even more important in the second aspect of statistical analysis. This is the fact that most climatic series, with a few notable exceptions (e.g. the ice ages), are dominated by largely random fluctuations. This means that the series is principally noise, and if there is any identifiable physical cause for some significant change in the series, this signal is often small. Unless we have a clear understanding of why the noise should take a given form, any attempt to use particularly sophisticated forms of statistical analysis to attribute particular significance to some small part of the variance must be treated with great caution. It may be reasonable to attach importance to a simple feature like the linear trend in a series. Using incredibly powerful spectral techniques, which are designed to enhance high signal-to-noise data that has been degraded for some practical reason, may, however, do no more than enable us to examine the noise in evermore excruciating detail. This is no benefit. What we want from statistics is techniques that help us to identify the physical causes of change, and, better still, assist in the production of useful predictions of future change.

QUESTIONS

1. Generate a series of random numbers by taking the last two digits from a set of consecutive numbers in a telephone directory and then subject them to statistical smoothing using a simple running mean (this is most easily done using a standard PC statistical package such as Excel or Lotus 123). What does the resulting curve tell you about the problems of interpreting the smoothing of noisy data?

2. Taking an example of a meteorological time series of, say, annual temperature or rainfall statistics found in many climatology textbooks (e.g. Hulme & Barrow (1997), or Lamb (1972)) and, using a standard PC statistical package, calculate the linear trend, the variance and the correlation coefficient for the series. Repeat the analysis for the first half and the second half of the series, and then compare the three sets of results. What do the differences in the various figures tell you about the series you have examined?

FURTHER READING

A complete reference list is available at the end of the book but the following is a selection of the best books or articles to follow up particular topics within this chapter. Full details of each reference are to be found in the Bibliography.

Burroughs (2003): This book provides a basic guide to the statistical methods that are used to extract information from meteorological time series.

von Storch and Zwiers (1999): A comprehensive and thorough presentation on many aspects of the statistical analysis of climatic data. Many readers will find the detailed mathematical treatment of the analysis intimidating, but the technique of splitting the work into a large number of bite-sized pieces makes it easier to grapple with specific aspects of the detailed analysis.

Jolliffe (1986): A useful introduction to principal component analysis.

Kendall (1976): A thorough analysis of the mathematical techniques for analysing the nature and information content of time series.

6 The natural causes of climate change

Lucky is he who could understand the causes of things.   **Virgil, 70–19** BC

By now it should be obvious that there are a lot of things that can contribute to changes in the climate. So assessing the causes of these fluctuations opens up a huge variety of physical processes. To keep this analysis manageable we must concentrate on the most obvious factors, and then focus on how these may contribute to future climatic trends. This means we will consider both short- and long-term processes in seeking explanations of natural variability. Then the analysis will narrowed down to the causes of more rapid fluctuations when the question of the impact of human activities is brought into the debate. So, the principal objective will be to identify those aspects of climate change that provide the most insight into how the global climate may change in the foreseeable future.

Particular attention will be paid to mechanisms for climate change which have cyclic properties. This emphasis is not based on the fact that the evidence of periodic behaviour is stronger than other forms of change but because the linking of cause and effect is easier. This is a consequence of both the ability to attribute cyclic variations to a specific cause when an *a priori* reason (see Section 5.1) is identified for the periodicity to occur and then to examine a physical link between observed fluctuations and their postulated cause. So, to the extent they can be identified, they offer the best insights into many of the forces driving climate change. In addition there is the fact that, whether or not they are real, cycles are the subject of immense and sustained speculation.

6.1 Autovariance and non-linearity

The description of the principal elements of the global climate in Chapter 3 introduced the concept that they can interact with one another in a complex manner to generate change. These processes are often termed *autovariance*, because they can be regarded as an internal part of the climate and so are essential to understanding the causes of change. Only when we know how the climate can fluctuate of its own accord will it be possible to separate out the impact of external influences (e.g. solar activity and astronomical tides) and human activities (e.g. the emission of greenhouse gases or deforestation). So, although these external factors will interact with the internally generated variations, it is preferable to address these issues sequentially.

Before doing so, there is, however, one further complication to consider. This is the question on *non-linearity* (see Box 6.1). As the consequences of this phenomenon can be exaggerated, erratic or even contradictory, their implications for understanding climatic change are profound. At the simplest level it is reflected in how errors in our knowledge of the initial state of the atmosphere will grow in any numerical weather prediction making the

Box 6.1 Is the climate chaotic?

A chaotic system is one whose behaviour is so sensitive to the initial conditions from which it started that precise future prediction is not possible. Even quite simple systems can exhibit chaos under some conditions. A condition for chaotic behaviour is that the relationship between quantities, which govern the motion of the system, be non-linear, in other words a description of the relationship on a graph would be a curve rather than a straight line. Since the physical relationships governing the atmosphere are non-linear, it can be expected to show chaotic behaviour. The performance of numerical weather forecasting shows a strong dependence on the quality of the initial data, confirming the atmosphere is a chaotic system.

Although the atmosphere is chaotic, the same need not apply to the climate as a whole. We know full well that the climate in any particular part of the world at any given time of the year sticks within prescribed limits: the temperature hardly ever rises above $-20\,°C$ at the South Pole, or falls below $20\,°C$ in Singapore. Computer models of the climate provide a reasonable description of these global patterns and their variation with the seasons. They can also be combined with what we know about the longer-term behaviour of the oceans to make useful predictions of seasonal weather in the tropics. In addition, the much longer variations associated with the ice ages can be largely explained in terms of changes in the Earth's orbital parameters. These results imply that some features of the climate are largely predictable. So while there appear to have been circumstances when the climate has behaved in a chaotic way in the past, for the most part it is not strongly chaotic.

detailed behaviour of the atmosphere more than a few days ahead unpredictable (see Section 10.2). In the longer-term analysis of how the climate may change, this behaviour may well set fundamental limits to our ability to predict the future.

In considering the basic aspects of non-linear processes our starting point is how the global climate system might behave when exhibiting some regular or approximately regular fluctuations. These may be the result of internal fluctuations or regular external influences. The most obvious effect of non-linear systems when subjected to a variety of cyclic forces is *harmonic generation*. This means that when forced to oscillate at a given frequency it will produce higher harmonics of the fundamental frequency. Moreover, if the system is excited by two or more frequencies, it will produce sums and differences of these frequencies. Simple harmonic generation produces multiples of the fundamental frequency. The amplitude of the higher harmonics will depend on the non-linearity of the system, but will, in general, decrease rapidly with increasing frequency. The sum and difference effects are best described in terms of two frequencies (ν_1 and ν_2). Non-linear systems acted upon by two such periodic input will generate not only harmonics of these frequencies but a whole range of combinations given by the general expression $m\nu_1 \pm n\nu_2$, where m and n are integers. This process produces not only high harmonics but different frequencies known as *subharmonics*, such as $\nu_1 - \nu_2$, $2\nu_1 - \nu_2$, $\nu_1 - 2\nu_2$, and so on, which produce low-frequency oscillations in the system.

Another interesting frequency response is known as '*entrainment*'. If a system, which has a natural self-excitation frequency ν_1, is subjected to an input of a slightly different frequency ν_2, the system may not behave in the way described above. Instead of both ν_1 and ν_2 and the frequency $\nu_1 - \nu_2$ being present, the whole system may oscillate at ν_2 with the original self-excitation oscillation effectively entrained by the imposed frequency. The range of frequencies over which this phenomenon can occur depends on the properties of the system and is known as the zone of synchronisation. A related but more unlikely effect is that in some non-linear systems it is possible either to start or stop an oscillation by the starting up on an entirely different frequency. This excitation or quenching is an entirely arbitrary consequence of the system, and usually termed *asynchronous* to reflect its unpredictable nature.

These somewhat abstract concepts may seem of little relevance to the problems of climate change, but in practice they have the potential to provide useful insights. An example may help to show this. As we will see, one of the most enigmatic features of the climate is a periodic feature of about 20 years duration. To the extent that this periodicity is accepted as real, it is variously ascribed to solar activity (the 22-year double sunspot cycle), lunar tides (18.6 years) and the inherent natural variability of certain parts of the climate, notably the ocean–atmosphere interactions in the Pacific Ocean, which appear to fluctuate on a 20-year timescale (see Section 3.7). If a natural resonance of this ocean basin happens to have a periodicity of around 20 years and both solar and lunar fluctuations on this timescale are capable of having some influence on the climate then how they combine will be a complicated process. In particular, at different times, we could expect either of the external periodicities to dominate or, alternatively the natural frequency could take over for a while. So, over time, any of the three frequencies could be observed with varying amplitude. Switches between each periodicity could occur at random and there could well be periods when the three processes effectively cancelled out and there are no significant fluctuations in the 20-year range, all of which will make the record of changes hard to interpret.

One final form of behaviour worth mentioning is the differing response of systems to self-excitation. Some require only small oscillations from equilibrium to build up. This is known as 'soft excitation'. Other systems require much greater perturbations before they will break into oscillation. This 'hard excitation' then appears with a sudden jump. Conversely, it will exhibit hysteresis in that as the oscillation decays, the system will continue to oscillate at a lower amplitude than the original threshold needed originally to get it going. This variable or erratic response of non-linear systems to both self-excitation and forced oscillation is yet another indication of the unpredictable nature of such systems.

The basic observations about the properties of simple non-linear systems have considerable implications for something as complex and non-linear as the global climate. Any propensity for parts of the system to oscillate at their own given frequencies will add or subtract to one another to produce a wide variety of fluctuations whose period and amplitude will vary with time. Furthermore, the properties of entrainment and excitation mean that, if by chance, certain parts of the global climate start to oscillate at some frequency, the onset of this resonant behaviour will not be predicted and its breakdown will equally well be sudden and unexpected, irrespective of how long the apparently cyclic episode has lasted.

6.2   Atmosphere–ocean interactions

The consequences of non-linear phenomena are central to considering how the atmosphere and the oceans interact to produce longer-term fluctuations. This interaction is, however, a classic example of the 'chicken and the egg' in that there are no obvious starting points in analysing the circular nature of the processes involved. So, it can be argued that the capacity of the atmosphere to sustain circulation patterns, which can produce changes in ocean temperatures, which then reinforce these anomalous patterns, means that the atmosphere is in the driving seat. Conversely, the alternative argument is that the massive thermal inertia of the oceans suppresses the wilder fluctuations of the atmosphere and so dictates how the climate behaves over periods of a year or more.

The simple 'linear' answer to this question is that the atmosphere controls the short term, while the oceans define the longer term. This view is, in part, supported by the observed spectrum of climatic fluctuations (*red noise* – see Section 5.2). While this can explain some of the quasi-cyclic behaviour of the climate, it cannot, however, resolve the issue of more sudden, but lasting shifts that are so important in understanding the non-linearity of the climate.

Before considering more erratic behaviour, however, the quasi-cyclic behaviour of atmosphere–ocean interactions is best explored in terms of modelling the ENSO (see Section 3.7). The major events in the 1980s led to the development of a series of computer models that appeared to provide accurate forecasts of the progress of ENSO events a year or more ahead. The major 1997 event caught almost everyone napping until it was well in train, and the sudden arrival of a modest event in 2006 was not clearly identified until two months before it got up steam. So the physical arguments underlying the forecasting models are being reappraised. But these models are able to simulate a quasi-periodic oscillation between warm and cold events, which shows they can capture some of the important features of the atmosphere–ocean interactions involved. In particular, the ability to represent the behaviour of the surface layer, so as to bring about a switch from warm to cold events or vice versa, is reassuring. This overcomes the basic problem of the model being trapped by a combination of SSTs and atmosphere circulation that maintain either warm or cold conditions permanently.

How the models achieve this important result depends on their treatment of slow-moving undulations in the thickness of the ocean surface layer, which slosh back and forth across the tropical Pacific. There are two types of wave. Close to the equator eastward-moving changes in the depth of the thermocline, known as Kelvin waves (see Section 3.7), take two to three months to cross the Pacific. Westward-moving changes in the depth of the thermocline are known as Rossby waves, which exhibit the same meandering properties as the long waves in the upper atmosphere (see Section 3.2). Because of the different physical properties of the ocean they are, however, much slower moving. On the equator they can take as little as three months to travel westwards across the Pacific, but, they, unlike Kelvin waves, are affected by the Coriolis force and are much slower moving at higher latitudes (at 30° N and S they take 10 years to cross the Pacific). When both these types of waves reach the edges of the Pacific they tend to be reflected back in the direction they have come from, but they switch type, so Kelvin waves make the return as Rossby waves and vice versa. Computer models of these processes, when combined with realistic representations of atmospheric circulation patterns, exhibit the property of switching back and forth between El Niño and non-El Niño (La Niña) conditions every three to five years.

The other feature of some of the models is that they are more likely to produce quasi-cyclic oscillations after there has been a significant perturbation of conditions in the equatorial Pacific. But, when any oscillations die down the system becomes more chaotic. This response is precisely what would be expected from the study of simpler non-linear systems in terms of those that exhibit the property of requiring hard excitation before oscillating. Furthermore, it reflects the experience of the early 1990s when less extreme El Niño (warm) conditions prevailed for nearly five years, whereas the models repeatedly predicted an earlier switch to colder conditions.

These problems were compounded by the difficulties experienced with forecasting the extreme event in 1997. One component of these problems was the part played by what are known as intraseasonal variations in the tropical weather. These are pulses of strong winds and rain that travel eastward around the equator separated by 30 to 60 days. Known as the Madden–Julian Oscillation (MJO), after the scientists who first identified it, this phenomenon is at its strongest from December to May, when it is one of the most intense weather systems in the tropics, pumping huge amounts of heat into the atmosphere. More relevant is the fact that if one of these bouts of activity, with its strong convection and westerly winds, happens to hit the western Pacific just as an ENSO warming event is about to break, it can stimulate its rapid development. This is what appears to have happened in early 1997. It is, however, in the nature of these MJO oscillations that they are difficult to forecast, being as chaotic as other medium-term aspects of atmospheric circulation. So, it appears that with the ENSO, at some times, the chaotic nature of the atmosphere may be in the driving seat, but, like the proverbial supertanker, once the ocean is heading in a given direction it takes a long time to turn it around.

The quasi-periodic variations of the NAO and the incidence of blocking in winter again raises the question of whether the atmosphere or ocean is

making the running (see Section 3.7). As in the Pacific the observed behaviour involves a similar set of feedback processes across the North Atlantic,
the adjacent continents, and wider global atmosphere–ocean interactions.
There is evidence that the position of the Gulf Stream is linked to the NAO.
Measurements of the 'north wall' of the current between 1966 and 1996
show that 60% of the variance of the position could be predicted in terms of
the NAO. Moreover, much of the remaining variance could be linked with
fluctuations in the Southern Oscillation. In addition, it may be part the
tripole, which appears to be a controlling factor in the climate in the
tropical and North Atlantic (see Section 3.7).

One possible explanation for these apparent links rests on the fact that
during the positive phase of the NAO the cooling in the Labrador Sea leads to
creation of more deep water. This flows southwards down the east coast of
North America where it could influence the strength and direction of the
Gulf Stream. This, in turn, could lead to changes in the temperature of the
eastern North Atlantic, which in due course could tip the NAO back into its
negative phase. But, so far, no convincing measurements have yet been
obtained which substantiate this explanation of why the NAO switches
back and forth, or why it should do so at any particular frequency.

6.3   Ocean currents

The major part played by ocean currents in the transport of energy to high
latitudes (see Section 3.6) means that any changes in this pattern could have
substantial climatic implications. In addition, it is apparent that different
ocean circulation patterns associated with earlier distributions of the continents were a major factor in the radically different climatic patterns that
existed for much of the geological past (see Section 8.1). What is more,
changes during and around the end of the last ice age (see Section 8.5)
suggest that the circulation in the North Atlantic can undergo sudden and
substantial shifts. These could lead to the global climate being able to exist
in distinctly different regimes even though the overall energy balance of the
system had not changed appreciably. So understanding how the large-scale
motions of the oceans can shift as a result of both their own natural
variability and external perturbations is central to unravelling the causes
of climate change.

While there is no doubt about the importance of different patterns of ocean
circulation in establishing past climatic regimes, the real issue is whether
sudden changes are relevant to the current debate on the impact of human

activities. This revolves around the question of how sensitive the Great Ocean Conveyor (GOC) (see Section 3.8) is to changes in the amount of freshwater entering the northern North Atlantic. Modelling work suggests that circulation patterns are extremely sensitive to run-off from the continents, the number of icebergs calved off Greenland and the amount of precipitation from low-pressure systems tracking northeastwards past Iceland and into the Norwegian Sea. Small changes in the total input may be able to trigger sudden switches to alternative patterns that carry warm surface water less far north before it sinks and returns southwards. This pattern would reduce SSTs around southern Greenland and Iceland by 5 °C or more. This would have a drastic impact on the climate of Europe and completely alter the atmospheric circulation patterns of the northern hemisphere.

The real issue is whether the changes produced in the models simply by altering the balance between evaporation and precipitation across the northern North Atlantic are climatically realistic. While the behaviour of the model can be tuned to produce chaotic shifts in circulation, the fact that, with one short-lived exception (see Section 8.5), such large shifts have not occurred in the last 10 kyr can be seen as evidence that bigger perturbations are required to get the GOC to switch to a different mode. One explanation is that changes seen both at the end of the last ice age and during its various fluctuations were the result of the periodic partial collapse of the ice sheet over North America which produced a surge of icebergs to flood out into the North Atlantic. These Heinrich events, which can be seen in the records of ocean sediments (see Fig. 8.14), would have a much greater impact on the GOC, so, at the moment the potential for the current climate to undergo such sudden shifts is an unresolved issue. However, the fact that human activities may be leading to greater and more rapid changes than anything seen in the last 10 kyr may be sufficient reason to worry about the potential of the GOC to 'flip' into a different mode (see Section 11.5).

The longer-term effect of different ocean-circulation regimes is bound up with the question of continental drift. The changes that have occurred not only in the distribution of the continents, but also in the seaways that opened and closed at various times (e.g. the opening of the Drake Passage around 25 to 30 Mya, or the closing of the Panama Isthmus around 3 Mya) had a profound effect on ocean circulation and hence global climate (see Section 8.3). In terms of causes of climate change, it is, however, a moot point as to whether these changes can be regarded as principally matters of ocean circulation. What is beyond doubt is that the different circulation regimes that existed before and after these tectonic developments were capable of maintaining radically different climatic states.

6.4   Volcanoes

It was Benjamin Franklin who first identified the potential of volcanoes to alter the climate. He suggested that the bitter winter of 1783–84 in northern Europe was caused by the dust cloud produced by the huge eruption of Laki in Iceland in July 1783, which dimmed the sun in Paris for months on end. This eruption, and others at low latitudes (e.g. Tambora) are plain to see in the Greenland ice cores (see Fig. 4.10).

Explosive volcanic eruptions can inject vast amounts of dust, and more significantly, sulphur dioxide into the upper atmosphere where this gas is converted into sulphuric acid aerosols. At altitudes of 15 to 30 km, where there is no significant vertical motion, these minute particles remain suspended for up to several years and are spread round the entire globe. A dust veil in the upper atmosphere absorbs sunlight. This heats the stratosphere but causes compensating cooling at lower levels, as less solar radiation reaches the Earth's surface. Analysis of past eruptions suggested that these physical processes did have a significant impact on the climate. There was, however, considerable doubt about just how big the impact on the global climate was. This uncertainty was the product of the fact that any cooling is accompanied by shifts in global weather patterns. Because the analysis, prior to the late nineteenth century, was based largely on observations of climate in middle latitudes of the northern hemisphere, parts of which experience disproportionate cooling, it was hard to be certain that observed changes were representative of global changes. These problems were compounded by the fact that following Krakatau in 1883 there was no truly significant eruption until Agung in Bali in 1963. What was missing was an adequate global picture of the direct impact of a major eruption on the temperature of the atmosphere.

In the 1980s things changed. First, there was the eruption of Mount St. Helens in 1980. Contrary to popular belief, however, this had no appreciable impact on the climate because it blew out sideways and did not inject much of its dust high in to the stratosphere. More significantly, its plume was low in sulphur compounds, so the limited cooling effect provided climatologists with valuable insights in to the essential role of sulphur in altering the climate. The eruption of El Chichón in Mexico in 1982, although a relatively small volcano, produced emissions that were very high in sulphur. Satellite measurements of the resultant stratospheric aerosol cloud provided confirmation of the climatic importance of sulphur. The massive eruption of Mount Pinatubo in the Philippines in 1991 offered a second chance to test these hypotheses (Fig. 6.1). It injected around 20 million tonnes of sulphur

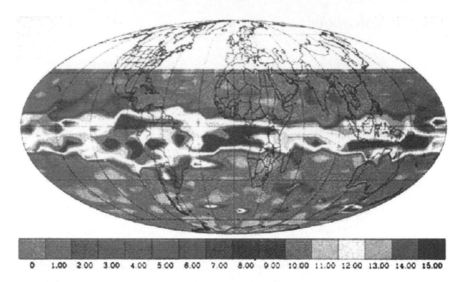

0   1.00   2.00   3.00   4.00   5.00   6.00   7.00   8.00   9.00   10.00   11.00   12.00   13.00   14.00   15.00

Fig. 6.1 Satellite measurements of the spread of the dust cloud from the eruption of Pinatubo in June 1991, showing that within five weeks the dust girdled the globe. (From Burroughs, 2005, Fig. 5.11.)

compounds in to the stratosphere, and was by far the most climatically important eruption of the twentieth century.

The eruptions of El Chichón and Pinatubo have been used to test theories of their impact on the climate, both by direct measurements and by the use of computer models of the climate. They are particularly valuable as the scale of the maximum perturbation caused by Pinatubo was estimated to be equivalent to a global reduction in solar energy reaching the Earth's surface of 3 to 4 $W m^{-2}$. This change is of the same scale as the radiative forcing due to the equivalent of doubling of CO_2 in the atmosphere ($4 W m^{-2}$ – see Section 2.1.3.). Accurate measurements of the temperature of the stratosphere (Fig. 6.2) using microwave radiometers on weather satellites (see Section 4.2) clearly show how the dust clouds from these eruptions warmed the upper atmosphere. The corresponding cooling of the lower atmosphere was not only observed in the surface-temperature record, but was also accurately predicted by computer models of the climate (see Section 10.2, Fig. 10.1).

What these observations confirm is that major volcanoes do cool the climate at ground level and there is a compensating and somewhat greater warming of the stratosphere. The effects of a single eruption last for around two to three years. This supports Benjamin Franklin's original hypothesis and the widespread assumption that the massive eruption of Tambora in

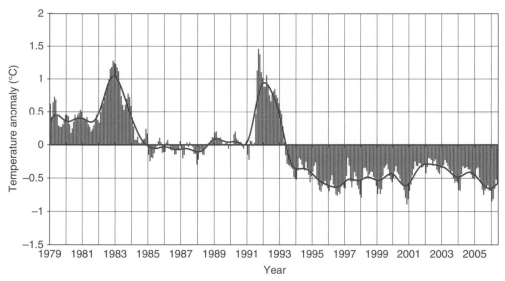

Fig. 6.2 Warming of the stratosphere between altitudes of 16 and 21 km observed by satellites showing monthly temperature anomalies smoothed with a 25-term binomial filter (thick black line). (Data from: http://vortex.nsstc.uah.edu/public/msu/.)

Indonesia in 1815, which injected five to ten times more material into the stratosphere than Pinatubo, was responsible, in 1816, for the 'year without a summer' when exceptionally late frosts destroyed crops in New England, and French vineyards experienced their latest wine harvest in at least the last five centuries. This cooling may also have been a contributory factor towards the consistently low temperatures in the 1810s (see Section 8.9). But Tambora is, at most, only part of the story, as the cooling of this decade was well and truly underway before the eruption occurred, and it has been postulated that an earlier, as yet unidentified, eruption in 1809, which shows up in Antarctic ice cores, may have set the cooling in motion.

This uncertainty about when and where earlier eruptions took place is a major limitation in establishing the role of volcanoes in past climate change. Recent studies of tree rings at mid-to-high latitudes around the northern hemisphere have been combined with ice-core data to produce an improved analysis of the impact of eruptions on the climate. This confirms that major eruptions produce a substantial drop in the summer temperatures for two or three years. If anything, the effect on winter temperatures is a warming, due to stronger westerly circulation at mid-to-high latitudes. This analysis has identified 1601 as the coldest summer in the last 600 years, possibly due to the eruption of a volcano in southern Peru the year before; 1816 is the second-coldest year in this period.

As for longer-term effects on the climate, the short-lived impact of volcanoes suggests they can only trigger lasting change if they coincide with other perturbations which are stimulating cooling as well. An interesting example of this is the fact that the largest eruption in the last million years – Toba in Sumatra – occurred at around 74 kya. Toba was gigantic, at least thirty times the size of Tambora. The caldera that resulted from this eruption is 100 km long and 60 km wide. It ejected about 3000 km^3 of material. In the central Indian Ocean, some 2500 km downwind of Toba, a 35 cm thick layer of ash was deposited. Furthermore, the eruption was rich in sulphur, which results in the formation of long-lasting sulphuric-acid-aerosol clouds in the stratosphere. This would have increased its climatic impact; the dust from the eruption would have dropped out of the atmosphere in a matter of months whereas these sulphuric-acid aerosols would have remained aloft for several years.

As for the precise timing, volcanic ash in sediment cores from the Arabian Sea provides the answer. They show clearly that this eruption coincided with the start of the sharp cooling that is seen in the GISP 2 ice core at the end of interstadial 20 (see Table 8.2 and Fig. 8.14). This eruption could have produced sufficient cooling for perennial snow cover to form for a number of years in, say, northern Canada. At a time when the Earth was already slipping into a cold phase of the last ice age, this snow cover with the additional cooling effect of reflecting more sunlight into space may well have tipped the balance.

More generally, the impact of Toba is estimated to have reduced the global temperature by about 5 °C. In summer the drop may have been as much as 15 °C in the temperate to high latitudes within a year or so and lasting for several more years. The effects on the growth of plants, and to life in the oceans of such a dramatic temperature drop would have been catastrophic. In many places the dust veil from the volcano would have effectively blotted out the sun. The cooling would have led to unseasonable frosts in many parts of the world and the disruption of growing seasons. The longer-term impact of Toba is more difficult to establish. It is in the nature of volcanic eruptions that they disrupt the climate temporarily. Furthermore, efforts to demonstrate that Toba was linked directly to subsequent changes in either monsoon rainfall in the Arabian Sea, or temperature trends in Greenland have been less convincing.

In geological history, there were much greater volcanic eruptions, usually known as *traps*, which spewed vast lava flows out over the course of thousands to hundreds of thousands of years. These massive events must have had an impact on the climate, but detecting the effects is made difficult by the blurring of the geological records. The most important consequence of these examples of volcanic activity is in connection with past mass extinctions (see Section 9.3), which show up so distinctly in the fossil record.

There may be another explanation for coincidences between volcanic eruptions and climate change, which reverses the chain of events. This is that they can be triggered by sudden sustained changes in atmospheric circulation. These alter the stress on the Earth's crust and are even detected in tiny changes in the length of the day (i.e. how fast the Earth rotates). An alternative mechanism, which is supported by the timing of some major eruptions, is that changes in sea level (see Section 8.3) alter the crustal loading in the vicinity of volcanoes close to the shoreline and this triggers some eruptions. So, as in many other proposals for causes of climate change, an appreciation of the complex web of feedback mechanisms between cause and effect are central to forming a balanced understanding of the processes at work.

6.5   Sunspots and solar activity

When compared to the Earth's climate, the behaviour of the Sun would seem to be a relatively simple matter. As a massive ball of ionised gas, it is comparatively homogeneous. It is, however, subject to colossal gravitational, electrical and magnetic forces. So it is a seething mass of convective and circulatory motions, which are all capable of oscillatory motions that produce a variety of features capable of having climatic consequences here on Earth. This means that its behaviour is probably every bit as complex as the Earth's climate, and we can only view it from afar.

There are three regions of the Sun that are of particular relevance to weather cycles. The first is the visible surface (the *photosphere*), which has a radiative temperature of around 5700 K and is the source of the bulk of the energy reaching the Earth. The energy output of this thin shell, some 100 km thick (the Sun's radius is 700 000 km), is affected by two principal features: transient dark areas known as *sunspots*, and brighter regions known as *faculae*. The second region is an irregular layer above the photosphere, known as the *chromosphere*. The importance of this layer is, as far as the Earth's weather is concerned, that observations of its behaviour can provide valuable insights into the Sun's magnetic field. In addition, changes in the chromosphere are a measure of fluctuations in the ultraviolet (UV) and shorter wavelengths, which have a disproportionate impact on the properties of the Earth's upper atmosphere. Chromospheric measurements include studying the bright boundaries surrounding sunspots (known as *plage*) that are associated with concentrations of magnetic fields in these areas and provide a means of linking sunspot activity with solar magnetic fields.

The third region is the outer atmosphere of the Sun: the *corona*. This region is in many ways a mystery. Although tenuous, it has an effective temperature of around 1 000 000 K. Its shape changes with the sunspot cycle, and in terms of impact on the Earth's weather, its most important features are *coronal holes*. Often found near the Sun's poles, these dark regions were first detected by X-ray equipment on Skylab in the early 1970s, and have been monitored since. They cover about 20% of the Sun's area when the solar activity cycle is at a minimum, and, as activity increases, they are replaced by smaller scale open-field regions scattered over the solar surface. The total area of coronal holes is closely associated with the high speed solar wind and this appears to be a good physical proxy of the global-scale 22-year solar magnetic cycle and shorter-term aspects of the modulation of the flux of cosmic rays entering the Earth's atmosphere.

Any discussion of solar variability must start with sunspots, as these have played a central role in the search for connections between solar variability and climate change. As noted in Section 2.2, sunspot activity has been monitored since the seventeenth century. The average number of spots and their mean area fluctuate over time in a more or less regular manner with a mean period of about 11.2 years (see Fig. 2.10). During this fluctuation the rate of increase in their number exceeds the rate of decrease, the period varies between 7.5 and 16 years, and the amplitude varies by about ± 50%. The variation in number during each period ranges from virtually no spots during the minimum in solar activity to just over 200 in the most active cycle that peaked in 1957. The instrumental record has now accumulated reliable data since around 1750 and is now in its twenty-third cycle of activity, which peaked at around 120 spots in 2000. Each cycle begins when the spots show up in both the northern and southern hemispheres some 35° away from the solar equator. As the cycle develops, the older spots fade away and new more numerous spots appear at lower latitudes (Fig. 6.3). Towards the end of each cycle the number decreases and the spots are concentrated at latitudes some 5° from the equator. This cycle of activity does not necessarily fall to zero at the minima, because a new cycle will start at high latitudes before the old one has died away at low latitudes. This overlap can exceed two years.

Faculae are closely associated with sunspots, and are most easily observed near the solar limb. Their output is linked with the incidence of sunspots

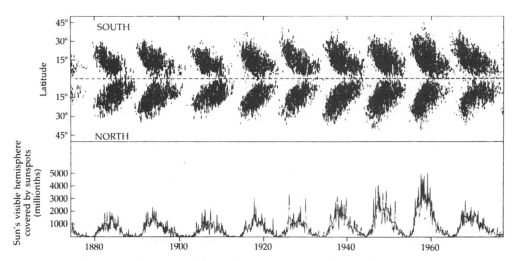

Fig. 6.3 Observed variations of the number of sunspots between 1875 and 1975 together with the 'Maunder butterfly' diagram showing the distribution of sunspots with heliographic latitude and their general movement in each hemisphere during successive sunspot cycles. (From Giovanelli, 1984. Data supplied by SERC, RGO. From Burroughs, 2003, Fig. 6.2.)

and it is now clear that they are a more important factor in explaining how changes in solar activity could affect the weather, as their increased brightness is the dominant factor in changing solar output rather than sunspot darkening. In addition, the surface of the Sun is affected by a whole range of shorter-term disturbances, but in terms of their scale and duration they are of less consequence to weather variations from year to year.

There have been many efforts to demonstrate that sunspots affect the amount of energy reaching the Earth from the Sun. Ground-based observations were, however, unable to provide convincing evidence as perturbations due to scintillations caused by the atmosphere swamped any small changes in the Sun's output. The advent of accurate satellite-borne instruments has, however, changed the situation radically. Starting in late 1978 a series of satellites have made measurements that have produced unequivocal observations of how the Sun's output varies with the 11-year cycle in solar activity (see Fig. 2.11). What these results show is that the level of total solar irradiance (TSI) is greatest around the solar maxima. At the same time, TSI variability is greatest, declining to a minimum when the Sun is least active. The TSI variation is slightly less than 0.1% over the 11-year cycle.

The fact that the TSI rises with sunspot number was, initially, a complication. Because sunspots are areas of low solar luminosity it was presumed that the output would decline with sunspot number, as the effect would be to block some of the Sun's output and so reduce the overall energy flux. Clearly this did not explain the longer-term observed changes in solar output. On the contrary, this observation supported the longstanding hypothesis that the cold period known as the Little Ice Age, and more particularly the colder weather of the late seventeenth century, was the result of an almost complete absence of sunspots during this period, known as the *Maunder Minimum* after the astronomer who brought this absence to the attention of the astronomical community.

The accepted explanation for the inverse relationship with sunspot number is that an associated increase in the brightness of the faculae outweighs the effect of sunspots. Judith Lean, now at the Naval Research Laboratory, Washington DC, and P. Foukal of Cambridge Research and Information, Massachusetts, first proposed a model that combined the changes in sunspot number and faculae brightness based on microwave observations. The relationship between sunspot darkening and faculae brightening is not simply a matter of periods of highest sunspot frequency being those when the Sun is radiating the most energy. For instance, the model estimated that the Sun's radiance was higher at the peak of solar activity in 1980 than it was during 1959 when the sunspot number was higher. Nevertheless, this model explained the observed changes in solar output since 1980, and supported the hypothesis that the colder weather of the seventeenth century was the result of the Maunder Minimum.

The small change in the TSI (less than 0.1%) during the last two solar cycles is, however, on its own, an order of magnitude too small to explain observed correlation between solar activity and global temperature trends since the late nineteenth century. This has led to a number of models being developed to predict changes in TSI since the seventeenth century. These have drawn on various proxies for solar activity. The models produce a variety of interesting results because, unlike sunspot numbers, which return to essentially zero at each solar minimum, the other indicators show longer-term modulation.

Judith Lean's recent analysis of the TSI (Fig. 6.4) estimates a 0.20% increase ($2.8 \, \mathrm{W \, m^{-2}}$) between the Maunder Minimum and the mean level in the late twentieth century. In addition, this work has calculated the variation of output over different spectral bands. Faculae dominate the solar irradiance variations at UV wavelengths: at 200 nanometres (nm) sunspots are a factor of two darker than they are at 500 nm while faculae are two orders of magnitude brighter. Using satellite measurements and modelling work Lean estimates that over the 11-year cycle solar radiance varies by 14% at

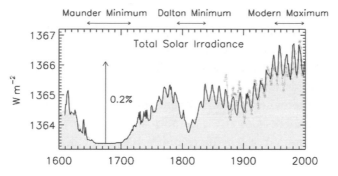

Fig. 6.4 An estimate of the annual total irradiance from the Sun since 1600 (from Lean, 2000). The shading identifies the 11-year running mean and the arrow shows the percentage increase to the mean of cycle 22 (1986 to 1996). The light grey line plus square symbols is the total irradiance (scaled by 0.999) determined independently by Lockwood & Stamper (1999).

around 150 nm and 8.3% at around 200 nm, but by less than 1% at wavelengths longer than 300 nm. What this means is that since the Maunder Minimum solar irradiance is estimated to have increased by 0.7% in the wavelength range 120 to 400 nm, by 0.2% in the range 400 to 1000 nm, but by only 0.07% at longer wavelengths.

The fact that the TSI varies so much with wavelength is probably the key to the close statistical link between solar activity and observed changes in the climate since the seventeenth century. So, it is possible that solar variability could be amplified in the Earth's atmosphere. The significance of changes in the UV region is that these wavelengths are largely absorbed in the atmosphere and this could enhance the impact on the climate. In particular, wavelengths shorter than about 300 nm are strongly absorbed by oxygen and ozone in the stratosphere (see Box 2.2), which affects the temperature at these levels. These temperature effects have an impact on circulation patterns in the equatorial stratosphere, which are transmitted to polar regions in winter months.

The more contentious issue is to explain how the impact of solar variability in the stratosphere influences the weather at lower levels. Clearly, the amount of solar radiation entering the lower atmosphere varies with solar activity. Alterations in stratospheric ozone concentrations caused by the changing UV flux suggest there is an amplification process at work. This reduces the amount of solar energy reaching the lower atmosphere in middle and high latitudes in winter during the active part of the solar cycle. These changes could have a significant impact on global circulation as preliminary calculations suggest that increased solar UV radiance in the lower tropical stratosphere will expand the Hadley circulation (see Section 3.1) leading to a poleward shift of the subtropical westerly jet and the mid-latitude storm track.

Another photochemical consequence of changing UV fluxes reaching the lower atmosphere is to affect the formation of free radicals in the lower atmosphere, notably the hydroxyl radical (OH). This alters the production of condensation nuclei and hence the formation of clouds. In effect more UV radiation reaching the troposphere will increase the concentration of condensation nuclei and hence make it cloudier – a process that would amplify any fluctuations due to varying solar activity.

The nature of the Sun's magnetic field is also an essential part of understanding possible links between solar variability and the Earth's weather. In general terms, at the beginning of a sunspot cycle the solar magnetic field resembles a dipole that is aligned with the Sun's rotation axis. This means that at low latitudes the field lines are closed, whereas at higher latitudes they are open. As the cycle builds up to a maximum this simple pattern breaks down into a disordered state. During the latter part of the cycle the dipole field is re-established.

In addition, the polarity of sunspots alternates between positive and negative in successive 11-year cycles. Sunspots tend to travel in pairs or groups of opposite polarity as if they are the ends of a horseshoe magnet poking through the surface of the Sun. During one 11-year cycle, as the spots traverse the face of the Sun in an east–west direction, the leading spots in each group in the northern hemisphere will generally have positive polarity while the trailing spots will be negative. In the southern hemisphere the reverse situation occurs with the leading spots being negative. It is this pattern that reverses in successive cycles. Known as the Hale magnetic cycle, this 22-year cycle could be the key to the amplification process, as it determines the solar-induced interplanetary magnetic-field direction, which is one of the controlling factors in the solar-wind interaction with the Earth's magnetosphere.

As a general observation the 20- to 22-year cycle has been more prevalent in climatic data than the more obvious 11-year sunspot cycle. So, any magnetic process, which amplifies the impact of solar variability on the weather, will have played part in climate change. One possibility is that the intensity of the solar magnetic field affects the quantity of energetic particles emitted from the Sun. A second is that it alters the Earth's magnetic field and so influences the amount of cosmic rays (energetic particles from both the Sun and from elsewhere in the universe) that are funnelled down into the atmosphere. These effects may alter the properties of the upper atmosphere.

Cosmic rays come in a wide variety of forms. They consist mainly of protons with smaller amounts of helium and heavier nuclei. Cosmic rays of low energy have their origin in the Sun and are absorbed high in the atmosphere. It is these particles that are the origin of aurora during periods of high solar activity active. Galactic cosmic rays (GCRs) are of much higher energy and have an appreciable impact on the troposphere. When the Sun is more active GCRs are less able to reach the Earth and so their impact on the lower atmosphere is inversely related to solar activity.

Cosmic rays produce various chemical species such as NO, OH and NO_3 that can catalyse chemical reactions. This leads to changes in the atmospheric concentrations of radiatively active molecules such as ozone (O_3), nitrogen dioxide (NO_2), nitrous oxide (N_2O) and methane (CH_4). These species are most likely to be seen in the stratosphere where their impact will be similar, but less significant to the changes caused by solar UV variations. In addition, the formation of ions will affect the behaviour of aerosols and cirrus clouds that have a direct radiative impact and also alter the amount of water vapour throughout the atmosphere. These changes could lead to shifts in the radiative balance of both the stratosphere and troposphere and so produce long-term fluctuations in the temperature.

The most controversial aspect of these changes is the question of whether GCRs alter cloud cover at lower levels. Studies by Henrik Svensmark, of the Danish Space Research Institute (DSRI), Copenhagen, Denmark and co-workers demonstrated a correlation between total cloud cover, from satellite studies, and cosmic ray flux between 1984 and 1991. Their analysis of total cloud using data over the oceans between 60 °S and 60 °N from geostationary satellites found an increase in cloudiness of 3 to 4% from solar maximum to minimum. They proposed that increased GCR flux causes total cloud amounts to rise and this cools the climate. More recently, however, the correlation has been far less impressive. On longer time-scales Svensmark also demonstrated that northern-hemisphere surface temperatures between 1937 and 1994 follow variations in cosmic-ray flux and solar-cycle length more closely than total irradiance or sunspot number.

The interannual variations in cloudiness are, however, difficult to distinguish from parallel changes caused by warm and cold ENSO events. Also the correlation with cosmic-ray flux tends to be reduced if high-latitude data are included. This would not be expected if cosmic rays directly induced increases in cloudiness, as cosmic-ray flux is greatest at high latitudes.

Moreover, a mechanism whereby cosmic rays resulted in greater cloud cover would be most likely to affect high cloud as ionisation is greatest at these altitudes. But, if high cloud does respond to cosmic rays, it is not clear that this would cause global cooling, as for thin high cloud the long-wave warming effects dominate the short-wave cooling effect. Clearly more work is needed to unscramble the range of impacts that solar particles and GCRs may have on the atmosphere at different altitudes and latitudes.

Ralph Markson at the Massachusetts Institute of Technology identified another possible influence of the changing flux in cosmic rays on the climate. He proposed that the modulation of cosmic rays by the Sun leads to changes in the Earth's electric field and hence thunderstorm activity. This mechanism has three attractions. First, it requires no significant change in solar energy to alter the state of the Earth's magnetic field and stratospheric conductivity, while offering the possibility of releasing and redistributing large amounts of energy already present in the troposphere. Secondly, it does not require strong links between the upper and lower atmosphere as the electric-field variations encompass the whole atmosphere from the ionosphere to the Earth's surface. Thirdly, the response of the electrical field to change in the magnetic field is almost instantaneous and so can explain how the weather responds within a day or to certain changes in solar activity.

Markson postulates that world-wide thunderstorms play a central role in maintaining a global electric circuit, so changing of the conductivity of the upper atmosphere can alter the incidence of thunderstorm activity. Greater stratospheric ionisation could lead to increased thunderstorm electrification either locally or as a consequence of global changes in the Earth's electric field, and this may alter thunderstorm development. Changes in the Sun's magnetic field will also alter the number of energetic particles it emits. This will have complicated effects on the Earth's atmosphere, which could include affecting the cloudiness caused by thunderstorms. At high latitudes the effects of the flux of solar protons, which is directly related to solar activity, will predominate. At low latitudes the magnetic-field variations will modulate galactic cosmic rays and produce an effect that is out of phase with solar activity. This may explain why there is evidence of a high positive correlation between the sunspot cycle and high-latitude thunderstorm activity whereas at low latitudes it is either non-existent or negative.

Although this theory was not pursued at the time, it has stimulated work on the possibility that charged particles were more efficient than uncharged ones in acting as cloud condensation nuclei (CCN). In particular, Brian Tinsley at the University of Texas developed a detailed mechanism for a link between cosmic rays and cloudiness, in which aerosols ionised by GCRs are more effective as ice nuclei and cause freezing of supercooled water in clouds. The consequent increased release of latent heat in clouds enhances convection that promotes cyclonic development and hence increased storminess.

The IPCC TAR concluded that these proposals required more research to establish whether or not it could be of sufficient magnitude to result in the claimed effects. Recent results obtained by Frank Arnold's group at the Atmospheric Physics Division, at the Max-Planck-Institut of Nuclear Physics, Heidelberg, Germany, have provided the first observational evidence of cosmic ray-induced aerosol formation in the upper troposphere. The link with Tinsley's work may be that, while ions cannot act as efficient condensation sites for water vapour, they can act as condensation sites for sulphuric acid vapour which then with water vapour grow into aerosols, and then CCN. In addition, the electrical charging of water droplets and aerosols increases their collision efficiency. In theory, a higher the level of charge carried by an aerosol leads to greater the collection efficiency, which may enhance the capture of those rare aerosols suitable as ice nuclei.

Unlike Markson' hypothesis, which focused on changes in thunderstorms, this approach is of more relevance to weakly electrified, non-thunderstorm clouds, such as marine stratocumulus, or nimbostratus. These are of much greater geographic and temporal extent, and are affected by changes in the ionosphere–earth current density and cosmic ray flux. By changing precipitation rates or radiative balance, the changes in the clouds then affect atmospheric dynamics and temperature. But, until we have confirmation that changes in atmospheric electricity lead to changes in patterns of cloudiness, we have to treat all these theories with caution.

There are two additional features of magnetic-field changes that need to be considered here. The first is that they will not be symmetric about the Earth's axis and this will affect how cosmic rays modify the upper atmosphere (Fig. 6.5). Because the geomagnetic poles are not coincident with the geographic poles, the perturbations of the magnetic field will be off axis. The circulation pattern in the northern hemisphere shows a similar off-axis form (see Section 5.1). This also ties in with the link between the QBO and the 11-year sunspot cycle (Section 3.2), which is reflected in both the circulation over North America centred on the geomagnetic pole and the latitude of the winter storm track across the North Atlantic.

The second consequence of magnetic-field changes is the long-term decline in the fair-weather potential gradient in the twentieth century. This measure of the strength of the global electrical circuit appears to be

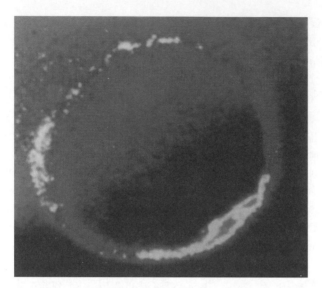

Fig. 6.5 A satellite ultraviolet image of the auroral oval. This shows that the auroral activity encircles the geomagnetic poles, but that the maximum activity is centred near local midnight and local noon. (From Lundin, Eliasson & Murphree, 1991.)

linked to the parallel decline in cosmic-ray flux over the same period. So to the extent that this is a measure of how solar activity could influence global cloudiness it provides additional insight into the processes that may be leading to longer periodic fluctuations in the Earth's climate.

If thunderstorms play a part in all these processes, then the key to monitoring their activity may lie in an unexpected area. This involves an intriguing property of the global atmosphere. The thin shell formed by the atmosphere between the Earth's surface and the ionosphere, at an altitude of some 80 km, forms a waveguide for very low-frequency electromagnetic radiation. In particular, a frequency of around 8 Hz has a wavelength of 40 000 km (the circumference of the Earth) can ring round the world with virtually no loss. Known as *Schumann resonance* after the German scientist who first proposed this phenomenon in 1952, this signal can be measured by electromagnetic detectors. The dominant source of such radiation is lightning in thunderstorms. So, at any time, the size of the 8-Hz signal is a direct measure of worldwide thunderstorm activity.

The potential importance of this phenomenon is that not only is it a measure of the global level of thunderstorm activity but also this activity appears to be peculiarly sensitive to the mean global temperature – a 1 °C rise will lead to a 10% rise in the number of thunderstorms. So the measurement of Schumann resonances offer the prospect of both checking whether the

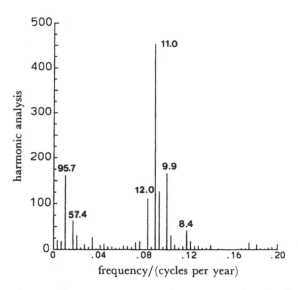

Fig. 6.6 The power spectrum of sunspot numbers for the period 1700 to 1986. (From Pecker and Runcorn, 1990, with permission of the Royal Society. From Burroughs, 2003, Fig. 6.7a.)

level of global thunderstorm activity is affected by solar activity and also providing a different means of monitoring global warming.

Thus far we have only considered the basic 11-year sunspot cycle and the double Hale cycle. As is obvious from Figure 2.10, the variation in the size of the peaks in successive cycles also shows evidence of a periodicity of around 90–100 years – often termed the Gleissberg cycle. This cycle is also associated with the change in the period between successive peaks of solar activity, which lengthens as the peak levels decline. When sunspot series are subjected to detailed spectral analysis two important features emerge. First, the principal feature at around 11 years (Fig. 6.6) is split into two main peaks at 11.1 and 10.0 years. Second, about 20% of the total variance in sunspot numbers is associated with the 90-year periodicity, which may be a difference frequency between the two features at 10 and 11.1 years. In addition, there is considerable evidence of a 200-year cycle in sunspot activity, including human observations, notably from China.

The importance of the longer cycles in solar activity is the marked parallelism between changes in global temperatures and both sunspot numbers and the length of the 11-year sunspot cycle. Recent work by Judith Lean and colleagues, building on the model described earlier, have reconstructed variations in solar UV radiation since the beginning of the seventeenth century, and its implications for climatic change. They conclude that the correlation between changes in solar UV and northern hemisphere temperatures is 0.86 and solar forcing has been responsible for half of the warming between 1860 and 1970, and one-third of the rapid increase between 1970 and 1990. In statistical terms, the reconstruction accounts for 74% of the variance in northern hemisphere temperature in the period 1610 to 1800 and 56% from 1800 to 1990. These figures are highly significant (see Box 5.2).

In considering the possible impact of longer-term variations of solar activity, the best source of information is tree-ring data. Because the strength of the solar magnetic field modulates the flux of cosmic rays entering the Earth's atmosphere, it affects the production of carbon-14 (^{14}C) (see Section 4.6). When solar activity is high, the magnetic shield is strong and the amount of ^{14}C formed, and hence incorporated in tree rings, is relatively low. Conversely, at times of low solar activity, the shield is weak and more ^{14}C is formed. By measuring the amount of ^{14}C in tree-ring series, and comparing it with what would be present if it had been produced at a constant rate, it is possible to build up a measure of past solar activity (see Box 4.4).

Studies of the spectra of ^{14}C fluctuations of tree-ring records going back some 9000 years may provide evidence of periodic fluctuations in solar activity. Studies have been conducted on tree-ring data from the White Mountains in California, and from fossilised oaks in Europe. This analysis identified five strong periodic features in the data. These occurred at about 2300, 500, 355, 204 and 154 years. The fact that these fluctuations have almost exactly the same periodicities and power densities, and are present in tree rings from widely separated parts of the globe, suggests they are due to real physical effects. What is not clear is the extent to which they are a product of solar variability as opposed to other factors that could influence the amount of radiocarbon in the atmosphere. Alternative explanations include changes in the magnetic field of the Earth, or changes in the size and exchange rates of the carbon reservoirs on the surface of the Earth, including the biosphere and the oceans.

The existence of the periodicity at around 200 years is intriguing, as it is close to observed fluctuations in sunspot numbers. Over the last 1000 years there appear to have been major lulls in solar activity at around AD 1280, AD 1480 and AD 1680 (the Maunder Minimum). These periods of low solar activity appear to coincide with periods of a cooler climate in the northern hemisphere, notably the period at the end of the seventeenth century (see Section 8.9). Since 1700 this pattern has been less evident, with the most marked minimum in the 1810s, and a broad reduction in the late nineteenth century (see Fig. 2.10).

6.6 Tidal forces

The obvious link between tidal forces and the climate is through the direct alteration of the movement of the atmosphere, the oceans, and even the Earth's crust. The nature of these links varies in complexity. Tidal effects in the atmosphere are relatively predictable and measurable but tiny compared with normal atmospheric fluctuations. In the oceans the broad effects can be calculated, but estimating changes in the major currents is far more difficult. When it comes to the movements of the Earth's crust, the problems are compounded by possible links with solar activity. The direct influence of the tides could influence the release of tectonic energy in the form of volcanism. Since there is evidence that major volcanic eruptions have triggered periods of climatic cooling (see Section 6.4), this would enable small extraterrestrial effects to be amplified to produce more significant climatic fluctuations. In addition, there is evidence of intense bursts of solar activity interacting with the Earth's magnetic field to produce measurable changes in the length of the day. Such sudden, albeit tiny changes in the rate of rotation of the Earth, could also trigger the release of tectonic activity (see Section 9.1).

The gravitational forces acting on the Earth as it orbits the Sun can be divided into four categories. First, there are the tides resulting from the combined pull of the Moon and the Sun. These tidal forces affect the movement of both the atmosphere and the oceans and also exert stress on the Earth's crust. Secondly, the forces exerted on the Earth by the changing positions of the other planets will play a similar but much smaller role. Thirdly, there is the possibility of the same tidal forces due to planetary motions affecting the Sun's circulation and with it solar activity. Finally, there are the orbital effects of these motions. Because these can cause the Earth to speed up and slow down in its orbit and also lead to small movements of the Sun about the centre of mass of the solar system, there is the potential for small periodic influences on the Earth's climate.

Clearly, all these tidal effects are interlinked. But, as a first step, we need to consider each potential influence separately before trying to make observations about what their combined effects may be. So, the obvious place to start is with the semi-diurnal tides in the atmosphere and the oceans. These are caused by the gravitational attraction of both the Sun and the Moon. On the nearside of the Earth the atmosphere and oceans are attracted towards both bodies, as is the Earth itself, which pulls away from its fluid envelope on the far side. Because of the Earth' rotation and the Moon's orbital motion, any particular place on the Earth's surface experiences two complete cycles of high and low tidal stress every 25 hours. On average the Sun's pull is roughly half that of the Moon. This means that there is a threefold variation between when the Sun and the Moon are on the same side of the Earth and pulling together, and when they are on opposite sides so that their effects are partially cancelled out.

In terms of longer-term climate change, the dissipation of tidal energy in the oceans may be an essential part of the mixing of abyssal water in the oceans and the maintenance meridional overturning circulation (MOC) (see Section 3.8). Walter Munk at the Scripps Institute of Oceanography, and Carl Wunsch of the Massachusetts Institute of Technology, first proposed this idea in 1998. They argued that without a mechanical source of energy thermohaline processes were insufficient to maintain vertical mixing. There are only two candidates for such a source: winds and tides. They concluded that half the power needed to return deep water to the surface was coming from the dissipation of tidal energy. Topex/Poseidon satellite altimetry studies have come up with some interesting results that may shed new light on this proposal. Prior to this work it was widely assumed that this energy, representing the effect the Moon receding from the Earth at a rate of some 4 cms a year, was dissipated in the shallow waters of the continental shelves around the world. It is now reckoned that about half this energy is fed into deeper water where it exerts a significant effect on the strength of the major ocean currents and hence the transport of energy from the tropics to polar regions (see Section 7.4). So we now need to look more closely at evidence of whether, say, the 18.61-year cycle can be detected in the strength of the principal ocean currents, because, if so, this would be a potentially important way for lunar tides to affect the climate.

When it comes to movements of the Earth's crust, the problems are compounded by possible links with solar activity. The direct influence of the tides could influence the release of tectonic energy in the form of volcanism. Since there is considerable evidence that major volcanic eruptions have triggered periods of climatic cooling, this would enable small extraterrestrial effects to be amplified to produce more significant climatic fluctuations. In addition, there is evidence that intense bursts of solar activity interact with the Earth's magnetic field to produce measurable changes in the length of the day. Such sudden tiny changes in the rate of rotation of the Earth could also trigger volcanic activity. It should be noted that, while there is no evidence that their occurrence was in any way related to this effect, the three climatically important volcanoes in the second half of the twentieth century (Agung in 1963, El Chichon in 1982 and Pinatubo in 1991) were spaced in such a way as to have a confusing impact on the interpretation of any solar or tidal effects. So this is yet one more complication that must be considered when seeking to identify the cause of oscillations in the climate.

The most convincing evidence of a periodicity in volcanic activity has emerged from recently published analysis of ice cores from high latitudes in both hemispheres. The work identified the 61 largest tropical volcanic eruptions in the last 600 years on the basis of those eruptions that produced an identifiable sulphate signal in at least one ice core in each hemisphere. This approach is likely to identify all the eruptions which had a global impact on the climate during this period, although it is possible that effectively simultaneous eruptions in both hemispheres could also be part of this analysis, but the chances of this having happened is small. What the study show is a remarkably strong 76-year cycle that appears to show a three-fold difference in the incidence of major tropical eruption between the peaks and the troughs of the cycle (rising and falling between one eruption every five years down to one every 15 years). There is, however, one difficulty with this periodicity as far as tidal effects are concerned. There is no obvious reason why the tides should be the cause. Indeed the nearest possibility is the 90-year solar cycle, and linking that with the release of tectonic activity poses a much greater challenge.

The longer-term fluctuations in the tides, which may affect the climate, depend on the nature of the Moon's orbit around the Earth. If the Moon completed an exact number of orbits of the Earth in the time it takes the Earth to complete a circuit of the Sun, the pattern of tidal forces would be relatively simple and reproducible. But this is not the case. The nearest it comes to this is that 13 tidal months amount to 355 days, which is sometimes referred to as the 'tidal year'. So although the Earth and the Moon return to approximately the same position after about a year, it takes much longer for more precise repetitions of alignment to occur.

This brings into play the relative motion of the Earth's perihelion (the point on its orbit when it is closest to the Sun), and the Moon's perigee (the point on its orbit when it is closest to the Earth). While the changing distance between these three bodies will exert a continual influence on the tidal forces, the relative position of the perihelion and the perigee play an important part in the longer-term periodicities that may influence the weather.

There are two periods of particular interest. The first is the 8.85-year period in the advance of the longitude of the Moon's perigee that

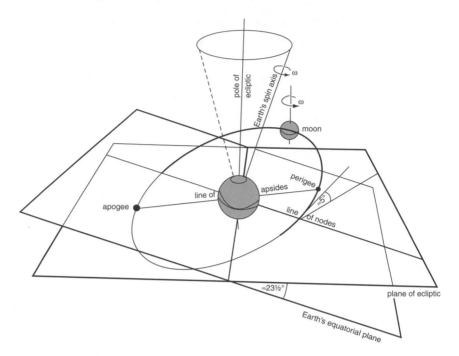

Fig. 6.7 The orbital geometry of the Earth–Moon system. The Earth's average orbital motion around the Sun defines the ecliptic and the Earth's spin axis rotates about the pole of the ecliptic once every 26 000 years because of the precession. The Earth's equinox, the intersection of the equator and the ecliptic, moves along the ecliptic at the same rate. The lunar orbit intercepts the ecliptic along the line of the nodes that moves around the ecliptic because of the solar attraction. For the same reason the line of the apsides precesses in the orbital plane. The Moon's spin axis remains normal to the ecliptic. (After Smith, 1982. From Burroughs, 2003, Fig. 6.9.)

determines the times of the alignment of the perigee with the Earth's perihelion (Fig. 6.7). The second is the 18.61-year period in the regression of the longitude of the node – the line joining the points where the Moon's orbit crosses the ecliptic. This period defines the exactness of the alignment of the Moon's perigee and the Earth's perihelion.

The 18.61-year cycle is the most widely studied aspect of the tidal stress. This is because the regression of the node defines how the angle of the Moon's orbit to the Earth's equatorial plane combines with or partially cancels out the tilt of the Earth's axis. This has the effect of altering the variation in the maximum declination of the Moon from the ecliptic. Because of the tilt of the Earth's axis, the equatorial plane is at an angle of 23° 27′ from the ecliptic. This combines with the angle of the Moon's orbit to the equatorial plane so that the maximum declination to the ecliptic is

28° 40′ N and S. But this extreme value occurs only every 18.6 years; at the opposite extreme, halfway between when the maximum declination is the difference between the tilt of the Earth's axis and the angle of the Moon's orbit from the equatorial plane, the value ranges only between 18° 21′ N and S. The significance of this variation is that when the declination is greatest, the tidal forces at high latitudes are greatest. Recent peaks in these forces have occurred in 1913, 1931, 1950, 1969, 1988 and late 2005.

The importance of the 8.85-year cycle in the alignment of the Moon's perigee and the Earth's perihelion is not its direct impact on tidal forces but how it combines with the 18.61-year cycle. Calculations of tidal stress at high latitudes since AD 1100 show that there is a tidal resonance of about 179.3 years. The significance of this period is that it is yet another candidate for the general group of periodicities that have been identified are around 180 to 200 years.

Looking beyond the tidal effects of the Sun, Moon and Earth, there is the influence of the other planets in the solar system. Of these, the motions of the giant planets Jupiter, Saturn, Uranus and Neptune are potentially the most interesting. These have orbital periods of 11.86, 29.5, 84 and 165 years respectively. So, singly or in combination, they could influence the tidal forces on the Earth. In practice, because of its mass (318 times that of the Earth) and its relative proximity to the Earth, Jupiter is by far and away the most important factor in these planetary tidal forces. Moreover, the fact that its orbital period is close to the 11-year cycle observed in the sunspot number means that it could either be a confusing factor in identifying the cause of periodicities in the weather or be directly linked with the sunspot cycle itself.

In respect of the tidal influences of the planets on the Earth's atmosphere and oceans, detailed calculations show that the scale of these perturbations are small compared with the variation of the tidal forces due to the motions of the Earth and Moon around the Sun. So unless there is a good physical reason for the much smaller gravitational influences of the planets to have a proportionately greater impact, they are unlikely to have a significant effect on the climate. One such possibility is that they can produce potentially important perturbations on the Earth's orbit. When the Earth is on one side of the Sun and all the other planets grouped in a tight arc on the other side, there is evidence that this configuration has reproducible effects in Chinese climatic records. Known as a synod, this particular alignment of the planets occurs every 179 years or so, although every five or six cycles it can be as short as 140 years. The fundamental rhythm of these synods is defined by the approximate alignment of Jupiter, Saturn, Uranus and Neptune. The movements of Mercury, Venus, Earth and Mars define the time of year when the synod occurs.

A study of Chinese records examined those occasions when the remaining planets were grouped within an arc subtending less than 90° at the Earth. After 1600 BC in China, when the synod occurred in the summer half of the year, the subsequent few decades tended to feature warm summers. Conversely, after winter synods there were more frequent cold winters. Moreover, where the grouping was narrower the effects tended to be more pronounced. The physical explanation for these observations appears to be in the way in which the planetary configuration causes the Earth to speed up or slow down in its orbit. While the period of the orbit (365.24 days) remains unaltered, when the Earth is travelling towards the grouping of giant planets it speeds up and when travelling away from them it slows down. This means that it spends less time in the half of the orbit when it is closest to the conjunction of planets and more time on the far side of the Sun. So if the synod occurs in the summer half of the year this period will be lengthened slightly and vice versa. In the extreme example in 1665 when the planets subtended an angle of 45° to the Earth, it is estimated that the winter half of the year was increased by almost two days with a corresponding shortening of the summer half of the year. This is potentially a significant physical shift and may, in part, explain why this synod marked the onset of the coldest period of the Little Ice Age in the northern hemisphere (see Section 8.9).

Another possibility is that the planetary motions influence the observed cyclic behaviour of sunspot numbers. The periods of the orbits of Jupiter, Uranus and Neptune roughly coincide with the 11-, 90- and 180-year cycles in the sunspot records and this has led to attempts to produce a planetary theory of sunspots. Of particular interest are the effects of the planets on the motion of the Sun around the centre of mass of the solar system. Calculations show that this complicated motion is dominated by the orbits of Jupiter and Saturn, and, in particular, the time taken for Jupiter to lap Saturn – 19.9 years. But over the last 1200 years this period of the Sun's motion has varied between 15 and 26 years. The other important cycle is 177.9 years that is a product of the near coincidence of 15 Jupiter orbits and 6 Saturn orbits. The consequence of the Sun's motion is to affect its oblateness, diameter and rate of spin, all of which could influence the sunspot mechanism. So we cannot consider solar variability and tidal forces in isolation.

6.7   Orbital variations

The Earth's orbit around the Sun is also influenced by the gravitational interactions with the Moon and other planets on much longer timescales.

The resulting perturbations give rise to cyclical variations in orbital eccentricity, obliquity and precession with periods of 413, 100, 41, 19 and 23 kyr respectively (see Section 2.1.4). These variations are climatically important as they control the seasonal and latitudinal distribution of solar radiation. This theory of the climatic effects is usually attributed to the Yugoslav geophysicist Milutin Milankovitch, who transformed the earlier semi-quantitative work by James Croll into the mathematical framework of an astronomical theory of climate. This theory has been refined since the 1960s to provide an explanation of the observed waxing and waning of the ice ages over the last million years.

As noted in Section 2.1.4, the key to explaining variations in the orbital parameters can trigger ice ages is the amount of solar radiation received at high latitudes during the summer. This is critical to the growth and decay of ice sheets. At 65° N this quantity has varied by more than 9% during the last 800 000 years. Fluctuations of this order are sufficient to trigger a significant climatic response. But, in explaining observed climatic change (see Section 8.4), there is a major snag. It is now accepted that latitudinal and seasonal variations in incident solar radiation due to the precession of the equinoxes (the 19- and 23-kyr periodicities) and the variations in the tilt of the Earth's axis (the 41-kyr periodicity) are sufficient to drive the significant climatic changes observed on these timescales. By comparison the 100-kyr-eccentricity periodicity is weakest of the orbital effects. This causes considerable difficulties, as the 100-kyr ice-age cycle is the strongest feature in the climatic record in the last 800 000 years. So it is necessary to consider separately the explanations of the observed 19-, 23- and 41-kyr cycles and the dominant 100-kyr cycle.

John Imbrie and John Z. Imbrie of Brown University, Rhode Island adopted a direct approach to possible models of the ice ages in 1980. Instead of using numerical models to test the astronomical theory, they used the geological record as the yardstick against which to judge the performance of various physical models. These models, which have become more sophisticated over the years, fall into two broad categories. The first group adopted an equilibrium approach to the changes in solar radiation. This involved calculating the climatic conditions that should occur for various combinations of orbital parameters. This produced reasonably realistic changes in temperature patterns, but was unable to reflect the inertia of the climatic system, notably the characteristic timescales of the growth and decay of the ice sheets, which appear to be of the same order of magnitude as those of orbital forcing.

The alternative approach is to use a differential model in which the rate of change of the climate is a function of both the orbital forcing and the current state of the climate. Not only is this a more realistic representation of climatic behaviour but it also contains a non-linear relationship between the input and the output that has important physical consequences. But it remains controversial, as there is no agreement as to which climatic factors should be given particular emphasis and what values should be attached to them. For this reason the simple empirical approach used in the model developed by the Imbries will be considered first, as it provides valuable insights into the processes involved.

The Imbrie model considered only the link between the orbital forcing function and the land-ice volume. This approach reflected the fact that the change in ice volume as recorded in the oxygen isotope records from deep-sea cores is the most accurately defined climatic parameter over the last million years or so, and also that the cryosphere is the part of the climatic system whose characteristic timescales of response closely match the dominant 100-kyr period of the orbital forcing. Indeed the common feature to emerge from all the modelling work in recent years is that only when the northern ice sheets exceeded a critical size did the 100-kyr cycle take over in the climate equation. This is seen as the key to explaining why before around 800 kya this cycle did not feature strongly in the climate records (see Fig. 8.9). The earlier interval of the Cenozoic ice ages, from, 3 to 0.9 Mya, was almost completely dominated by the 41-kyr-tilt cycle. Prior to this the 19- and 23-kyr cycles were more important. The advent of the 41-kyr cycle around 2.4 Mya ago is linked with the start of major northern-hemisphere glaciation. So the lengthening period of the major climatic variations during the last five million years or so appears to be due to the increasing size of the ice sheets.

If the size of the northern ice sheets is critical, the important factor is the sensitivity of the ice volume to the time constants for the growth and decay of the ice sheets and the lag between the changes in the solar radiation falling in summer at high latitudes of the northern hemisphere. Once the Imbrie model had achieved a reasonable representation of this long-term behaviour, the more rapid response of the rest of the global climate to the other orbital parameters, which have a linear impact, could be added in to build up a better picture of the progression of the ice ages.

In spite of developments in recent years, the Imbrie model remains a good start to considering the physical processes that may be at work in the 100-kyr periodicity. In this model the parameters were tuned to achieve the best fit between the calculated ice-volume changes and the oxygen-isotope record. The most important features of the model include, first, the orbital forcing is fixed by the changes in the tilt of the Earth's axis and the

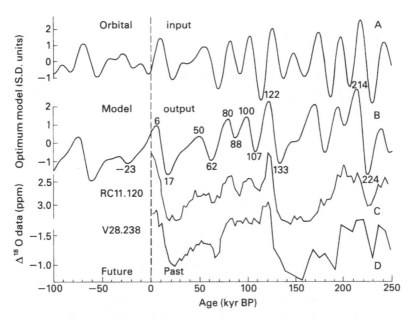

Fig. 6.8 The combined orbital effects shown in Fig. 2.6 can be used as the input (A) to a model whose output (B) shows a marked similarity to the oxygen isotope variations observed in deep-sea cores from the southern Indian Ocean (C) and the Pacific Ocean (D). (After Imbrie & Imbrie, 1980. From Burroughs, 2003, Fig. 7.2.)

precession of the perihelion of the orbit. The changes in eccentricity (i.e. the 100-kyr and 413-kyr periodicities) do not exert a significant influence on the seasonal and latitudinal variations of the radiation input. So the orbital forcing used in the model contained only the 19-, 23- and 41-kyr cycles, in spite of the fact that the ice-volume curve over the last 800 kyr is dominated by the 100-kyr cycle. The second essential feature is that the time constants of growth and decay of the ice sheets are markedly different. This reflects the evidence that the ice sheets built up slowly, but collapsed dramatically at the end of each ice age (see Fig. 8.12).

The best results were obtained with a set of parameters that included time constants of growth and decay of the ice sheets of 42.5 and 16 kyr, respectively, and a lag of 2 kyr between the orbital forcing and the response of the climatic state. This produced a good fit over the last 150 kyr (Fig. 6.8), although prior to this the match was less good. More importantly, the calculated changes in ice volume included 100- and 413-kyr periodicities, although the relative strengths were wrong, with the former being too weak and the latter too strong. But the fact that these essential periodicities were present at all highlights an important aspect of non-linear models. This is that the output spectrum has major features that are absent in the input-forcing function. This is an example of the phenomena described in Section 6.1 where a simple non-linear system can generate sum and difference frequencies. The important feature here is that the choice of the time constants of ice growth and decay plus the non-linearity of the model combines to produce the required longer periodicities. The difference between the 19- and 23-kyr will produce a 110-kyr periodicity and that between the 23-kyr and 41-kyr will produce a 52-kyr periodicity, and the difference between these two is a 100-kyr cycle. Tuning the model and highlighting a given frequency is both rewarding and also underlines the limitations of the approach adopted. The dependence on empirically derived time constants, which have only the broadest links with the physical behaviour of the ice sheets, is a major limitation.

The process driving the 100-kyr cycle remains the key challenge in improving ice-age models. An overall assessment produced in 1993 of the many models developed to address this issue confirmed that the non-linear response of the global climate to the northern ice sheets exceeding a critical value was the explanation of the 100-kyr cycle. All that was required was that the build-up of the ice sheets introduced a lag of some 15 kyr into the system for the climate to cease to be dominated by the 23 and 41-kyr cycles and switch into the 100-kyr mode. Components of the atmosphere–ocean–ice system could be responsible for generating this response. The eccentricity cycle need play no part in these changes: they could be nothing more than a natural response of the climate to the variations in the size of the ice sheets.

In 1998 Didier Paillard, at CEA/DSM in France, proposed an interesting refinement to this type of thinking. He has produced models which include the possibility if the climate system switching between three distinct climate regimes (e.g. interglacial, mild glacial and full glacial). These different states in the climate system could well be the product of different modes of global ocean thermohaline circulation (see Section 6.3). The switch between these regimes is controlled by a combination of changes in insolation and/or ice-sheet volume. By defining the conditions for the transition between the three regimes the model is capable of reproducing with remarkable accuracy the changes in ice-sheet volume over the last million years. So, yet again, the capacity of the global climate to move between markedly different, but relatively stable states appears to be an essential feature of explaining bigger climate changes.

Going beyond the interactions between the atmosphere, oceans and ice sheets, and the Earth's orbital parameters, other factors have been invoked to explain why the climate has become more susceptible to glaciation in recent geological history. In particular, it has been argued that the rapid tectonic uplift of the Himalayas and parts of western North America during the past few million years has increased the sensitivity of the global climate to orbital forcing. The link between the meandering of the jet stream, these mountainous areas (see Section 3.4) and the regions where the northern ice sheets developed may be the key to the current pattern of ice ages.

6.8 Continental drift

When it comes to climatic changes over millions of years we have to add in the effects of plate tectonics (*continental drift*). Most of these changes are best presented in terms of the changes that occurred at the time (see Chapter 8). There are, however, a number of general features that need to be considered alongside other causes of climate change.

On the longest timescales the amount of land at high latitudes is the most profound consequence of continental drift. When the proportion increases, it is easier for ice sheets to form in polar regions. This happened during the Permo-Carboniferous glaciation from 330 to 250 Mya, with the formation of the supercontinent, Pangaea. All the Earth's land masses came together, and stretched from the equator to the South Pole, with what are now Antarctica and India at high latitudes forming the centre of glaciation (see Section 8.1). The other occasion when a large proportion of high latitudes of the northern hemisphere became land-covered is during the last 3 Myr or so. This made it easier for ice sheets to form.

Another consequence of continental drift was the formation of the Himalayas and the Tibetan Plateau following the collision of the Indian subcontinent with Eurasia some 50 Mya. This process was gradual, becoming a major climatic factor at the beginning of the Miocene. At around the same time, the uplift of the Western Cordillera in both North and South America further altered the global atmospheric circulation patterns (see Section 3.2).

These sweeping climatic consequences of continental drift have been reinforced by more subtle changes in opening and closing of ocean gateways. In the North Atlantic the Iceland–Faeroe sill subsided below the sea around 38 Mya. Then Greenland and Svalbard separated around 36 Mya letting high-latitude water enter the North Atlantic.

Perhaps the most significant change in ocean circulation occurred with the separation of Antarctica from Australia and then South America. With the movement of Australia northward, the South Tasman Rise subsided to produce a shallow-water connection between the Indian and Pacific Oceans. The deep-water Tasmania–Antarctica passage opened around 34 Mya. About the same time, a deep-water channel developed to form the Drake Passage between South America and Antarctica. Although a shallow channel may have existed since the mid-Eocene, it was this event that allowed the circumpolar current to develop and isolated Antarctica climatically. Then, much later, the elevation of the Panamanian Isthmus closed the link between the Atlantic and Pacific Oceans around 3 Mya, radically altering ocean circulation patterns in both oceans. The final change may have been an important additional factor in creating the right conditions for permanent ice sheets to start forming in the northern hemisphere.

6.9   Changes in atmospheric composition

Against the background of current concern about greenhouse-gas emissions it is easy to lose sight of just how much the levels of these gases have varied throughout the history of the Earth. Since pre-Cambrian times the level of carbon dioxide (CO_2) in the atmosphere has changed massively. The scale of these changes is still the subject of some scientific debate. The balance between CO_2 and oxygen (O_2) is a consequence of the biotic cycle, which includes the interaction of these gases with living matter and its waste products. Oxygen in the biological cycle is produced by photosynthesis of plants and consumed in the oxidation of organic materials to form CO_2. The consumption of CO_2 by plants in the process of photosynthesis is the other half of this cycle. So changes in the productivity of the Earth's biosphere (see Section 3.5) have altered the amount of CO_2 in the atmosphere. Also, where large quantities of organic material have been deposited in the form of coal, oil and natural gas, it depleted CO_2 levels.

On the timescale of tens of millions of years, changes in the level of CO_2 in the atmosphere are, however, controlled by two other processes. The first is the atmospheric input by volcanoes. The second is the weathering of exposed silicate rocks, which depletes the atmosphere of CO_2. These two processes are linked to plate tectonics. When combined with changes in the level of plant life these processes have led to striking changes in the amount of CO_2 in the atmosphere on geological timescales. Prior to the Cambrian period, variations in CO_2 levels are the subject of great uncertainty and may have been anything from five to several hundred times current levels. These changes are, however, of great climatic importance as they were probably a central factor in the relative stability of the Earth's temperature for much of

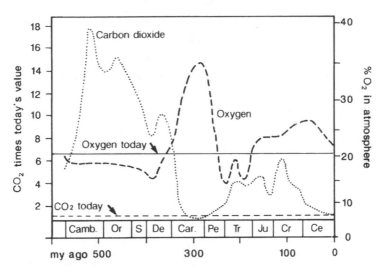

Fig. 6.9 The precise levels of atmosphere constituents carbon dioxide and oxygen depends on interpreting the complex fossil record in terms of the many reservoirs, fluxes and interactions which could lock up or release carbon and oxygen from rocks. The most comprehensive analysis by Robert Berner of Yale University contains two dramatic features. First, the high levels of CO_2 in the Cambrian and Ordovician, and second, the sharp reduction in CO_2 and the build up of oxygen in the late Carboniferous which are linked with the laying down of huge deposits of coal at the time. (From van Andel, 1994, Fig. 14.6.)

pre-Cambrian times. The lower solar constant at the time (see Section 2.2) would have led to lower temperatures but for the enhanced greenhouse effect (see Box 2.1) of the high levels of CO_2 at the time. This is why the evidence of much colder episodes (the Snowball Earth – see Section 8.1) and fluctuations of CO_2 levels are central to understanding longer-term climate change.

Over the last 600 Myr, although CO_2 levels appear to have fluctuated less, they remained high compared with more recent changes. In particular, there is some evidence that the huge deposition of organic material during the Carboniferous period led to major perturbations in both CO_2 and O_2 levels (Fig. 6.9). Thereafter, CO_2 levels rose, and during the Cretaceous it is estimated they were four to ten times current levels. After the mid-Cretaceous CO_2 levels declined possibly because the sluggish circulation of the oceans in a warm world created anoxic conditions at depth, which allowed large quantities of carbon to be deposited in the form of organic-rich shales, which contain high concentrations of hydrocarbons. Levels of atmospheric CO_2 declined around the end of the Cretaceous, and apart from some evidence of an increase in the Eocene around 45 Mya, have remained at relatively low levels since then.

A remarkable set of independent measurements of CO_2 levels has come from fossil leaves. The density of pores (*stomata*) on the leaf surface, through which cell metabolism draws in the CO_2 and transpires its products, including oxygen, has evolved in response to changing CO_2 levels in the atmosphere. When CO_2 levels are low, more pores are needed, and vice versa. Where it is possible to find fossils of a species that has a long evolutionary history, the pore density of ancient leaves provides clues into CO_2 environment in which the plant. One such tree is the *Ginkgo biloba* (the Maidenhair Tree). There are four genera of plants that are closely related to Ginkgo whose fossil record stretches back 300 Myr.

Gregory Retallack of the University of Oregon has measured the stomatal index of the Ginkgo leaves and those of related genera to assess CO_2 levels since the early Permian. The general pattern tallies well with the results of other measurements. In particular, CO_2 levels were high levels between 275 and 250 Mya and from 130 to 90 Mya. Periods of low CO_2 prevailed between about 296 and 275 Mya, between 30 and 20 Mya, and during the past 8 Myr. The periods of low CO_2 coincide with the periods of cool (*icehouse*) modes of Earth's climate (see Section 8.2), as inferred from sedimentary evidence such as the extent of glacial deposits at high latitudes.

The stomatal index record is incomplete, and its resolution is rather coarse (about 5 Myr). Nonetheless, compared with previous reconstructions of CO_2, Retallack's record shows many more maxima and minima that roughly correspond to the warmer and cooler periods seen in the marine temperature record. All in all, this seems the data point to a long-term coupling between CO_2 and temperature. Moreover, the data seem to resolve the paradox of a cool climate and high levels of CO_2 at times during the Jurassic and Cretaceous periods (between 210 and 120 Mya). Retallack's results indicate that CO_2 levels were not as high as has been suggested by other geological data, which raises interesting questions about the extent to which CO_2 levels controlled climate change at this time.

The changes during the Pleistocene are possibly of greater interest. Analysis of air trapped in bubbles in the ice in Antarctica (see Section 4.4.2.) show changes in the concentration of trace gases (e.g. CO_2 and CH_4) that closely track changes in temperature. The correlation between CO_2 levels and temperature over the last 220 kyr is 0.81 (the figure for CH_4 is 0.76). Both these figures are statistically highly significant and are clear evidence that variations in the greenhouse gases played a direct part of the observed change in the climate. Although these changes are small

Fig. 6.10 The meteorite crater in Arizona which was formed some 50 000 years ago. The diameter of the crater is 1.2 km. (From Smith, 1982. Fig. 2.13.)

compared with changes in CO_2 over geological time, they have important implications for explaining the physical processes driving both the last ice age and for predicting the consequences of the present-day build-up of greenhouse gases in the atmosphere (see Section 11.2).

Although vitally important in terms of evolution, changes in oxygen levels during the Phanerozoic have been of little direct consequence to climate change. Oxygen first appeared in atmosphere in very low levels around 2000 Mya and rose to some 10% of current levels at about 700 Mya. During the Cambrian it reached concentrations comparable with levels now. The fluctuations over the last 600 Myr shown in Figure 6.10 remain the subject of scientific dispute and investigation.

These changes in atmospheric constituents are bound to have exerted some influence on the climates of the past. The higher levels of CO_2 in the atmosphere during, say, the Cretaceous must have been an important factor in the maintenance of the benign climate of that period (see Section 8.1). The more difficult question is just how important a factor was it when compared to the issues of continental distribution and ocean circulation and what part the long-term decline in CO_2 over the last 50 million years or so has played in the general cooling of the climate. The broad conclusion of various computer-model studies (see Chapter 10) is that the location of the continents and the oceans are not a sufficient explanation, and that the long-term cooling over this period may be associated with a gradual decrease in atmospheric CO_2.

6.10   A belch from the deep

There is one aspect of changing atmospheric conditions that requires separate consideration. This is the question of methane locked up in what are known as *marine clathrates*. In the right circumstances, methane and water can form an ice-like substance called a clathrate, or gas hydrate, at temperatures above normal freezing point. Ice has an open molecular structure and under pressure this openness enables it to accommodate gas molecules. These, in turn, lend support to ice crystals that would otherwise melt. The pressure at the bottom of the oceans often provides the right conditions for the formation of clathrates.

The stability of clathrates is limited by temperature and pressure: they are stable at low temperatures and/or high pressures. These requirements are combined with the need for relatively large amounts of organic matter for bacterial action to produce methane in the sediments. As a result, clathrates are mainly restricted to two regions (high latitudes, and along the continental margins in the oceans). In polar regions they are often linked to permafrost occurrence onshore and on the continental shelves. On the outer continental margins, they are found where the supply of organic material is high enough to generate enough methane, and water temperatures are close to freezing. It is estimated that globally some two to eight trillion tons of methane are locked up in clathrates deposits.

The climatic importance of clathrates is that at times of warming it is possible that conditions underground could alter sufficiently to release large amounts of methane. This 'belch from the deep' could lead to an additional abrupt warming. The most likely example of this type of warming event occurred in the late stages of the Palaeocene, around 55 Mya, close to the Paleocene–Eocene boundary (see Section 8.2). Measurements of ^{18}O isotope levels and Mg/Ca ratios suggest that average global temperatures rose by some 5 to 6 °C. The warmth lasted approximately 100 kyr. Known as the *Late Palaeocene Thermal Maximum (LPTM)*, measurement of carbon isotope ratios (see Box 4.1) at the time suggest that this aberrant climatic event was caused by a giant release of methane. This conclusion is supported by the fact that, at the same time, ^{13}C isotope records show that a large mass of carbon with low ^{13}C concentration must have been released. The mass of carbon was sufficiently large to lower the pH of the ocean and drive widespread dissolution of seafloor carbonates.

The importance of this event to our climate change studies is the similarity with the current release of carbon into the atmosphere by humans. The estimated magnitude of carbon release at the LPTM is on the order of one to two trillion tons, similar to the predicted release of greenhouse gases by humans during the coming century. Moreover, the period of recovery after the release ($\sim$100 kyr) is similar to that forecast for the time it will take to get over current human activities. So, the LPTM is an interesting example of the climatic impact of a massive carbon release.

6.11   Catastrophes and the 'nuclear winter'

Few subjects in climatic change excite greater passions than the possibility of the Earth being hit by an extraterrestrial object. Such a catastrophe, whether as the result of a collision with a large meteorite, asteroid or fragments of a comet, is regarded by many as being in the realms of science fiction. Part of the reason for this reaction is that it is sometimes seen as undermining the principle of uniformatorianism (see Section 8.1) that was originally conceived to deal with earlier explanations of the geological record. These had placed great emphasis on catastrophes being the explanation for many features of the landscape. But, as presented here, the occasional impact from an asteroid is not seen as undermining the principle of uniformatorianism in any real sense.

The simple fact of the matter is that there can be little doubt that an object greater than, say, a kilometre across would have a massive temporary impact on the climate. Equally well, there can be no doubt that objects of this size and greater have struck the Earth in the past. What is at issue is whether these events can explain past changes in the climate and what is the probability of a significant event occurring in the foreseeable future.

The most recent example of the Earth being hit by a sizeable object was on 30 June 1908 in Siberia. This was either a meteorite or a fragment of a comet some 50 m across. It burnt out at an altitude of around 10 km over the Tunguska River, about 1000 km north-north-west of Lake Baikal, with an explosive force equivalent to a 10–20-megaton bomb. It devastated

$2000 \, km^2$ of forest and generated a seismic shock of about 5 on the Richter scale. While the climatic impact of this event was small, it clearly demonstrates the potential of collision of the Earth with objects in space. Equally impressive is the impact of the fragments of the comet Shoemaker-Levy 9 with Jupiter in July 1994 providing a graphic illustration of the consequences of such an event. The biggest fragment was about 3 to 5 km across ejected superheated gas to a height of 1300 km and created a circular cloud larger than the entire Earth. Although there are a variety of physical arguments about the relative impact of different types of objects (e.g. meteorites with a variety of compositions, or cometary fragments – the generic term *bolide* is used to describe these objects), the message is simple – any body a kilometre or more across would do immense damage and have a profound and immediate impact on the climate.

The possibility of bolide collisions being the cause of some or all of the major mass extinctions is a matter of vigorous controversy (see Section 9.3). The most comprehensively analysed event relates to the end of the Cretaceous, and the passing of the Age of the Dinosaurs. These arguments about the scale, pace and nature of the extinction that occurred at around 65 Mya will continue to rage for a long time. But, what we need to consider is the climatic consequences of the proposed bolide collision. The accepted view is that a crater on the Yucatan peninsula in Mexico is the site of impact. A body some 10 km across hit this part of the world making a crater between 200 and 300 km across. The immediate consequences of this impact were cataclysmic. The combination of massive earthquakes, huge tidal waves, widespread forest fires and vast amounts of dust and debris hurled high into the atmosphere would have done untold damage. What is more pertinent is whether this produced lasting climatic effects that could have precipitated the extinction of the dinosaurs?

The answer may lie in work that was done for entirely different reasons. During the late 1970s a lot of studies were conducted into the climatic impact of nuclear warfare. These considered the range of environmental consequences of a large-scale nuclear exchange between the major powers involving the detonation of the equivalent of some 6000 megatons of nuclear devices. They estimated the direct consequences of nuclear explosions, the subsequent fires and smoke generation and the impact of this smoke on the global climate. This combination was predicted to produce a sudden and dramatic cooling that was widely known as the *nuclear winter*. Many features of this analysis can be translated into the catastrophic consequences of a bolide impact. This is because there is some comparability between the events. A kilometre-diameter bolide would have an impact energy of 100 000 megatons, while a 10 km object, comparable to the one that hit 65 Mya, would be equivalent to 100 million megatons. Although the predicted impact of a large bolide is much greater, its impact is concentrated in one spot whereas a nuclear conflict would many much smaller explosion spread over a much wider area. But such an impact would eject huge amounts of material high above the Earth. This would then rain down like molten meteorites over a vast area, setting fire to forests and grasslands. So the analogy between a nuclear winter and these prehistoric catastrophes may be a useful approximation. Indeed there is evidence of global fires in the form of large amounts of soot present in sediment marking the Cretaceous–Tertiary extinction boundary.

There is another feature of this boundary that fits in well with the nuclear winter hypothesis. This is the sudden drop in pollen from flowering plants and a jump in fern spores. This is known as the *fern spike*. In a few millimetres of the sedimentary sequence, the fern spore content goes from 25 to 99%. This change is reminiscent of sudden shifts of vegetation often seen after a forest fire today – for a few years a lush forest is replaced by an opportunistic flora dominated by ferns.

The broad conclusion is that a nuclear winter would be a dramatic, but short-term event. If it occurred in the summer half of the year in the Northern Hemisphere it would cause a drop in continental land surface temperature of between 20 and 40 °C within a few days. Thereafter, the smoke clouds would stabilise and might be held aloft for as much as a year, thereby maintaining winter-like conditions over the northern continents for at least one growing season and possibly longer. This more general cooling would have a global impact, causing frosts in regions that normally never experienced them. So the overall consequences on all forms of land-based life would be catastrophic, with only the most adaptable surviving.

In terms of climate change such an event can be regarded as both exceedingly dramatic but also transient. Unless other parts of the climate system experienced some more profound shift, it would soon return to normal as the atmosphere cleared of smoke and debris. One such change might be a coincident sustained period of volcanism. There have been a number of occasions in geological history when such events have occurred. Lasting hundreds of thousands of years, they could have a huge impact on the atmosphere and maintain an atmospheric dust veil (see Section 8.4) for long enough to precipitate a longer cold episode with the expansion of the polar ice-sheets. There is evidence of massive volcanism at around the end of the Cretaceous, which led to the formation of the Deccan traps in western India. This geological episode, which involved a series of events many of which led to the spewing out of several thousand km^3 of lava and added up to a total of several million km^3 of lava. On their own, or in conjunction with a bolide impact, they could have led to change the climate. Indeed, there is no reason why either the bolide impact or volcanism alone has to be the sole explanation of the extinction; they could both have contributed to the death of the dinosaurs.

On the more general question of the incidence of impacts and their distribution over time the arguments rage equally fiercely. Two issues matter. The first is to establish just how often impacts of different sizes can occur. The second is to decide whether there are good reasons for there being any regularity in the spacing of the biggest impacts. The frequency of impacts is no easy matter as most objects burn up in the atmosphere and leave no detectable evidence of their arrival. Even the Tunguska impact left no clearly identifiable trace of its make-up after its explosive arrival. So usually only the biggest objects produce craters. The notable exception is iron meteorites which often reach the surface and if sizeable produce distinctive craters (Fig. 6.10). Furthermore, identifying major impacts in geological history is difficult, as many features have been covered up by sediment, eroded away or lie beneath the oceans. For this reason, the estimates of the current rate of impacts of different sizes is based on the analysis of new small craters on the Moon (Fig. 6.11), which show up because of the bright rays of ejecta around them. This suggests a Tunguska event might happen every few hundred years, while a climatically significant event of, say, about 1000 megatons could occur every 10 000 years or so.

The question of whether collisions will be random or exhibit some regularity depends on how debris is distributed in the vicinity of the solar system, in general, and in Earth-crossing orbits in particular. Asteroids and meteorites can probably be regarded as randomly distributed, although the variations in the frequency of micrometeorite impacts (*shooting stars*) throughout the year suggests that there are regions of higher density in the vicinity of the Earth's orbit. Cometary debris is more likely to be located in distinctive patterns related to the orbits of massive comets that have broken up in the past. The whole issue of where these orbits are, in respect of the Earth, and how frequently our planet will run into this type of debris is the subject of controversy. Suffice it to say; if there are regions in space where such objects are more common, when the Earth moves through these regions the risk of impact will rise.

Leaving aside the questions about the more distant prospect of mass extinctions whose frequency is measured in tens to hundreds of millions of years, there is the question of whether any events in recorded history

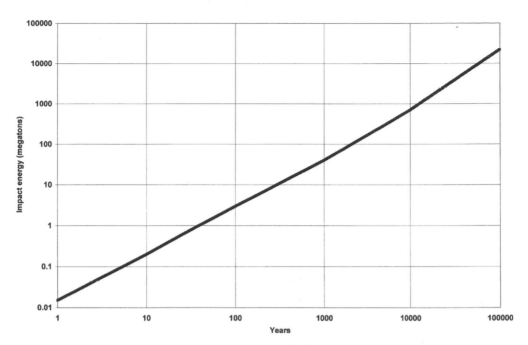

Fig. 6.11 An estimate of the incidence of bolide impacts of given energies on the Earth. (Data from Shoemaker, 1983.)

could be attributed to smaller impacts. Perhaps the best candidate for such an event is the 'mystery cloud' of 536 AD. It is recorded by chroniclers from Rome to China that the Sun dimmed dramatically for up to 18 months and widespread crop failures occurred. These contemporary observations are supported by clear evidence of climatic deterioration in tree-ring widths (Fig. 6.12). This event shows up much more strikingly in tree-ring data than the impact of the volcano Tambora in 1815 (see Section 6.4).

The most commonly assumed explanation is that this event was the result of a major volcanic eruption, possibly Rabaul in New Guinea, which created a global dust veil in the upper atmosphere that led to a cooling of the Earth's surface (see Section 6.4). But the evidence of the eruption in Greenland ice cores (see Section 6.4) is, at best, ambiguous; an event at around this time does not match up to either Tambora, or an unidentified eruption in 1259 AD. In the absence of confirmation of an exceptional eruption around 536 AD, it can be argued an alternative explanation is needed and a collision with an extraterrestrial object is a plausible option. This debate will only be resolved by new measurements in, say, ice cores that point unequivocally in one direction or other. In the meantime, all that can be said is there is no

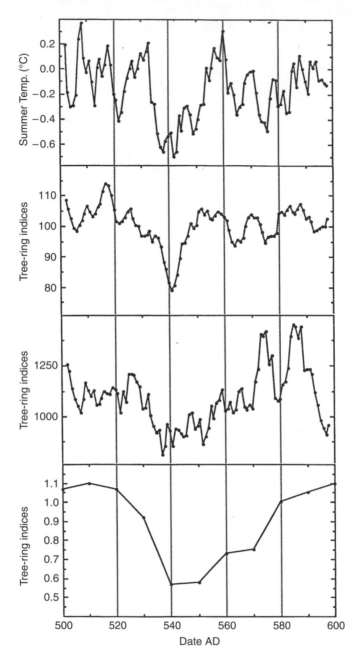

Fig. 6.12 Tree ring observations from various parts of the northern hemisphere showing a consistent picture of markedly cold summers around AD 540. The curves from top to bottom are the five-year running means of Fennoscandian summer temperatures, European oak width indices, and bristlecone width indices from the White Mountains in California, together with decadal foxtail pine width indices from the Sierra Nevada Mountains in California. (From Baillie, 1995, Fig. 6.4.)

doubt there was a cataclysmic event in 536 AD, which had far-reaching historical consequences (see Section 9.7), and it is possible it was caused by an extraterrestrial object.

6.12   Summary

This overview of the principal causes of climate variability and climate change provides a flavour of the challenges facing anyone wishing either to explain past events or to predict future developments. The combination of the wide variety of physical processes that could produce fluctuations and the fact that many of these are difficult to distinguish from one another means it is not easy to decide which effects matter most. But we must take a view on the past and make plans for the future. This requires decisions, as it is not feasible to try to include all the possible causes of change in any analysis. So the message is to draw out the leading factors from the areas covered in this review.

Two priorities emerge from this review. First the most pressing problem is to get a better fix on the internal variability of the climate. Finding out more about how the oceans and the atmosphere interact on timescales of decades and centuries is central to deciding to what extent changes during the twentieth century are the product of climatic autovariance as opposed to human activities. In addition, providing an explanation as to why the climate of the last 10 kyr has been relatively stable and whether this stability could be disturbed by natural causes or human activities is equally important.

The second, and related priority is to establish a balanced and more thorough understanding of how human activities can affect the climate. This is not just a matter of increasing levels of greenhouse gases, but also the direct and indirect effect of particulates and the changes in land use including desertification and deforestation. At the same time more research must be conducted into the impact of other factors (e.g. volcanoes, solar activity, and tidal effects). But, these are subordinate issues, although a growing knowledge of how these factors alter the climate will assist in the primary objectives of understanding the nature of autovariance and the impact of human activities.

Progress on these priority areas will depend on both improved measurements of the type described in Chapter 4, and better computer models. But, first we should consider the specific question of which human activities are climatically the most important.

QUESTIONS

1. Given the non-linear properties of the climate, explain why we need to consider the dominant nature of the annual cycle in our analysis of weather cycles. What periodicities could be generated between this cycle and the QBO (see Fig. 3.9), a 3 to 5-year ENSO signal, and the 11 and 22-year solar cycles? What do your calculations tell you about the value of putative cycles in forecasting the future climate.

2. What are the differences between high clouds and volcanic dust injected into the upper atmosphere that mean that high clouds warm the climate while volcanic dust has a cooling effect?

3. What are the arguments for and against a policy that proposes that many of the climatic consequences of human activities may compensate each other, and hence, we do not need to take drastic action to prevent climate change?

FURTHER READING

A complete reference list is available at the end of the book but the following is a selection of the best books or articles to follow up particular topics within this chapter. Full details of each reference are to be found in the Bibliography.

Bigg (2003): Particularly valuable in providing analysis of the growing understanding of how the atmosphere and oceans combine to govern many aspects of the climate.

Bryant (1997): Well worth reading as it adopts a somewhat different position to, say, the IPCC on the relative importance of various causes of climate change and our ability to estimate their impact.

Courtillot (1999): A balanced and highly readable account of how the major mass extinctions are attributable to periods of immense volcanic activity and this is a much more persuasive explanation than collisions with asteroids.

Diaz and Magraff (2000): A compilation of papers providing a useful review of the large-scale climatic fluctuations in the tropical Pacific and its impact on many economic and social aspects of life around the world.

Gray *et al.* (2005): A comprehensive review prepared for the Hadley Centre, which provides a balanced picture concerning the various physical mechanisms that could link solar variability to variations in the Earth's climate.

Imbrie *et al.* (1992) and (1993): These papers review the vast amount of work that has been done on modelling the response of the Earth's climate to orbital forcing and provide a good assessment of our current state of knowledge.

Open University Oceanography Series (2001): The ocean-circulation volume provides a particularly clear introduction to the basic aspects of ocean dynamics and how the oceans and the atmosphere interact.

Paillard (1998). An original and influential paper that produces model for explaining the changes in ice sheet volume during the last million years.

Pecker and Runcorn (1990): A collection of papers, which provides a particularly comprehensive review of the nature and origin of solar variability and a set of interesting observations about how this behaviour may be linked to climatic change.

Shoemaker (1983): The definitive analysis of the likely incidence of significant bolide impacts with the Earth.

Thomas and Middleton (1994): A stimulating and controversial analysis of the causes of desertification that is required reading for anyone who wishes to understand the complexity of this difficult subject.

7 Human activities

I have striven not to laugh at human actions,
not to weep at them, nor to hate them,
but to understand them.   **Baruch Spinoza, 1632–1677.**

How human activities can alter the global climate is best considered separately from natural causes of climatic change. When it comes to predicting how the climate change in the future, in Chapters 10 and 11, we will then need to consider how the combination of natural and human impacts can be analysed together. Now, however, in terms of current perceptions of the causes of climatic change, human activities take priority. Encapsulated in the expression 'global warming', it is assumed that these activities are leading to changes are one of the greatest threats to the well-being of humankind in the twenty-first century.

At the local level there is no doubt that human activities alter the climate. As discussed in Section 4.1, correcting for the effects of urbanisation is a major factor in obtaining a reliable measure of temperature trends. Apart from the warming effect of cities, they also reduce wind speeds, reduce visibility by the formation of particulates and photochemical smogs, and, in certain circumstances, increase the chances of heavy precipitation. These effects are, however, restricted to a very small part of the Earth's surface and so far, in terms of global climate change, their consequences are negligible. But they do provide an indication of how human activities can produce a complex web of climatological impacts and some insight of what a warmer world will be like.

7.1   Greenhouse gas emissions

On a global scale, the most closely studied aspect of human activities is the emissions of various radiatively active gases that are leading to an increase in the greenhouse effect (see Section 2.1.3.). Pride of place goes to CO_2. Its atmospheric concentration has been monitored closely since 1958 at Mauna Loa in Hawaii (Fig. 7.1) and, for nearly as long, at a variety of

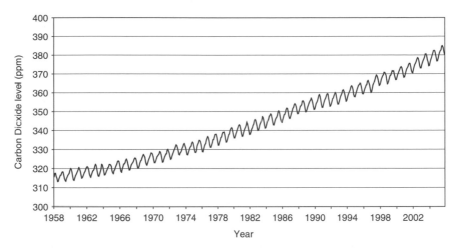

Fig. 7.1 CO_2 concentrations measured at Mauna Loa, Hawaii, since 1958 showing trends and seasonal cycle. (Data from NOAA, http://www.cmdl.noaa.gov/ccgg/trends/.)

other sites, including the South Pole. When combined with ice-core data the conclusion is that CO_2 levels have risen from about 280 parts per million by volume (ppmv) in the pre-industrial period (defined as the average of several centuries before 1750) to around 380 ppmv in 2005 – the precise figure varies around the globe and with the time of year, as the growing season in the northern hemisphere has a dominant influence on the annual cycle (Fig. 7.2). This rise appears inexorable. A closer examination of the Mauna Loa data shows that the rate of growth has fluctuated appreciably with marked peaks and troughs on a steady increase (Fig. 7.3). These fluctuations have not been explained but suggest complicated feedback mechanisms between short-term climatic variations (e.g. the ENSO) and the up-take of carbon in the biosphere.

Analysis of CH_4 concentrations shows a similar pattern, having risen from a pre-industrial level of around 700 parts per billion by volume (ppbv) to about 1780 ppbv at the end of 2005. Again the rate of growth has fluctuated in recent years having slowed from 15 ppbv in 1980 to about 10 ppbv in 1990 and still more since then, although fluctuating considerably from year to year. Other important greenhouse gases, which are increasing in the atmosphere, are oxides of nitrogen, notably nitrous oxide (N_2O). Halocarbons are powerful greenhouse gases, which include the CFCs and other chlorine and bromine containing compounds. These have, however, been the subject of effective international action (see Section 7.4).

The overall effect of the build-up of these gases is to produce a radiative forcing (see Box 2.1) of about 2.5 W m^{-2} between 1850 and now. CO_2 has

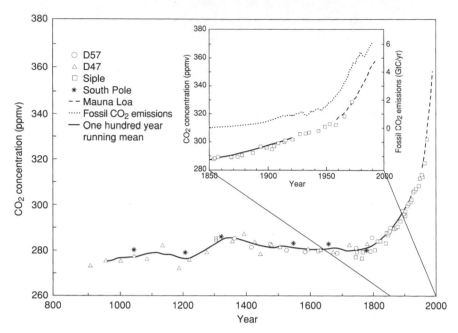

Fig. 7.2 CO_2 concentrations over the last 1000 years from ice-core records (D47, D57, Siple and South Pole – all in Antarctica) and (since 1958) Mauna Loa, Hawaii, measurement site. The smooth curve is based on a 100-year running mean. The rapid increase in CO_2 concentration since the onset of industrialisation is evident and has followed closely the increase in CO_2 emissions from fossil fuels (see inset of period from 1850 onwards). (From IPCC, 1995, Fig. 1.a.)

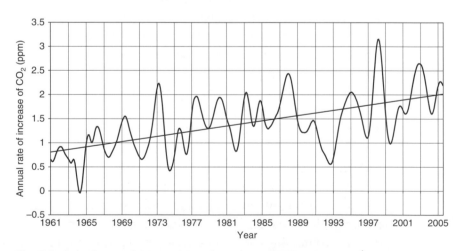

Fig. 7.3 Growth rate of CO_2 concentrations since 1958 in ppm year^{-1} at Mauna Loa, Hawaii, together with a smoothed curve to show variations on timescales longer than around 2 years. (Data from NOAA, http://www.cmdl.noaa.gov/ccgg/trends/.)

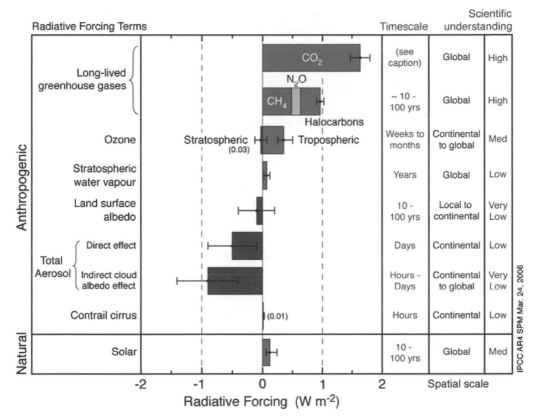

Fig. 7.4 Global-mean radiative forcings and their 65% uncertainty range for various agents and mechanisms. Columns on the right-hand side specify: (timescale) the approximate duration of the variation of the agent; (spatial scale) typical geographical extent of the forcing; (Scientific understanding) a measure of the scientific confidence level.

contributed some 60% of this figure and CH_4 about 25%, and N_2O and halocarbons the remainder (Fig. 7.4). As noted in Section 2.1.3, the radiative forcing due to the equivalent of doubling CO_2 from pre-industrial levels is estimated to be $4\,W\,m^{-2}$. The impact of changes in radiative forcing due to the build-up of greenhouse gases resulting from anthropogenic activities is considered in detail in Chapter 11.

7.2   Dust and aerosols

The climatic impact of other human activities is less well understood. For instance, the formation of atmospheric particulates and how these affect the extent and nature of cloudiness is a murky business. As Section 3.3

makes clear we do not have adequate measures of global cloudiness and so getting to grips with how human activities affect cloudiness and atmospheric haze is difficult. The impact of aerosols, dust and other particulates on the radiative balance of both incoming solar radiation and outgoing terrestrial radiation depends on their absorptivity, shape and size. It also depends on how they interact with water vapour and water droplets in the atmosphere. This means whether particulates absorb, reflect or scatter radiation, and how much, is a complex function of their properties.

A great deal of work has been conducted to establish the physical impact of particulates on the microphysical properties of clouds and how this affects the size shape and longevity of clouds. A good example of this type of study is the formation of sulphate particulates as a result of the combustion of sulphur-containing compounds in fossil fuels. Emissions of sulphur dioxide are converted to sulphuric acid in the atmosphere, which then forms sulphate aerosols that can both absorb and reflect sunlight in their own right. Their net effect will depend on the underlying surface. Over low albedo surfaces (e.g. the oceans and forests) they will probably increase the amount of sunlight reflected to space. Over high albedo surfaces (e.g. snow and deserts) they may be net absorbers. Overall, it is estimated that these effects have delayed the warming effect anticipated with the build-up of greenhouse gases in the atmosphere during the twentieth century (see Section 11.2).

Another important form of particulates results from the combustion of both fossil and bio fuels in developing countries. These form high concentrations of sooty particles, which are efficient absorbers of sunlight. Combined with the widespread use of slash-and-burn agriculture, the net effect is to alter radically the radiative properties of the lower atmosphere in many parts of the tropics. The net effect is to reduce the amount of sunlight reaching the surface and warming the lower atmosphere.

In addition, vast quantities of particulate emissions are the direct result of altering the surface of the land. Where agriculture has removed forests and exposed the soil there is an increased chance of dust being swept up into the atmosphere, especially during times of drought. It is estimated that the overall impact of altering the Earth's surface and injecting extra dust into the atmosphere has had a significant cooling effect. It could even be sufficient to cancel out much of the impact of the build-up of greenhouse gases in the last century. Conversely, increased atmospheric dustiness at the end of the last ice age caused by strengthening global circulation could have accelerated the warming process as the dust would have absorbed more sunlight than the ice sheets beneath it. So, as with so many other aspects of the climate, the response is non-linear and depends on the state of other climatic factors.

The complexity of the impact of particulates and dust on the climate is compounded by the fact that they also affect the properties of clouds. While the impacts described above are often termed as the *direct* effects, there are also indirect effects, which alter the droplet size in clouds, reduce precipitation rates and lengthen the lifetime of clouds. In general, these effects increase the amount of sunlight that is reflected to space and overall have a cooling effect on the climate. At present, what measurements there are suggest that the effect on marine stratocumulus and cumulus has a significant cooling effect. In the case of other clouds, especially over land, there is considerable doubt about the scale of this cooling.

The physical consequences of all these particulates has been detected in the significant decrease in the amount of incident shortwave reaching the land surfaces between 1960 and 1990. Known as *global dimming*, the rate of decline varied around the world but is on average estimated at around 2–3% per decade. There is doubt about the exact figure because of the difficulty in accurately calibrating the instruments and the problem of spatial coverage. Nonetheless the effect is almost certainly real. This dimming has occurred despite the fact that the land temperature has increased by 0.4 °C over the same period (see Fig. 6.1). The largest reductions are found in the northern-hemisphere mid-latitudes.

In terms of surface energy balance, the impact of dimming on surface temperature is counterintuitive, as less energy reaching the surface would be expected to lead to cooling. It can only be explained if there is an increase in the downward longwave radiation from clouds or aerosols in the atmosphere, which outweighs the decreased insolation, or a decrease of upward heat flux in the form of convection plus a decline in surface evaporation. The latter reduces evaporative surface cooling. It is suggested that the long-wave effect is not large enough. So, it is argued that less solar energy is available for evaporation. This could explain the fact that standard observations of the amount of evaporation from open pans has decreased over the last 50 years. As evaporation has to equal precipitation on the global scale, a reduction in the latent heat flux leads to a reduction in precipitation.

A more alarming possibility is that the warming due to the build-up of greenhouse gases more than counteracted the effects of dimming during the period 1960 to 1990. This means that as planned reductions in the amount of particulates produced by human activities take effect the rate of warming will rise steeply. The fact that recent surface observations suggest that the *dimming* trend has reversed since about 1990 adds fuel to this debate. It is likely that at least some of this change, particularly over Europe, is due to decreases in pollution, as most governments have done more to reduce aerosols released into the atmosphere than to reduce emissions of greehouse gases.

The phenomenon underlying global dimming may also have regional effects. While most of the earth has warmed, the regions that are downwind from major sources of air pollution (specifically sulphur dioxide emissions) have generally cooled. This may explain the cooling of the eastern United States relative to the warming in the western part of the country. All of this indicates that we do not know enough about the direct and indirect effects of particulates in the atmosphere, and on cloudiness in particular; a major gap in our knowledge.

7.3   Desertification and deforestation

Human activities have, since the dawn of history, had a major impact on the land surface of the globe. The domestication of grazing animals and the clearance of forests to establish agriculture have had a substantial impact on the surface of the land. In particular, these processes have led to the more permanent formation of desert areas (*desertification*). This became a major political issue in the 1970s because of events in sub-Saharan Africa (the *Sahel*). The drought in the Sahel started in earnest in 1968 and reached a nadir in 1972. It then abated, only to return in the late 1970s and worsened during the 1980s before easing off in the 1990s (Fig. 7.5). Its most dramatic

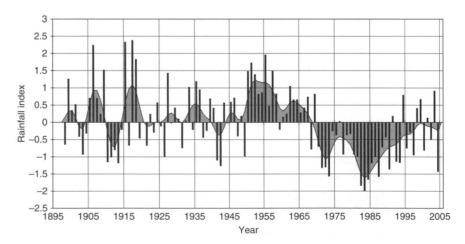

Fig. 7.5 Standardised precipitation anomaly in Sahel Region (20–8 N, 20W-10E) showing the sharp reduction in rainfall from the late 1950s onwards. (Data from http://tao.atmos.washington.edu/data_sets/sahel.)

social consequences were during the initial phase when over 100 000 people died and the pastoral economy of the more arid regions was effectively destroyed. These awful events had a dramatic impact on thinking in respect of both the provision of international aid and the consequences of both the natural variability of the climate and impact of human activities on arid regions.

The first consequence was that the issue of desertification became established as an environmental issue of vital importance. Terrifying figures entered environmental mythology. More than 20 million hectares (an area well over half the size of the British Isles or the size of Kansas) of once-productive soil was being reduced to unproductive desert each year. The image of the Sahara marching inexorably southwards at up to 50 km a year galvanised many aid agencies into action. This concern culminated in the UN Conference on Desertification, held in Nairobi in 1977. It launched a plan of action, which funded projects amounting to some US$ 6 billion over the subsequent 15 years to prevent desertification. During the same period increasing doubts have arisen as to whether the whole concept of desertification was misconceived and what was really needed were better measures of the changes that were actually occurring. In particular, there was no adequate distinction between degradation due to human activities (e.g. overgrazing by pastoralists' herds, collection of firewood, and inappropriate farming) and the effects of drought (see Section 4.2).

The underlying physical issue remains whether the increased albedo associated with the removal of vegetation had led to a permanent expansion of the desert. To the extent that nomadic pastoralists with their flocks were responsible for the removal of the vegetation, it was argued the desertification at the time was the result of human activity. Subsequent satellite observations showed that the extent of the Sahara Desert was closely linked to fluctuations in rainfall, and vegetation cover rapidly regenerated in wetter years. Furthermore, even within seasons, analysis of satellite images and rainfall between 1982 and 1999 shows that 20% more rain falls when the land is covered with greenery. Now, it is recognised that desertification is a complex process in which climatic change is probably the dominant factor and human activities play a relatively minor part (see Further reading). Nevertheless, recent local projects have demonstrated that planting and protecting trees in Niger has had a dramatic effect on re-establishing agriculture.

Whether or not climatic events in the Sahel are due to human activities, another aspect of the drought in the region is the amount of dust that is injected into the atmosphere. A significant increase in long-range atmospheric dust transport due to drought modifies the radiative budget and the water cycle over both Africa and the tropical Atlantic. Surface measurements at Barbados between 1965 and 2000 and satellite observations of dust optical thickness between 1979 and 2000, suggest that background dust loads over the Atlantic have doubled since the mid 1960s. So, alongside evidence of global dimming (see Section 7.2), these measurements provide additional evidence that increases in atmospheric dust load has a potentially significant global consequence.

The related question of the impact of other human activities modifying the land surface has concentrated on the impact of deforestation. The subject of the destruction of tropical forests has attracted most attention. The arguments for this leading to appreciable global climate change are, however, not compelling as the impact of the reduction in albedo may be compensated by a decline in cloudiness as the land dries out. By comparison, the impact of removing northern forests may be more substantial. The reason is all to do with snow (see Table 2.1). Snow-covered forests absorb about two-thirds of the sunlight falling on them, whereas snow-covered croplands and grasslands absorb only about a third of the sunlight falling on them. So, the effect of widespread deforestation would be to produce a dramatic cooling across the snowy higher latitudes of the northern hemisphere, especially at the end of winter and in early spring. It is possible that the effect on global mean radiative forcing of land cover change is comparable to that due to aerosols, ozone, solar variability and minor greenhouse gases. Model simulations suggest that land-use change may have had a cooling influence on climate, leading to a cooling of 1–2 °C in winter and spring over the major agricultural regions of North America and Eurasia.

7.4   The ozone hole

In the 1980s, the discovery of the *ozone hole* high over Antarctica in spring provided dramatic evidence of the global impact of human activities on the atmosphere. This discovery had a profound influence on both scientific and political thinking. It led immediately to large-scale international scientific programmes to provide answers to questions about the cause of the rapid decline in ozone over Antarctica and also to international action.

The cause of plummeting ozone levels in the stratosphere each September and October over Antarctica was *chlorofluorocarbons* (CFCs), which are used as aerosol propellants, in industrial cleaning fluids and in refrigeration equipment. These pollutants were involved in a complex set of chemical reactions, which combined with the return of the Sun, following the sunless Antarctic winter, led to a set of photochemical reactions that destroyed ozone in the stratosphere. The October values of total ozone in the atmosphere over parts of Antarctica declined by over 70% between 1979 and 2000, and the size of the ozone hole had grown from nothing to a mighty 16.7 million km^2 (twice the size of Antarctica). In the lower stratosphere, at altitudes of between around 15 and 20 km where the concentration of ozone is normally greatest, disappeared completely over the South Pole in October.

The explanation of these changes is based on how ozone is created and destroyed in the atmosphere. It is formed by the action of ultraviolet solar radiation that breaks oxygen molecules into two oxygen atoms (see Box 2.2). These can combine with oxygen molecules to form ozone molecules that contain three oxygen atoms. But ozone is itself reactive and in the presence of certain other chemicals in the atmosphere it can combine with them and revert to oxygen. The balance between all the features of this chemical cycle produces a sharp peak in the concentration of ozone in the stratosphere at altitudes of around 15–20 km – the ozone layer.

CFCs can interfere with these normal processes because they break down in the upper atmosphere to form highly reactive chlorine compounds. But this is not enough to explain what is happening over Antarctica. The peculiar conditions that develop in the Antarctic winter vortex provide the additional ingredient. The intense cold, often below –90 °C at altitudes around 15 km, produces clouds of ice crystals which accelerate the depletion of ozone through complex chemical processes on their surfaces.

The dramatic difference between the chemistry of air effectively trapped in the polar vortex, and that in regions outside it, suggest that to a large extent the effects of CFCs on ozone will be confined to Antarctica. But the chemical models have had to be refined to take account of other factors. In particular, the larger-scale dynamics of the stratosphere, which affect how much the air over Antarctica is isolated in winter from the rest of the global atmosphere, need to be incorporated. The effects of changes in solar activity also have to be included, as there is evidence that the 11-year cycle in this activity affects the rate of ozone production in the upper atmosphere (see Section 6.5). What is more, volcanic activity may increase the destruction as, following the eruption of Mount Pinatubo in 1991, there was an even greater loss of ozone.

Even more important was the immediate response of the international community. By 1985 the Vienna Convention for the Protection of the Ozone Layer had been agreed, and the 1987 Montreal Protocol and subsequent

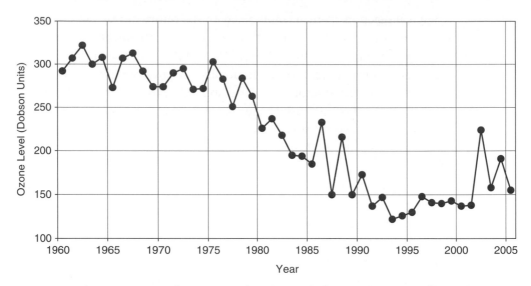

Fig. 7.6 Mean total ozone in October above British Antarctic Survey Halley Station since 1960. (Data from http://www.antarctica.ac.uk/met/jds/ozone/data/Z0Z5699.DAT.)

amendments to eliminate certain CFCs from industrial production followed. As a result of this rapid action the global consumption of the most active gases fell by 40% within five years and the levels of certain chlorine-containing chemical in the atmosphere started to decline. This prompt action shows that, where there is unequivocal evidence of significant global impact and the cause is accurately identified, it is possible to take effective international action.

Although the climatic impact of the ozone hole is limited, the lessons learnt from the events of the 1980s and 1990s provide guidance for current efforts to address the challenge of anthropogenic emissions of greenhouse gases and their build-up in the atmosphere. Here the challenges are much greater and the pressures to duck the issues are politically correspondingly persuasive.

7.5   Summary

Clearly there is a wide variety of ways in which human activities can affect the climate. The most important challenge is to establish how the scale of these compares with natural variations of the climate. At the same time, it is essential that we ensure that we do not concentrate too much on those natural effects and human activities that are better understood, while effectively sweeping the more intractable processes under the carpet. This challenge can only be addressed by the combination of improvements in measuring the nature of climate change and better models of the climate.

FURTHER READING

A complete reference list is available at the end of the book but the following is a selection of the best books or articles to follow up particular topics within this chapter. Full details of each reference are to be found in the Bibliography.

IPCC (2007) and the earlier reviews: In terms of obtaining a comprehensive picture of where the debate on the science of climate change and the nature of the impact of human activities on the climate has got to, this latest review is the definitive statement of the consensus view. It provides both a carefully balanced analysis, and an exhaustive presentation of the competing arguments. As such, it is not an easy read and may appear evasive, if not confusing, until you have got a grip of the basic issues. Nevertheless, it provides a truly comprehensive review of the many features of human activities and the extent to which they may be altering the climate.

8 Evidence of climate change

Time, which antiquates antiquities, and hath an art to make dust of all things, hath yet spared these minor monuments. **Sir Thomas Browne, 1605–82**

Turning to the evidence of climate change, much of what we have to work with is circumstantial and fragmentary; in effect only a few pieces of a jigsaw. Differing interpretations can be put on what actually could have occurred and what the causes might have been for the proposed changes. This analysis requires continual re-evaluation as more evidence becomes available or improved measurement techniques enable the existing evidence to be reappraised in a more critical light. For example, the holding of Frost Fairs on the frozen River Thames in London during the seventeenth and eighteenth centuries have long been seen as confirmation of the winters being much colder then. But, how much can we read into a few extremely cold seasons? Moreover, how do we take account of the fact that the old London Bridge, which was removed in 1831, acted as a weir to slow down the flow of the river? Combined with the absence of embankments, which meant the river was much wider, and the lack of waste heat from industrial plants, ensured that the river froze much more readily in cold weather. Only with the careful examination of instrumental records (see Section 4.1) did it become clear that winters were on average $1\,°C$ colder; a smaller difference than might be inferred by relying on the records of Frost Fairs alone.

Similarly, evidence of ancient civilisations thriving in areas which are now desert have to be examined closely to ensure that other factors, which could have contributed to their collapse, have been adequately analysed. In the same way, the evidence that mountain glaciers were more extensive in the past shows clearly that periods of colder, wetter weather led to more snowfall at high levels. But, how precisely this happened and whether successive expansions and contractions of the glaciers erased all evidence of previous surges makes it difficult to construct a clear record of climate change over time. The same applies to the evidence of the formation of huge ice sheets over large areas of the northern hemisphere which shows beyond a shadow of doubt that there have been a series of ice ages during the last million years.

Ideally, evidence should be based on an accurate measure of a specific meteorological parameter at a given spot at a known time. With enough measurements it is possible to build up a picture of how the climate has changed with time. In practice measurements rarely achieve this goal. Even with modern instrumental observations there are gaps, and these grow rapidly as we go back in time. But for almost all of the Earth's history we must rely on indirect *(proxy)* measurements, such as tree-rings, ice cores, and lake and ocean sediments. But even with the exploitation of a wide variety of sources, there are vast gaps in the record.

Exploring the evidence of climatic variations over the complete history of the Earth covers processes that have taken an immense time to occur (e.g. continental drift). In terms of current concerns about future changes in the climate these changes seem immeasurably slow and hence of little relevance to contemporary issues. This view is short-sighted. An understanding of longer-term changes not only sets current events in context, but also identifies the importance of different components of the global climate (e.g. changes in ocean circulation). So the more we know about what has happened in the past, and why events occurred, the easier it may be to appreciate the questions that need to be addressed today.

This catholic approach means we need to be clear what we trying to do in this chapter. The objectives are to identify the principal evidence of climate change and to expose the limitations in our knowledge of what changes have occurred. This approach will prepare the ground for considering the consequences and causes of climate change, and whether we can take a view on how it may change in the future.

8.1   Peering into the abyss of time

The science of geology is the key to understanding how the Earth's climate has undergone huge changes in the past. As geologists started to build up models of how the stratigraphy of the Earth had been laid down over the vastness of time, it became clear that huge changes had occurred in the climate. More recently, it was realised that the distribution of the continents had changed over time. So the interpretation of the climatic clues locked up in the rocks has been a painstaking business. This involved building up a picture of the conditions prevailing at the time the sediments were deposited and whereabouts on the globe these processes took place.

The geological evidence of fossilised plant and animal remains, organic material, pollen, and the shells of creatures contained in sedimentary rocks provide many clues. Working on the basic principle that some idea of past climate can be ascertained by comparing these rocks with similar sediments being deposited today, in early parts of the record, geologists draw on whatever sediments are available to build up a consistent picture. But, as we approach the current distribution of the continents the analysis of ocean sediments can be more discriminating because the topography and oceanic conditions when they were laid down are better known.

In particular the remains of various forms of phytoplankton are of great value. These occur widely in large numbers in ocean sediments. Because the different forms of plankton are readily identifiable and are known to have lived in certain conditions (i.e. warm or cold waters, and near the surface or at great depth) their distribution and physical properties (i.e. isotopic composition – see Section 4.5.3) provides detailed information about climatic conditions during their lifetimes. For example, a few groups of oceanic plankton living near the surface of the ocean can be used to subdivide the last 65 Myr into no fewer than 50 to 60 zones.

The nature of the rocks can provide a variety of climatic information. The remains of dried-out lakes and seas (*evaporites*) are signs of prolonged aridity. Large deposits of *carbonate rocks* indicate periods when the carbon dioxide in the atmosphere was drawn down. Glacial tills, which contain a poorly sorted mixture of soil and rocks, become consolidated to form *tillites*, while ice will transport boulders from identifiable sources across the landscape and leave them in unlikely places (*erratics*) or carry them out to sea before melting to leave *ice-rafted debris* in ocean sediments. Direct physical indications of the movement of ice in the form of *striations* on rocks provide evidence of the presence and extent of glaciers and ice sheets.

In building up a consistent explanation of past geological developments a number of basic principles apply. First, there is the underlying assumption that the physical processes involved in creating sedimentary layers have neither slowed down or speeded up over time (*uniformatorianism*). So the processes occurring now have been the same throughout geological time. In short – 'The present is the key to the past.' Second, successive layers of sediments are superimposed with the youngest at the top and the oldest at the bottom. Moreover, these layers exhibit lateral continuity and were initially horizontal, or nearly so. This means breaks in the record are due to either erosion or vertical movement. Finally, where there are, say, faults or intrusive dikes (due to volcanoes), these features are younger than the sedimentary layers they cut across.

In practice, the interpretation of stratigraphic layers is made much more complicated by the changes that disturb them over time. The combination of crustal movement, which leads to folding, faulting, uplifting and down-warping, together with the continual process of erosion has confused and destroyed much of the record. Estimates of the amount of sediments retained in the stratigraphic records suggest that between 90 and 99% of the material originally laid down has subsequently been eroded. So the resulting record has huge gaps in time and from place to place. This means that for two centuries geologists have been piecing together evidence from the abyss of time. In so doing, their views of such essential matters as the age of the Earth, continental drift, the rise and fall of sea level, and the nature of mass extinctions have changed radically over the years.

The defined geological timescale (Fig. 8.1) has been constructed on these principles. The important thing to concentrate on is the time covered by each period rather than the names which are a wonderful combination of Greek/Latin terms (e.g. *Palaeozoic*, meaning time of old life) and whimsical links with the regions where rocks belonging to given periods were first identified (e.g. *Ordovician* and *Silurian* are named after the ancient British tribes *Ordovices* and *Silures* which once inhabited the Welsh borders where these distinctive rocks were first identified). Many of the quoted dates are only accurate to a few per cent. This means that for the late Cenozoic the boundaries between various periods can be trusted to within 100 kyr or so. This uncertainty increases to around a million years in the Palaeocene and grows to several million years in the early Mesozoic. Nevertheless, for the purposes of maintaining a consistent chronology dates will be quoted with greater precision to enable adjacent events to be discriminated, as these changes can be resolved with greater precision than the absolute timescales can be defined.

There is little remaining to tell us how the climate changed for most of the first 90% of the Earth's lifetime. We do not know where the oceans and the continents were, or what precisely the constituents of the atmosphere were. The first sedimentary rocks were laid down some 3700 Mya, when it is believed that the climate may have been some 10 °C warmer than now. The first primitive forms of life may have appeared as early as around 3800 Mya but they provide little evidence of the climate. All we know is that between 2700 and 1800 Mya widespread glacial conditions were a feature of the climate for at least part of the time.

In Ontario and Wyoming, for example, there is evidence, in the form of preserved tillites, of three discrete glaciations between 2500 and 2200 Mya. Moreover, recent work suggested glaciers covered most of southern Africa

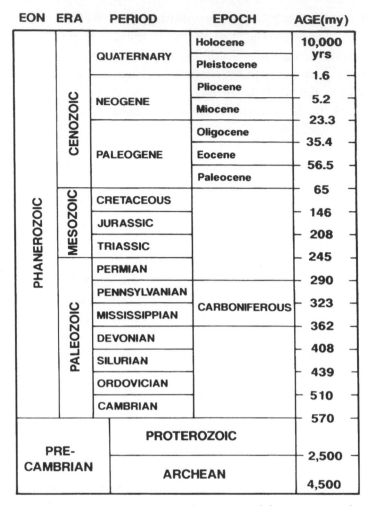

EON	ERA	PERIOD	EPOCH	AGE(my)
PHANEROZOIC	CENOZOIC	QUATERNARY	Holocene	10,000 yrs
			Pleistocene	
				1.6
		NEOGENE	Pliocene	
				5.2
			Miocene	
				23.3
		PALEOGENE	Oligocene	
				35.4
			Eocene	
				56.5
			Paleocene	
				65
	MESOZOIC	CRETACEOUS		
				146
		JURASSIC		
				208
		TRIASSIC		
				245
	PALEOZOIC	PERMIAN		
				290
		PENNSYLVANIAN	CARBONIFEROUS	
				323
		MISSISSIPPIAN		
				362
		DEVONIAN		
				408
		SILURIAN		
				439
		ORDOVICIAN		
				510
		CAMBRIAN		
				570
PRE-CAMBRIAN		PROTEROZOIC		
				2,500
		ARCHEAN		
				4,500

Fig. 8.1 The geological timescale. (From van Andel, 1994, Fig. 2.4.)

at this time when that region was close to the Equator. This raises the question that the Earth was largely covered with ice – sometimes referred to as *Snowball Earth* – a condition that might be expected to last forever, as it would have reflected so much of the Sun's energy back into space. Instead these glacial rocks are capped with either carbonates or volcanic magma, which can be interpreted as a catastrophic event (e.g. volcanic eruption, overturning of a stagnant ocean, or cometary impact – see Chapter 6) that could have pumped enough carbon dioxide back into the atmosphere to melt the planet's icy shell. Whatever the explanation thereafter, it seems the Earth remained warm and free of ice caps and glaciers for around a billion years.

The mists begin to clear around 1000 Mya. The late Precambrian period experienced a glacial period, which as best we can tell lasted some 200 Myr (see Fig. 8.1), sometimes termed the *Cryogenian* period of the late Precambrian. It appears there were at least two icy epochs between 800 and 590 Mya. These two glacial periods may have been made up of a number of shorter glaciations, which most geologists consider to have been among the coldest in the Earth's history. The older episode, called the *Sturtian* glaciation, occurred about 750 Mya, and the younger one, called the *Varanger* glaciation, took place roughly 590 Mya. Analysis of ancient glacial sediments suggest that continental-scale ice sheets extended to latitudes as low as 10°. An intense debate continues about the maximum extent of the snow and ice cover, with the two opposing camps supporting either the icy Snowball model (see above) or the less icy *Slushball model*. The former argues that the tropical oceans were also totally covered by sea ice, while the latter points to computer-modelling studies of conditions at the time that suggest some equatorial oceans remained ice-free.

The latest state of play in this debate is that measurements have been made of cosmic dust from outside the Earth that became trapped in the ice sheets long ago. When the ice sheets melted these deposits appear as a characteristic pulse of identifiable elements such as iridium in ocean sediments. Assuming the rate of deposition is not radically different from figures obtained from existing ice sheets, then it is possible to estimate the total time the ice sheets were in existence. The figure for one such melt-water pulse is that the ice sheets existed for some 12 Myr; an immense length of time compared to more recent glaciations, and a figure that tends to support the Snowball Earth hypothesis.

Towards the end of the Precambrian period dramatic evolutionary changes began to occur. Around 700 Mya there was a rapid diversification of higher plants and animals with soft bodies. Then, following the Varanger glacial, during the Cambrian there was a sudden dramatic acceleration in evolution starting around 565 Mya when all the basic designs for animal life, involving shells and hard skeletons, make their first appearance in the fossil record. This is often termed the *Cambrian explosion*. The scale of these changes and the improving fossil record means that for the purposes of this book our examination of climate change will now be restricted to events from the beginning of the Cambrian. These last 600 Myr are known as the *Phanerozoic* (the *Age of Visible Life*).

The interpretation of the geological evidence of the Phanerozoic is made easier in recent decades by the parallel development of theories of continental drift. Since measurements of paleomagnetism of ocean-floor

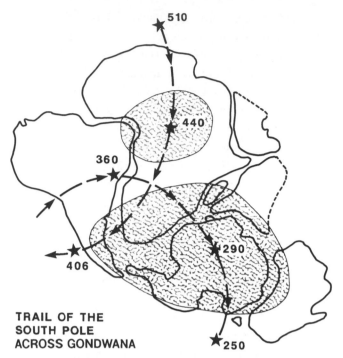

510

440

360

406

290

TRAIL OF THE
SOUTH POLE
ACROSS GONDWANA

250

Fig. 8.2 Between the late Precambrian and the Permian, the main continent was Gondwanaland, made up of Africa, South America, India, Antarctica and Australia, so its geographical position during this time is crucial to understanding climatic change. The South Pole traversed the continent at least twice (the numbers along its track mark its progress in My), and ice caps formed around the pole (shaded areas), but at other times the Polar Regions were free of ice. (From van Andel, 1994, Fig. 7.3.)

spreading in the 1960s confirmed earlier hypotheses about how the plates, which make up the Earth's crust, move, the history of the migration of the continents has been developed. During the first half of the Phanerozoic the main continent was *Gondwanaland*, which was made up of Africa, South America, India, Antarctica, and Australia (Fig. 8.2). Other continental fragments included parts of what are now North America and Eurasia. Mapping the movement of Gondwanaland has enabled the changes in geological record, which might be attributed to climate change, to be put into context.

During the Cambrian the climate warmed appreciably and remained relatively warm for most of the following 300 Myr. This era is known as the *Palaeozoic*. There was a relatively brief glacial period around the end of the Ordovician and beginning of the Silurian. The most extensive evidence of this cool period is now to be found in the Sahara (Fig. 8.3), which was close to the South Pole at this time (see Fig. 8.2). During the Carboniferous the

Fig. 8.3 Ordovician tillites, about 440 Myr old, from the Saharan desert. The striated boulders are clear indicators of a glacial origin. (From Sherratt, 1980, Fig. 23.23.)

temperature dropped, culminating in the long Permo-Carboniferous glaciation from 330 to 250 Mya. This icy epoch may well have been the coldest period in the last 600 Myr. It featured the expansion and contraction of a succession of major ice sheets, which at the greatest extent covered more than 50 million km^2 (four times the extent of the current Antarctic ice sheet). Its later stages, when all the Earth's land masses came together, Gondwanaland stretched from the equator to the South Pole. At various times, parts of what are now Antarctica, Australia, India, South America and South Africa were affected by massive ice sheets (see Fig. 8.2). For example, evidence suggests that around 280 Mya western Australia was covered by ice up to 5 km thick, although climatic conditions in other parts of the globe may have been rather warm. Also the interiors of the unglaciated continents became more arid.

Compiling a coherent explanation of these changes has been complicated by dramatic shifts in the fossil records. Certain cataclysmic events, including possibly asteroid impacts with the Earth, immense volcanic eruptions, or abrupt climatic change have led to mass extinctions of flora and fauna. The five most clearly established examples of these events (the *Big Five*) were Late Ordovician 440 Mya; Late Devonian, 365 Mya; Late Permian and Early Triassic, 255 Mya; Late Triassic, 210 Mya; and end-Cretaceous, 65 Mya. The consequences of this process are breathtaking. Over the last 600 Myr, 99.9% of all species that ever lived have perished. Nevertheless, biological

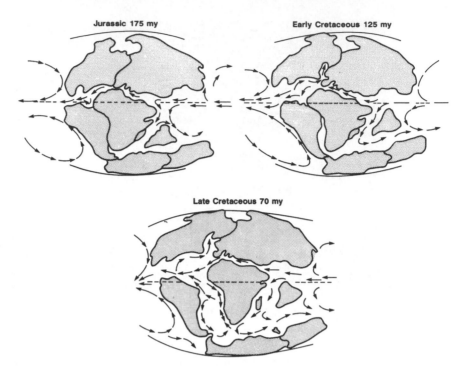

Fig. 8.4 During the Mesozoic the supercontinent Pangaea slowly broke up and the surface circulation of the oceans evolved from a simple pattern of a single continent in a single ocean to a more complicated pattern in the new oceans of the Cretaceous. The combination of the open circum-equatorial path and the absence of circum-polar currents during this period produced a more even temperature distribution than today. (From van Andel, 1994, Fig. 10.7.)

diversity has increased so there are now at least a million species of animals, of which insects make up three-quarters of the total (see Section 9.2). But extracting coherent record of climate change from the fossil record, against the background of such immense changes, is a daunting task.

During the early Mesozoic era, the continental masses amalgamated to form the supercontinent known as *Pangaea*. During the Jurassic and Cretaceous it drifted towards lower latitudes and broke up (Fig. 8.4). The standard view is that this was a period of generally warm climate with relatively little difference in temperatures between the poles and the tropics and insignificant seasonal variations. This equable period is usually associated with the *Age of the Dinosaurs*. There is now new evidence, however, of some fluctuations in climate throughout this era, including a cooler period with cold winters at high latitudes around the end of the Jurassic and the beginning of the Cretaceous. This includes examples of ice-rafted

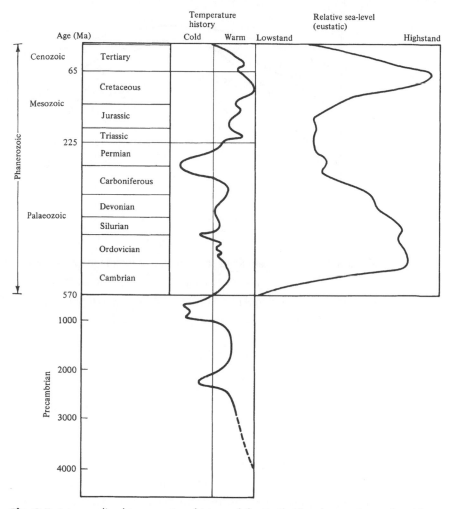

Fig. 8.5 A generalised temperature history of the Earth. The changes in sea level for the Phanerozoic are also shown. (From Brown, Hawkesworth & Wilson, 1992, Fig. 24.7.)

debris in Siberia, the Canadian Arctic, Spitzbergen (Svarlbard) and central Australia, all of which were at high latitudes at this time. There is also evidence of a sharp reduction in sea level around 128 and 126 Mya (see Section 8.3), which suggests glaciation of continental interiors at these latitudes. The greater seasonality of the climate at this time is confirmed by tree-ring studies of fossil wood, which show definite annual boundaries, indicating that growth ceased in the winter (see Section 4.5.1).

Following this cooler period, the mid-Cretaceous, around 100 Mya, was probably the warmest period in the Earth's history for which there is reasonable data (Fig. 8.5). Globally the average temperature is estimated to

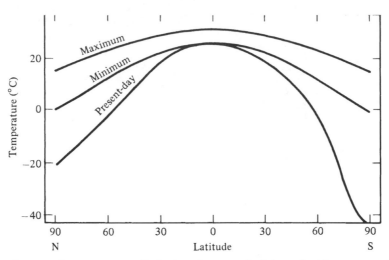

Fig. 8.6 The temperature limits (maximum and minimum) with respect to latitude for the mid-Cretaceous (~100 Mya) based on the full spectrum of biological, chemical and physical observations. The limits are compared with present-day values. (From Brown, Hawkesworth & Wilson, 1992, Fig. 24.8.)

have been 6 to 12 °C warmer than the present day. In the tropics it was 0 to 5 °C warmer, with the Arctic being 20 to 35 °C warmer and the Antarctic being at a similar temperature to the Arctic, as opposed to being markedly colder now (Fig. 8.6). The distribution of continents in the Cretaceous was probably the major factor in maintaining these balmy conditions. The existence of a circum-equatorial seaway and the absence of circum-polar currents may explain the warmth (Fig. 8.7). Measurements of sediments from the Arctic Ocean dated to around 70 Mya, suggest ice-free conditions and an average temperature as high as 15 °C. This would also have had a profound effect on the vertical circulation patterns in the oceans, which are now driven by the descent of cold waters in these regions (see Section 3.6), making the turnover far more sluggish. The possibility that changes in ocean circulation can have such a profound impact on the global climate is of particular interest in considering current events (see Section 6.3).

Fossil remains found the high Canadian Arctic (Axel Heiberg Island), which was well within the Arctic Circle in the mid- to late-Cretaceous, sum up the extreme warmth. Dated at around 90 Mya, they indicate that the mean annual temperature was at least 14 °C. The fossils include the bones of a large (2.4 m long) champsosaur, an extinct crocodile-like reptile with a long snout and razor-sharp teeth. These freshwater reptiles, which lived on fish and turtles, needed an extended warm period each summer to survive and reproduce.

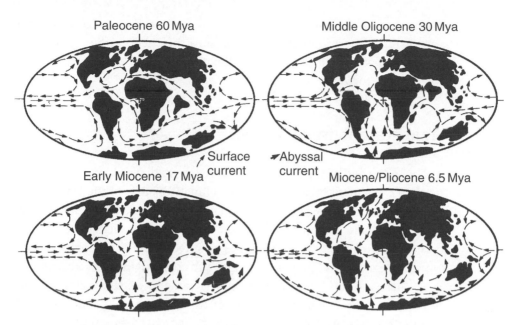

Fig. 8.7 During the Cenozoic the continents moved to the positions we recognise and the climate was strongly influenced by changes in ocean circulation. Two events matter most: the opening of the Antarctic circum-polar seaway (~25–30 Mya), and the closure of circum-equatorial seaway, which was completed in the Pliocene when the isthmus of Panama emerged. (From van Andel, 1994, Fig. 11.1.)

It is, however, a measure of the gaps in our knowledge that there is still dispute about what caused this warmth, as computer models of the climate (see Chapter 10) cannot reproduce the Arctic conditions of the mid-Cretaceous without having unrealistically high levels of CO_2 in the atmosphere. There are also doubts about whether the tropical oceans were as warm as is usually assumed, and arguments over the extent to which continental interiors would have experienced much colder winters than data based on ocean sediments would infer; yet another example of the problems of building up a reliable picture of past climates on the basis of fragmentary information. It is also another example of the value of understanding past climates. In considering the implications of possible global warming (see Section 11.2), the extreme warmth of the late Cretaceous has been used as an analogue for examining the most extreme future climatic conditions.

At the end of the Mesozoic era there was a sudden brief cooling. This coincided with the mass extinction that wiped out the dinosaurs. The termination of the Cretaceous is the best-known example of where a cataclysmic event appears to have caused a mass extinction. The favoured explanation is that this was caused by a collision with an asteroid (see Section 6.11). The arguments for and against this theory, which postulated a huge temporary dislocation of the climate, still rage some 25 years after it was first put forward.

8.2   From greenhouse to icehouse

Although the end of the Age of the Dinosaurs is an awful long time ago, it marks a convenient break point in the climatic history of the Earth. From now on we are considering a series of changes and specific events that has more immediate relevance to current concerns. The changes that have occurred during the Cenozoic era (the most recent era of geologic history), which covers the last 65 Myr of the Earth's history, fall into three categories. First, there are the long-term changes governed by plate tectonics, and possibly changes in the composition of the atmosphere (see Section 6.8). These effects have worked on timescales of hundreds of thousands to several million years. Then there are the increasingly important effects of the Earth's orbital parameters that operate on timescales from thousands to hundreds of thousands of years. Finally there are abrupt shorter-term events.

Throughout virtually all of the Cenozoic there has been a cooling trend (see Fig. 8.8). Initially, during the Palaeocene remained global temperatures

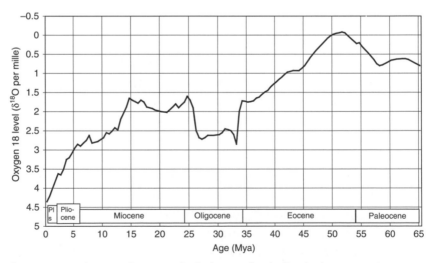

Fig. 8.8 From the greenhouse to the icehouse: the decline in the oxygen isotope ratio in foraminifers in a wide range of ocean sediments shows the long-term cooling of the global climate during the last 65 Myr. (After Zachos *et al.* 2001, Fig. 2.)

remained on a par with the late Cretaceous and even rose to a peak in the early Eocene. Moreover, what is not clear in the broad sweep of Fig. 8.8 is that at around 54 Mya there was a sudden rapid warming over less than 10 kyr of some 5 to 6 °C. Known as the *Late Palaeocene Thermal Maximum*, measurement of carbon-isotope ratios at the time suggest that this aberrant climatic event was caused by a giant release of methane (see Section 6.8). This event lasted around 100 kyr and the climate reverted to a gradual warming in the early Eocene reaching a climatic optimum between 52 and 50 Mya.

As the continents moved into the distribution we now recognise (see Fig. 8.7), things began to change. The real cooling started in earnest around 50 Mya, and was greatest at high latitudes. There was a relatively smooth decline until around 35 Mya, apart from a pause between 42 and 40 Mya. Glaciers first formed on Antarctic mountains around 50 Mya and by 36 Mya the first ephemeral ice sheets formed. The abrupt cooling around 34 Mya resulted in a 7 °C decline in ocean temperatures. This event, which appears to coincide with the Eocene–Oligocene boundary, lasted around 400 kyr. Although there was a partial rebound, this event marked the onset of a cooler period that lasted some 7 Myr, during which a permanent ice sheet covered parts of East Antarctica.

During the Oligocene significant changes occurred in ocean circulation with the separation of the Antarctic from Australia and then South America. The Tasmania–Antarctica passage opened around 34 Mya and the Drake Passage formed between South America and Antarctica around 29 Mya, which allowed the circumpolar current to develop and isolated Antarctica climatically, and provided the right conditions for the maintenance of the Antarctic ice sheet.

The climate warmed appreciably around 26 Mya, and there is evidence that as late as 25 Mya there were forested areas on Antarctica. At the beginning of the Miocene, there was a short-lived setback around 24 Mya, which lasted about 200 kyr. Thereafter the climate returned to warmer conditions and reached an optimum between 17 and 15 Mya. Then around 14 Mya a marked cooling trend set in and a permanent ice sheet formed over East Antarctica. At the same time mountain glaciers started to form in the northern hemisphere.

With only minor interruptions the cooling continued throughout the rest of the Miocene. The final stages of the formation of the Antarctic ice cap occurred during a sharp cooling between 6 and 5 Mya. At the same time, ice started to form in the central Arctic Ocean. The other major climatic event of this period was the isolation and subsequent desiccation of the

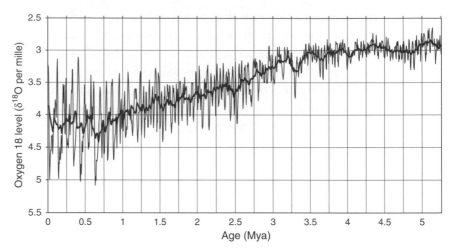

Fig. 8.9 A Pliocene–Pleistocene stack of 57 globally distributed benthic oxygen 18 records (δ^{18}O) showing how, alongside the long-term cooling of the last 5 Myr, there has been an increase in the amplitude and the period of the fluctuations, notably since around 1 Mya. (After Lisiecki & Raymo, 2005.)

Mediterranean around 5.8 Mya. Known as the *Messinian Salinity Crisis*, the deposition of huge quantities of salt as the sea dried up and the water vapour entered the hydrological cycle. This would have reduced the salinity of the rest of the world's oceans and caused major perturbations of the climate. Then around 5.3 Mya tectonic activity opened the Straits of Gibraltar and a vast torrent of water flooded back into the Mediterranean sending further shudders through the climatic system.

These changes, which coincide with the boundary between the Miocene and Pliocene, included a sharp increase in sea level and a marked warming. Thereafter, the cooling continued and accelerated around 3 Mya in the mid-Pliocene (Fig. 8.9). At the same time the fluctuations increased and by around 2.5 Mya there was a marked cooling and fall in sea level, which is thought to have resulted from the initial rapid build-up of northern hemisphere ice sheets. A feature of these developments is the establishment of a well-defined 40-kyr cycle in ice ages.

The Cenozoic era culminated in the Quaternary period which started around 1.8 Mya and runs up to the present time. This period of repeated glaciations, when up to 32% of the Earth's surface was covered by ice, is usually divided into the *Pleistocene*, covering the glaciations, and the *Holocene*: the warm period covering the last 10 kyr. The steady increase in the scale of climate change is indicative of increasing ice volume and climatic fluctuations, and declining average sea level, it is evident that this variability is periodic in nature. During the last 900 kyr the fluctuations have been larger and featured longer cold periods with greater ice volumes and lower average sea level (see Fig. 4.12). At the same time the period of the fluctuations shifted from about 40 kyr to nearer to 100 kyr.

Making an estimate of the temperature change over a period of 50 Myr is complicated, but by combining Mg/Ca ratio measurements with the oxygen-18 ($\delta^{18}O$) observations it is possible to construct a temperature record (see Box 4.2). This shows a consistent long-term decrease in deep-ocean temperature of roughly 12 °C over the past 50 Myr. The overall form of this curve is remarkably similar to the corresponding composite benthic $\delta^{18}O$ record.

The profound changes in the climate during the Cenozoic can be linked to a variety of causes that have been considered in Chapter 6. The increasing amount of land at high latitudes in the northern hemisphere made it easier for ice sheets to form. The opening and closing of ocean gateways, including the isolation of Antarctica during the Oligocene, and the Iceland–Faeroe sill subsiding in the late Eocene and then much later the forming of the Panamanian Isthmus closed the link between the Atlantic and Pacific Oceans around 3 Mya, altered ocean circulation patterns and allowed new currents to develop. The latter may have been crucial in creating the right conditions for ice sheets starting to form in the northern hemisphere.

The formation of the Himalayas and the Tibetan Plateau following the collision of the Indian subcontinent with Eurasia some 50 Mya, combined with the uplift of the Western Cordillera in both North and South America altered the global atmospheric circulation patterns (see Section 3.2). Meanwhile, the decline in the level in the concentration of carbon dioxide in the atmosphere (see Section 6.9) reinforced the cooling trend.

8.3   Sea-level fluctuations

Rises and falls in sea level are an integral part of climate change. During the ice ages a large amount of water was locked up in the ice sheets that covered the northern continents, so the level of the oceans dropped by more than 100 m. But for much of the history of the Earth there have been no ice caps, so the question arises as to whether evidence of sea-level variations contain significant climatic information. Over the last 600 Myr there have been substantial long-term variations. While absolute figures are hard to produce the scale of changes fall in the range from being as much as 200 m lower than the current level to being over 250 m above this. These fluctuations provide additional information about past climates but also highlight how climatic factors are tangled up with many other physical processes. So, care is needed to avoid jumping to conclusions about what part the climate played in these changes.

The most comprehensive analysis of changes of sea level have been prepared by geologists for the Exxon Corporation, and the resulting record is often known as the *Exxon curve*, or the *Vail curve* after Peter Vail, the leading geologist in the work. This analysis are the subject of fierce debate among geologists because of some of the conclusions reached, and because much of the data is commercially confidential and so cannot be the subject of independent scientific assessment. The reason for secrecy is that the analysis looks at the way sediments were laid down along the shores of ancient seas and how different layers show evidence of rising or falling sea levels (Fig. 8.10). Because these sedimentary layers may contain hydrocarbon reserves they are of great commercial value and oil companies are reluctant to share the data with other scientists.

In spite of the unresolved argument over the Exxon curve the broad features are trustworthy (Fig. 8.11). The first-order changes involve time-scales that are so long that they are unlikely to be the product of climatic factors. Although the lower sea level at the end of the Carboniferous and during the Permian may be partially due to the glacial epoch at this time, the most likely explanation for the long-term rise and fall is that it is the consequence of continental drift and changes in the rate of formation of mid-ocean ridges. Superimposed on this broad pattern is evidence of more rapid supercycles and cycles (see Fig. 8.11). More striking is the nature of the rises and falls, with the rises being relatively gradual and the falls being precipitate, giving the Exxon curve a characteristic 'saw-tooth' appearance. The intriguing question is whether these fluctuations, and in particular the sudden drops in sea level, are evidence of the formation of ice caps.

The timing and scale of the sudden substantial drops in sea level are hard to pin down. Little is known about the causes of the drops in sea level prior to around 300 Mya, although those at the end of the Silurian and during the Carboniferous may be linked to ice sheet formation. In the late Permian there was a sharp drop around 255 Mya, which may be linked to the mass extinction at this time (see Section 9.3). Then in the Triassic there was an event at about 232 Mya, but after that there are no striking examples during the next 100 Myr. The next two major falls are at about 128 and 126 Mya and coincide with the longer-term low sea level in the early Cretaceous. These

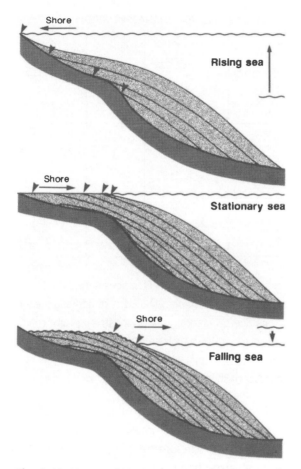

Fig. 8.10 Many estimates of past sea levels depend on the interpretation of seismic studies of identifiable layers of sediments laid down on the floors of ancient oceans. The highest point at which any layer is found (see arrow) will mark the highest sea level when it was formed. Depending on whether the sea was rising, stationary or falling, the extent of a sequence of layers formed near continental margins will provide information on changes in sea level. This method is known as *sequence analysis*. (From van Andel, 1994, Fig. 9.7.)

may be linked with continental glaciation (see Section 8.1). Thereafter, in spite of the broad rise in sea level, there were major short-term dips in the late Cretaceous at around 90 and 67 Mya.

During the Cenozoic the detailed picture becomes clearer. There were short sharp falls at about 58 and 49 Mya, followed by a series of falls and rises of levels between 39.5 and 35.5 Mya, and again between 30 and 25 Mya. The fall at around 30 Mya appears to have been particularly marked and

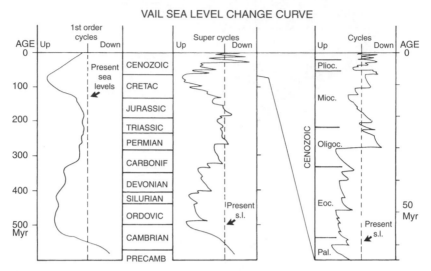

Fig. 8.11 Sequence analysis of global sea level changes during the Phanerozoic on three different scales. The first-order changes show the underlying megacycles of several hundred Myr in duration. Superimposed on these long-term changes are a series of 'supercycles' whose length varies from 20 to 50 Myr in the Palaeozoic to 4 to 15 Myr in the Mesozoic and Cenozoic, while more detailed analysis of the Cenozoic shows an increasing number of shorter 'cycles'. (From van Andel, 1994, Fig. 9.8.)

may be linked to the start of the build-up of the Antarctic ice sheet (see Section 8.1), but the scale of the drop suggests other wider tectonic effects were also involved. Moreover, the timing of these events does not tally closely with the temperature changes recorded in Figure 8.8. This may, however, reflect improvements in the more recent dating used for the temperature record. There were then two sharp dips around 16.5 and 15.5 Mya. Since around 10.5 Mya there have been a series of low levels (~ -100 m), punctuated with higher stands, with the lowest levels at 2.5, 1.5 and since 0.9 Mya. Changes during the last ice age are considered in Section 8.4, while those over the last 20 kya are discussed in Section 9.4.

Overall data on sea-level changes charted in the Exxon curve imply a more dramatic and variable global environment than is normally inferred from other geological records. As noted earlier, however, it is not possible to make a direct connection between rises and falls in sea level and climate change, because other factors could have been the driving force for both sets of variations. For instance, tectonic activity, including both volcanism and continental drift, plus fluctuations in ocean chemistry leading to changes in atmospheric composition, have to be considered in seeking to explain the observed changes (see Chapter 6).

8.4   The ice ages

Although there have been a number of epochs when large areas of the Earth were covered by ice sheets, the term Ice Age is most frequently applied to the Pleistocene glaciations. Moreover, how the discovery of these icy events has come to be associated with the Swiss naturalist Louis Agassiz is an excellent example of how geological evidence was put together to provide a coherent theory of climate change.

The possibility of ice sheets covering part of northern Europe had first been proposed by James Hutton, the founder of scientific geology in 1795, and reiterated by a Swiss civil engineer, Ignaz Venetz in 1821. In 1824 Jens Esmark, a Norwegian geologist offered the theory that Norway's mountains had been covered by ice, while in 1832 a German professor of forestry, Bernhardi, published a paper suggesting that a colossal ice sheet extended from the North Pole to the Alps. But these ideas received scant attention.

In the summer of 1836, while on a field trip in the Jura Mountains with Jean de Charpentier, a friend of Ignaz Venetz, Agassiz became convinced that blocks of granite had been transported at least 100 km from the Alps (Fig. 8.12). In 1837 he first coined the term Ice Age (*die Eiszeit*) and in 1840 his

Fig. 8.12 Glacially polished rocks and morainic debris at the edge of the Zermatt glacier. From Louis Agassiz's *Études sur les glaciers* (Neuchâtel, 1840). (From Smith, 1982, Fig. 1.12.)

proposals were published in an ground-breaking book. At first the geological community ridiculed the theory, but his passionate advocacy of the ice age was to prevail.

Agassiz travelled to Scotland where he saw more evidence of glaciation and then in 1846 arrived in Nova Scotia where again the evidence of ice was plain to see. In 1848 he joined the Harvard faculty and was active in many fields, notably marine science, but continued glacial research in New England and around the Great Lakes. Over the next few decades a variety of geological evidence made it clear that many features of the northern hemisphere could only be explained by ice ages, and Agassiz was vindicated, although in some quarters the subject remained controversial until the end of the nineteenth century.

Following the work of Agassiz an orderly view built up during the remainder of the nineteenth century about ice ages. This was that over the last 600 kyr to a 1 Myr there had been four glacial periods lasting around 50 kyr separated by warm interglacials ranging from 50 to 275 kyr in length. The present interglacial started about 25 kya and was destined to last at least as long as previous interglacials and possibly indefinitely (Fig. 8.13). The principal evidence of the stately progression was seen in glaciated landscapes of the northern hemisphere. The scoured U-shaped valleys, eroded mountains,

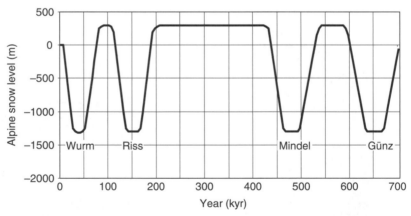

Fig. 8.13 Penck and Brucker model of past ice ages, which was based on studies of the geological evidence of fluctuations in the snow level in the European Alps and was published at the beginning of the twentieth century. The four main ice ages (Wurm, Riss, Mindel and Günz) were named after places in the Alps that showed clear evidence of each particular glacial period. In other parts of the world these ice ages are known by place names that were identified by good examples of this succession of events. (From Burroughs, 2005a, Fig. 2.4.)

Table 8.1 Correlation of the Pleistocene stages

Stages	Date (kya)*	Alps	Northern Europe	European Russia	North America
Post-Glacial	0–25	Post-Glacial	Flandrian	Post-Glacial	Post-Glacial
4th Glacial	25–75	Würm	Weichsel	Valdai	Wisconsin
Interglacial	75–125	Riss-Würm	Eem	Mikulino	Sangamon
3rd Glacial	125–175	Riss	Salle	Moskva	Illinois
Interglacial	175–450	Mindel-Riss	Holstein	Likhvin	Yarmouth
2nd Glacial	450–500	Mindel			Kansan
Interglacial	500–600	Günz-Mindel		Morozov	Afton
1st Glacial	600–700	Günz	Menap	Odessa	Nebraskan

*These dates and the correlations between the events identified in different parts of the world must be regarded as tentative in the light of more recent geological evidence. Nevertheless, these names are still often used to describe past climatic events and so it as well to know to what they refer.

drumlins (mounds of stiff boulder clay moulded under, and by, the creeping ice), *eskers* (ridges of gravel and sand formed by the meltwater flowing out of edges of the ice sheets), large glacial boulders (erratics), glacial tills, and terminal moraines are obvious landscape features. This view was encapsulated in the work of Penck and Bruckner published between 1901 and 1909. While some other geologists were more cautious about the chronology of the events, the broad picture of four major glaciations was the accepted interpretation of the geological evidence in both Europe and North America. The nomenclature for these ice ages and the associated interglacials, which reflects the local sites where the evidence of their existence was identified, together with the broad timing of these events is summarised in Table 8.1.

In the 1950s, the situation started to change. Caesari Emiliani, at the University of Chicago, published a set of papers on the properties of fossil shells of the tiny creatures found in the sediments of the tropical Atlantic and Caribbean. Using the reversal of the Earth's magnetic field 700 kya as a marker, he was able to show that there had been seven glacial periods since then, occurring every 100 kyr or so. This new picture of the most recent ice ages was not immediately accepted, but since the 1960s a growing body of evidence principally from ocean sediments (see Section 4.5.3.), but also from pollen records from part of Europe which had not been covered by ice, and Antarctic ice cores, has confirmed Emiliani's conclusions. It is now agreed that glacial periods occurred more frequently during the

Pleistocene than early theories suggested and, during the last 800 kyr, their period has been roughly every 100 kyr. These cold periods have been interspersed by shorter warm interglacials (see Fig. 4.9). Within each glacial period there were substantial fluctuations in the climate ranging from extreme cold to near-interglacial warmth. Statistical analysis of these fluctuations show that they are dominated by three major cycles of periods of around 21, 41 and 100 kyr (see Section 2.1.4).

When it came to a more detailed picture of events during the last ice age, it was the ice cores from Greenland that provide the greatest insights. A series of these have been extracted since the 1960s. These extend back into the last interglacial (the *Eemian*), which peaked around 125 kya, when the mean temperature in the northern hemisphere was around 2 °C above current levels and widespread Arctic melting contributed several metres to sea levels above today's. These conditions are regarded as useful comparison with the conditions that may prevail by the end of the twenty-first century (see Chapter 11), although the sea-level rise would take much longer to reach its full extent.

More striking is the evidence of shorter, frequent and dramatic fluctuations throughout the last ice age. In particular the isotopic temperature records show some 20 interstadials (known as *Dansgaard/Oeschger [DO] events*) (Table 8.2) after the scientists that first identified them; the precise number is the subject of slight variations from analysis to analysis depending on whether or not different climatological groups award the accolade of being an 'event' to the most transient warmings) events between 15 and 100 kya (Fig. 8.14). These events are probably linked to changes in the rate of formation of *North Atlantic Deep Water (NADW)* (see Section 3.8 & 6.3), which, in turn may be related to the amount of freshwater entering the ocean at high latitudes.

Typically these events start with an abrupt warming of Greenland of some 5 to 10 °C over a few decades or less. This warming is followed by a gradual cooling over several hundred years, and occasionally much longer. This cooling phase often ends with an abrupt final reduction of temperature back to cold (*stadial*) conditions. The spacing between these events is most often around 1.5 kyr, or, with decreasing probability near 3 or 4.5 kyr.

A second form of abrupt climate change is associated with what are known as *Heinrich layers* in the ocean sediments of the North Atlantic, after the scientist who first identified them. These events are associated with the surging of the Laurentide ice sheet over North America that released armadas of icebergs into the North Atlantic. They have variable spacing of several thousand years (see Fig. 8.14), and left telltale signs of debris in the ocean sediments and the accompanying freshwater was apparently sufficient to

Table 8.2 The chronology of events during the last ice age

$\delta^{18}O$ Stage*	Age (kya)	DO Events	Heinrich events (kya)	Comments (approximate date in kya)
1	0–15	1 (14.5)	H1 (16.5)	Ice sheets start to retreat (17)
2	15–29	2 (23.4)	H2 (23)	Maximum extent of ice sheets (25–18)
		3 (27.4)		
		4 (29.0)		
3	29–59	5 (32.3)	H3 (29)	Interstadial in North America and Europe (30)
		6 (33.6)		
		7 (35.3)		
		8 (38.0)	H4 (37)	Scandinavian ice sheet expanded (42–35)
		9 (40.1)		
		10 (41.1)		
		11 (42.5)		Partial amelioration in Europe (45–40)
		12 (45.5)		Much of Western Canada not glaciated (60–30)
		13 (47.5)		
		14 (52.0)	H5 (51)	
		15 (54.0)		
		16 (57.0)		
		17 (58.0)		
4	59–74	18 (62.0)		Interstadial in Europe (60)
		19 (70.5)	H6 (~70)	Considerable doubt about extent of glaciation in North America (79–65); cool dry conditions in northwest US.
		20 (74.0)		Expansion of Scandinavian ice sheet (80–75)
5a	74–85	21 (84.0)		Pronounced interstadial (85–80); return of cool mixed forest to parts of northern Europe. Warmer wetter summers northwest US than now.
5b	85–94		H7? (87.0)	Northern Europe descends into Taiga and Tundra Cool dry conditions in northwest US.
5c	94–107			Pronounced interstadial (105–94), forests return to northern Europe. Warmer wetter summers northwest US than now.
5d	107–117			Onset of glacial climate with northern Europe shifting from temperate forest to taiga. Cooler and wetter in northwest US than now.
5e	117–130			The last interglacial (the *Eemian*), generally warmer and drier in North America and Europe than now.

*The numbers denote alternately cold (even stages plus 5a, 5c and 5e) and warm periods (odd stages plus 5b and 5d).

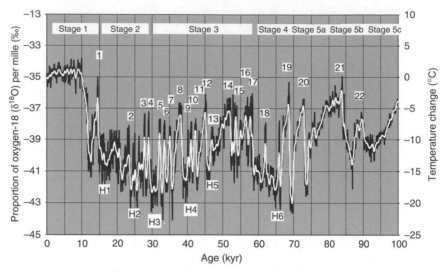

Fig. 8.14 The record of the proportion of oxygen-18 ($\delta^{18}O$) per mille (‰) in the GISP2 ice core (the black line is the data-average values for every 50 years and the white line is the 41-term binomial smoothing of this data). These curves show the 20 Dansgaard/ Oescheger warming events (labeled 1 to 20), six of which coincided with Heinrich events (labeled H1 to H6). The temperature range covered by these changes is reckoned to be about 20 °C between the coldest periods and the warmth of the last 10 kyr. The various marine isotope stages identified in the SPECMAP analysis are shown at the top of the diagram. (Data archived at the World Data Center for Paleoclimatology, Boulder, Colorado, USA.)

shut down the formation of NADW and particularly cold stadial conditions. These cold Heinrich events also appear to be associated with unusual warming in East Antarctica. These different modes of change fit in rather well with Didier Paillard's model of ice ages (see Section 6.7).

There is another aspect of ice core data that provides additional information about the climatic changes during the last ice age. This is the level of greenhouse gases in bubbles in the ice cores. When compared with the changes in temperature derived from the isotope measurements, it becomes clear that there are subtle differences between the changes in the temperature and the rise and fall in the levels of CO_2 and CH_4. This suggests that other climatic factors were the primary cause of the shift in the climate, and the changes in greenhouse gases may have been no more than a secondary factor. Most notable is the fact that, while CO_2 levels rose in line with rising temperature, they did not follow it down into the glacial conditions. Instead they remained at relatively elevated levels for several thousand years at the end of the last interglacial, and then descended in two stages to their lowest levels. One explanation for the intermediate level over the period from around 115 to 75 kya was that the oceans continued to maintain a high level of biological activity until around the end of Stage 4 (see Table 8.2). So, while the greenhouse gases acted as part of a positive feedback mechanism, which reinforced the changes during the last ice age, using these events as a useful guide to what may happen as a result of human emissions of these gases requires careful handling.

The product of all this work is an increasingly detailed picture of the waxing and waning of the ice sheets across the northern hemisphere during the last ice age. These fluctuations are summarised in Table 8.2. A key element in this presentation is the agreed $\delta^{18}O$ stages in the chronology from the oxygen isotope record in the ocean sediments, which has been built up from a large number of cores (see Section 4.5.3), which now extend back over 5 Myr (see Fig. 8.9). Table 8.2 covers only marine isotope stages 1 to 5, dating back to 130 kya. The description of conditions on the northern continents after the major build-up between 117 and 74 kya is based on fragmentary evidence. This reflects that as the ice sheets built up, and then fluctuated with changing climatic signals, their extent would have varied appreciably from place to place and time to time around the northern hemisphere. So there is considerable uncertainty about whether these changes were synchronous or were independent of one another. Table 8.2 is therefore designed to give some idea of where there is broad agreement about general developments.

The last glacial period reached its greatest extent around 21 kya, defined as the *Last Glacial Maximum (LGM)*. At this time, ice sheets up to 3 km thick covered most of North America, as far south as the Great Lakes, all of Scandinavia and extending to the northern half of the British Isles and the Urals. In the southern hemisphere much of Argentina, Chile and New Zealand were under ice, as were the Snowy Mountains of Australia and the Drakensbergs in South Africa. The total amount of ice locked up in these ice sheets has been estimated to be between 84 and 98 million km^3 as compared to the current figure of about 30 million km^3. This was sufficient to reduce the average global sea level by about 130 m (Fig. 8.15).

The global average temperature during the LGM was at least 5 °C lower than current values. Over the ice sheets of the northern hemisphere the cooling was around 12 to 14 °C. The position in the tropics is less clear. A major international study (CLIMAP) in the 1970s concluded that temperatures were much closer to current figures, especially over some of the tropical oceans. The largest inferred changes were in the mid-to-high latitudes, especially in the North Atlantic region bordered by large ice sheets.

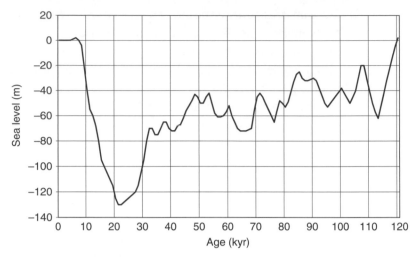

Fig. 8.15 Changes in sea level during the last 100 kyr, based on Mix, Bard & Schneider (2001) and Lambeck *et al.* (2003).

Near the equator in all oceans the changes were much less. Most striking was that the western Pacific temperatures appear to have stayed essentially the same as, and in some cases even warmed relative to, modern conditions. Overall, the inference was that Earth cooled surprisingly little during the ice age, except at high latitudes. This concept of *polar amplification* of climate change implies a feedback mechanism driven by increased albedo associated with expanded snow and ice cover. On land, forests shrank and deserts expanded. The question of just how thick the northern ice sheets were was the subject of much discussion. Sufficient to say, the fact that the sea level fell so far in the LGM (see Fig. 8.15), whatever their precise extent during the LGM, they exerted a massive influence on the climate at high latitudes.

During the last two decades the view of the conditions in the LGM has developed. The growing body of knowledge emphasises the complexity of the changes taking place during the LGM. The cause of these difficulties is two-fold. First, there is a basic measurement challenge of reconciling the results of different measurement techniques applied to a wide variety of proxy records. Secondly, there is no doubt that the lowest temperatures occurred at different times around the northern hemisphere. The rapid advance in the development of computer models of the climate has, however, made it much easier to explore the nature of the climatic conditions at time.

The important shifts in our picture of the LGM relate to the tropics. Although the general cooling of the oceans appears to have been no more than a degree or two, the tropical lowlands cooled by between 2.5 to 3 °C, and, at higher altitudes the cooling may have about 6 °C. This shift would have resulted in a weaker hydrologic cycle, which is consistent with the evidence of desiccation in many parts of the tropics. Nevertheless, the considerable uncertainties in the regional climate during the LGM pose a considerable challenge to the computer models used predict future climate change (see Chapter 11). What is certain is that around 15 kya a dramatic warming started, and with it, our perspective alters.

8.5   The end of the last ice age

So far the evidence of climate change has considered timescales that are beyond our current experience. The consequences of the huge, but gradual changes over geological time seem too remote to influence our lives. From now on, however, the emergence of the Earth's climate from the last ice age over the period of 15 to 10 kya is of immediate relevance to contemporary concerns about climate change. It is not just a matter of how the eventful transition between glacial and post-glacial climates prior to the settled period covering the last 10 kyr (the *Holocene*) affected the development of the earliest human societies. Of equal importance are the insights they provide about the stability of the climate.

The sporadic retreat of the ice sheets at the end of the last ice age and the associated advance of vegetation and forests left many clues on the landscape. The terminal moraines of the ice sheet together with the fluctuating levels of the lakes trapped behind the ice provided plenty of evidence for geologists to identify various stages of the retreat in both northern Europe and North America. At the same time botanists extracted information from both the vegetation and insects collected in peat deposits, plus pollen records from both peat beds and the sediments of lake beds. During the early part of this century this work produced detailed but somewhat separate chronologies for events in Europe and North America. The essence of this work is reviewed in Table 8.3 using ice-core data to define the timescale.

Going into a bit more detail, the first act in the unfolding drama at the end of the LGM was a collapse in part of the Laurentide ice sheet. This led to a surge of icebergs out into the North Atlantic and the last Heinrich event around 16.5 kya (H1 in Table 8.2). Pollen records for northern Europe show that after this last cold interval was followed by a sudden and profound warming (known as the *Bølling*) around 14.5 kya, which coincides with a rapid rise in sea level. This rise must be associated with a significant collapse of one of the major ice sheets. The warming with influx of freshwater is not consistent with influx of freshwater occurring in the North Atlantic, as this

Table 8.3 Changes at the end of the last ice age		
Time (kya)	Northern Europe	North America
17	Glaciers retreat in Alps	Major collapse of part of ice sheet
16		Wet period across southern US with lake levels from Great Basin to Florida reaching their highest levels around 15 kya.
15	Abrupt warming starts ~14.9 kya with evidence of forest expansion northwards (known as *Bølling* interstadial)	
14	Return of cold conditions around 14 kya for as much as 500 years (known as *Older Dryas*). Warming around 13.5 kya (known as *Allerød* interstadial) although some doubt as to whether as warm as Bølling.	Cordilleran ice sheet retreats from Puget Sound
13	Around 12.9 kya a sudden and dramatic cooling (known as *Younger Dryas*), forests retreated far south, and ice sheets expanded.	Drainage of meltwater from Laurentide ice sheet switches from Gulf of Mexico to St Lawrence. Together with surges of icebergs out of Hudson Bay these changes could have cooled the North Atlantic, thereby triggering the Younger Dryas.
12	Around 11.6 kya sharp and sustained warming, the initial stages of which are usually known as the *Pre-Boreal*.	Readvance of the Laurentide ice sheet may have redirected meltwater back to Gulf of Mexico and so switched off the Younger Dryas.
11	Forests started to spread across northern Europe.	Cordilleran ice sheet melts rapidly and disappears around 10 000 BP.
10	Forests became established at high latitudes, although the Fennoscandian ice sheet did not finally disappear until around 8.5 kya.	Laurentide ice sheet declines more slowly, and finally melts away around 7 kya.

would reinforce Heinrich Event 1. Current thinking is that this water came from the break-up of part of the Antarctic ice sheet. This surge of melting ice into the Southern Ocean led to a dramatic cooling and had the counterintuitive effect of switching on the thermohaline circulation in the North Atlantic (see Sections 3.8 and 6.3). This led to a rapid warming of the northern hemisphere.

The Bølling was relatively short-lived. Within a few hundred years it suffered a sharp interruption (known as the *Older Dryas*) before returning to a somewhat warmer level (this interstadial, the *Allerød*, is named after the place in Denmark where the first evidence of its existence was identified). Then around 12.9 kya the climate dropped back into near glacial conditions for over a thousand years, known as the *Younger Dryas*. It has been argued that the scale of this event justifies its being defined as an additional Heinrich event. There is, however, little evidence of it having been associated with a surge of icebergs and it is more likely that the release of a large amount of glacial meltwater from the lakes that formed behind the melting Laurentide ice sheet.

The Younger Dryas can be regarded as the last major paroxysm of the ice age and affected the entire northern hemisphere. The sudden drop in temperature, which lasted over a thousand years, was followed by an equally sudden warming. In Europe during this frigid spell summer temperatures were 5 to 8 °C below current values while in midwinter they were 10 to 12 °C lower. The extent to which the events recorded in the ice cores are matched by the changes observed in Southern Germany is striking (Fig. 8.16). This demonstrates how different proxy measurements can combine to provide a wider and more coherent picture of climate change. Then, following the end of the Younger Dryas, apart from a temporary interruption known as the *Preboreal Oscillation*, around 11.25 kya, temperatures rose until around 10 kya.

Around Antarctica there is little evidence of the dramatic ups and downs that marked the emergence of the northern hemisphere from the icy grip of the glacial. More striking is that the warming started two to three thousand years earlier in the southern hemisphere. The delay in the North may be a consequence of the huge thermal inertia of the ice sheets on the northern continents, which slowed things down at higher latitudes for several millennia. This is hardly surprising if you think about how long it would take for the ice to melt. Even allowing for the dramatic changes in atmospheric circulation that may have occurred in the northern hemisphere, together with the increased summer insolation at high latitudes, the time needed to supply the amount of energy needed to melt the ice sheets is measured in thousands of years.

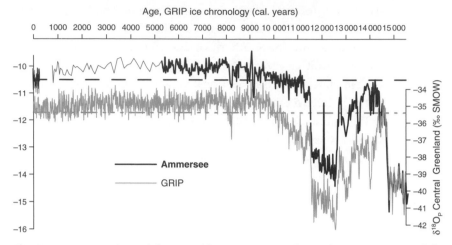

Fig. 8.16 A comparison of the record from Ammersee, in southern Germany, and the GISP2 ice core showing the close correlation between the Younger Dryas cold event between 12.9 and 11.6 kya at the two sites. (From von Grafenstein *et al.*, 1999, with kind permission from Uli von Grafenstein.)

How the great ice sheets in both hemispheres collapsed is central to explaining the climate changes of the time. Over northern Europe, the Fennoscandian ice sheet largely disappeared by around 8.5 kya after creating a series of meltwater lakes in the region of what is now the Baltic Sea, and there is little evidence of this having a cataclysmic impact on the climate.

The melting of the more massive Laurentide ice sheet was a different matter. It exerted a profound influence on the climate of the northern hemisphere. The principal meltwater lake that formed in central North America during retreat of the Laurentide Ice Sheet is known as *Lake Agassiz*. During its 4.5-kyr history the lake (around 13.7 to 8.2 kya) underwent major changes in volume. Its early stages had areas and volumes of up to about $170\,000\,km^2$ and $13\,000\,km^3$ respectively (about the volume of modern Lake Superior). The largest of the middle stages of the lake had areas and volumes of about $250\,000\,km^2$ and $23\,000\,km^3$ (about the total volume of the modern Great Lakes). In its final stages Lake Agassiz grew substantially as a result of merging with a second glacial lake, known as *Lake Ojibway*. Just prior to the final catastrophic release of its waters into the Hudson Bay, the area and volume of the combined Lake Agassiz–Ojibway had grown to $841\,000\,km^2$ and $163\,000\,km^3$; about seven times the total volume of the modern Great Lakes.

During its long history Lake Agassiz released a number of huge pulses of meltwater into the North Atlantic. These included events at 12.9 kya

(9500 km^3) and 11.3 kya (9300 km^3). These outbursts coincided with the start of the Younger Dryas and the Preboreal Oscillation. This suggests that outbursts from Lake Agassiz may have repeatedly influenced hemispheric climate by affecting ocean circulation and North Atlantic Deep Water production. This, in turn, altered the temperature of the surface of much of the northern North Atlantic, and with it the climate of the northern hemisphere.

The Laurentide ice sheet did not completely disappear until around 6 kya. Furthermore, well into the Holocene, in its death throes, it produced one last cataclysmic meltwater pulse of 163 000 km^3 around 8.2 kya. Although this was much larger pulse than the earlier ones, it occurred when the global climate had shifted into the Holocene mode. So, although it led to a sharp cooling of about 5 to 6 °C over Greenland and more like 1.5 to 3 °C around the North Atlantic region between about 8.2 and 8 kya, it was not on the scale of the Younger Dryas. It is assumed that this surge of freshwater out through the Hudson Strait and into the Labrador Sea could have disrupted the circulation of the North Atlantic and precipitated one last cold event.

The transition from the Younger Dryas to the Holocene marks a transition from the turbulent climate that characterised the ice age to something far more stable and hence benign. The timing of this change is a matter of definition. The temperature rose to something comparable to modern values by around 10 kya. The cooling between 8.2 and 8 kya can be regarded as the final relic of the last ice age. Apart from this hiccup, the ice-core temperature record suggests the Holocene was period of remarkable climatic stability.

Changes in the shorter-term variability of the climate are just as important. The GISP2 ice-core record since 60 kya has been sampled for 20-year intervals over its entire length (Fig. 8.17a). What we see, superimposed on the ups and downs during the last ice age and the subsequent warming, is the fuzz of shorter-term climatic variability. This fuzziness is a measure of the changes from decade to decade (i.e. well within a human lifetime). How this variability has changed over the last 60 kya can be extracted from the ice-core records by computing a measure of the variance of the detailed data (see Box 5.1). This is done by calculating the squares of the difference between the individual measurements (the black line in Fig. 8.17a) in the record and the smoothed version of the data (see Section 5.4) that removes periodicities shorter than about 200 years (the white line in Fig. 8.17a). This value of the variance provides an indication of the disruptive potential of shorter-term climatic fluctuations at any given time.

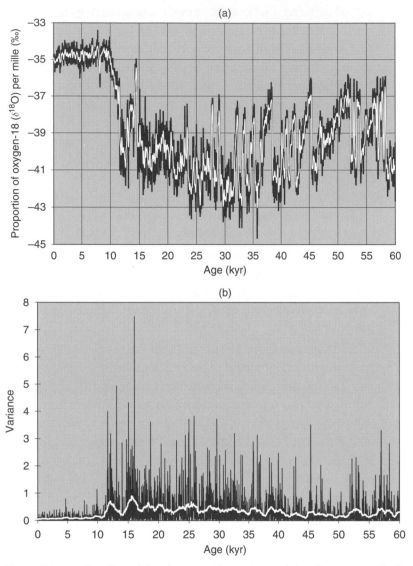

Fig. 8.17 An estimation of the change in the variance of the climate over the last 60 kyr. In (a) the GISP2 ice-core data for every 20 years (black curve) is overlain with the 21-term binomial running mean (white curve). In (b) the square of the difference between the two curves in (a) is presented. This presentation provides a measure of the variance of the climate in the vicinity of Greenland and the North Atlantic. (From Burroughs, 2005a, Fig. 2.9.)

This use of the variance to provide a measure of the impact of extreme weather events is not a statistical sleight of hand. In fact, it can be supported by a rather simple example. In considering the economic impact of tropical storms, it has been shown that the damage they cause when they strike land is proportional to the square of the wind speed. So in keeping a tally of trends in damaging storms a measure of the square of their strength is used to give a more accurate indication of their potential to cause economic damage. More generally, there is widespread evidence that bigger deviations in the weather have a disproportionately greater impact on many human activities.

The calculation of the variance of the short-term fluctuations from the longer-term trend in the GISP2 record shows how much greater the variability of the climate was prior to around 10 kya, and the relative stability since then (Fig. 8.17b). The striking feature of this calculation is the huge decline in the variance at the end of the last ice age. Virtually the entire period from 60 kya until around 12 kya was marked by extreme variability, apart from a preliminary lull during the Bølling. After the Younger Dryas the climate settled down into a much more quiescent pattern. Broadly speaking the variance dropped by a factor of five to ten. This change, which in statistical terms, can be defined as if the world emerged from the climatic 'long grass' into more hospitable conditions. This drop in variability was probably the vital factor in the ability of humans to develop more settled societies. The wild swings before the start of the Holocene must have been immeasurably more demanding than our present climate. They would have required an extraordinarily adaptable, flexible and migratory lifestyle to adjust to changing environmental conditions. At the simplest level, it is probably true to say that even now such a climate would make any form of agriculture, as we currently know it, virtually impossible (see Section 9.6).

8.6   The Holocene climatic optimum

By around 6 kya the post-glacial warming reached a peak. On the basis of evidence of tree cover the average summer temperature in mid latitudes of the northern hemisphere was 2 to 3 °C warmer than it is today. Not only had trees spread farther north than now but they also extended higher into upland areas; in Britain trees grew at levels 200 to 300 m above the current timberline. This phase of the climate is generally known as the *Atlantic* period, because it is assumed to have featured a strong westerly circulation which brought warm wet conditions to high latitudes of the northern hemisphere. As such it is sometimes seen as an example of what we might expect if the current global warming (see Section 8.10) were to continue.

At lower latitudes the warmer conditions showed up most noticeably in a stronger summer monsoon circulation. This brought heavier rainfall to many parts of the sub-tropics. This affected not only the Indian sub-continent but also the Middle East and across much of the Sahara. Ice cores from glaciers high in the Peruvian Andes confirm that the period from 8 to 5 kya marked the climatic optimum in the tropics. This warmer wetter period probably provided the right conditions for the spread of agriculture from its earliest beginnings in the *fertile crescent* of the Near East before 10 kya, but the evidence of the unusual stability of the climate during the Holocene period (see Fig. 8.17) has the most profound implications for assessing the potential consequences of current climate change. So, throughout the first half of the Holocene, with two notable exceptions, the changes have been small compared with the sudden shifts at the end of, and during, the last ice age. The first, as noted in the previous section, was the sudden cooling that occurred about 8.2 kya. The second was the abrupt desiccation of the Sahara around 5.5 kya (see Section 8.7).

Measures of dust and sea salt in the Greenland ice cores during the Holocene indicate that the strength of the mid-latitude westerlies varied considerably on millennial timescales. So it may be that these aspects of global weather patterns explain why in some parts of the world there were more dramatic shifts than seen in the Greenland temperature records. This takes us into the more subtle areas of latitudinal shifts in global weather patterns and the non-linear response of the climate to long-term changes in the mid-Holocene. At the same time as the Sahara began to dry out, southern Africa became warmer and moister and the Antarctic pack ice receded to higher latitudes. A sediment core from what is now close to the limit of pack-ice cover, showed that between 10 and 5.5 kya there was virtually no ice-rafted detritus in the core. Thereafter, pack ice returned to these lower latitudes.

Another interesting change was that ENSO (see Section 3.7) appears to have been largely absent during the early Holocene. A record of sea-surface temperatures (SSTs) obtained from magnesium/calcium ratios in foraminifera from seafloor sediments near the Galápagos Islands since the LGM provides insights into how ENSO fluctuations. The observed LGM cooling of just 1.2 °C implies a relaxation of tropical temperature gradients, and a southward shift of the location of the ITCZ. This pattern suggests a persistent El Niño pattern in the tropical Pacific. Temperatures then fluctuated over a range of about 1 °C from the end of the LGM to the early Holocene. During the mid-Holocene there was a cooling of nearly 1 °C, which suggests a La Niña pattern with enhanced SST gradients and strengthened trade winds.

8.7   Changes during times of recorded history

Around 5.5 kya the climate began to cool gradually and became drier. This trend coincided with the establishment of the ancient civilisations of Egypt and the Middle East, and so can be regarded as the beginning of *recorded history*. So from this point onwards the physical evidence of climate change can be combined with written records from various civilisations. Some of the greatest changes occurred in the Sahara, Middle East and farther east. The stronger summer monsoon circulation, which had brought heavier rainfall to many parts of the sub-tropics, weakened. These effects were most striking in the Sahara. Suddenly, around 5.5 kya, drier conditions set in across the Sahara and Arabia. The moister conditions, which had lasted for some 9 kyr, ceased. The decline in rainfall across the Sahara can be seen clearly in deep-sea sediment records off the west coast of North Africa, which show a sudden increase in the amount of dust being transported in the winds from the east (Fig. 8.18). At the drop of a hat the climate shifted to a drier form and the desert began to take over.

A general survey of conditions across the Sahara by 5 kya shows a general drying, although the distribution and timing of these changes shows interesting regional variations. In particular, the eastern Sahara aridification may have started as early as 6.5 kya. By 5.5 kya it was well advanced with the Libyan Desert extending southwards into what is now Sudan. Only in favoured wadis did the savannah remain a feature of regions that are now desert. This dramatic shift has to do with how the the cycles in the latitudinal-variation solar insolation (see Section 6.7) influence the position of the ITCZ. Any shifts in its position or intensity can have a crucial impact on how wet or dry different areas of the tropics are. Around 14.5 kya the level of summer insolation rose to a critical level at 65° N, sufficient to trigger a sudden shift northwards of the ITCZ into the Sahara during the summer. This brought relatively heavy rainfall to much of the Sahara and reduced the amount of dust blown out into the Atlantic (see Fig. 8.18). The level of summer insolation remained sufficient to maintain this pattern well into the Holocene, with the exception of during the Younger Dryas when the disruption of North Atlantic circulation was sufficient to override it. Then

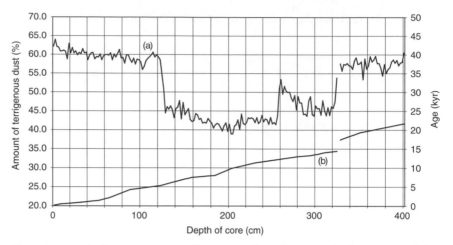

Fig. 8.18 Results from an ocean-sediment core from the tropical Atlantic west of the Sahara Desert. This data shows a sharp decline in the amount of mineral dust (defined as terrigenous) being transported from the Sahara between around 15 kya and 5 kya. (From Burroughs, 2005a, Fig. 2.11.)

around 5.5 kya the summer insolation at 65° N fell back below the critical level, and the ITCZ moved south and the Sahara rapidly became desert. This change was most rapid in the central and eastern Sahara, plus Arabia, where extreme aridity took over within a century of so. The fact that this sudden shift coincided with changes in the tropical Pacific emphasises the global nature of connections in the tropical climate.

In the mid-latitudes the storm belts appear to have moved a few degrees of latitude towards the Equator. Observations of changes in the amount of sodium ions (a measure of sea salt) and potassium ions (a measure of continental dust) in the Greenland ice cores suggest that both the Icelandic Low and Siberian High intensified between 5.8 and 5.3 kya. This indicates that the circulation in mid-latitudes of the northern hemisphere strengthened at this time. Around the same time, there was an abrupt increase in northern hemisphere sea-ice cover and widespread North American drought, which lasted for centuries. An ice core from high in the Bolivian Andes shows also that El Niño events returned around 5.5 kya. There was clear shift to more variable conditions that can be attributed to the return of the El Niño around this time. This change is supported by archaeological evidence of shells of mollusks along the coast of Peru, and by detailed cores taken from the bottom of Laguna Pallcococha in Southern Ecuador.

The next marked change is evidence of a decline in rainfall in the Middle East and North Africa setting in around 4.3 kya. At the same time the tree

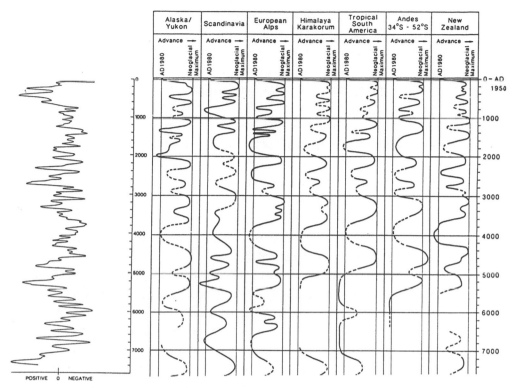

Fig. 8.19 An analysis of the fluctuations of glaciers in the northern and southern hemispheres during the last 7600 years, compared with radiocarbon production variations. Periods of negative radiocarbon production signal lower solar activity and are associated with glacial advance, which is a sign of a cooling climate. (From Burroughs, 2003, Fig. 4.12.)

line across the Canadian Arctic and Siberia started to recede and over subsequent millennia moved some 200 to 300 km southwards. This trend continued until near the end of the first millennium AD. What is less clear is how great the fluctuations on this broad trend were. The considerable evidence of mountain glaciers around the world expanding around 2500 BC, but then receding to high levels around 2000 BC (Fig. 8.19).

Explanations of this period of climatic deterioration have concentrated on either natural climatic variability or more specifically to a large volcanic eruption. Natural climatic variability has been attributed to an approximately 1500-year cycle in the climate of the North Atlantic, possibly of solar origin. A notable cooling event in the North Atlantic, which is associated with ice-rafted debris in ocean sediments in the vicinity of Iceland and Greenland, occurs around 4.5 to 4.2 kya. In addition there is evidence of a large volcanic eruption in ice cores in Greenland and Antarctica at 2354 BC, while tree rings also show evidence of climatic deterioration at this time. The event has been attributed to the eruption of Hekla in Iceland.

Tree-ring data across northern Europe suggests that the climate was warm and dry until around 1700 BC, and then cooled until the fifteenth century BC, at which point there was a marked warming for about 70 years, before an even colder wetter period set in from around 1400 BC to 1230 BC. This led to some glaciers in the Alps expanding to limits not matched since, and there is evidence of widespread abandonment of lakeside settlements around 1200 BC as water levels rose. Broad correlation with deep-sea sediment records suggests that the transition may reflect colder sea-surface temperatures in the North Atlantic. At the same time the eastern Mediterranean cooled markedly, initially in the form of colder winters, which could have led to an increasing incidence of drought in the region, which ushered in another Dark Age (see Section 9.5).

There is plenty of evidence of glaciers around the world expanding again around 200 BC and around AD 500–800. Tree-ring data from northern Eurasia and the White Mountains in California show cooler conditions prevailed during the latter period. Changes in temperature and rainfall at lower latitudes during these times are less easy to identify. Furthermore, attempts to link these and earlier changes to 'Dark Ages' when ancient civilisations went into decline is the subject of heated debate (see Section 9.4). For ease of analysis, a summary of the various changes during the period 8000 BC and AD 1000 is given in Table 8.4.

What is clearer is that there were shorter periods of climatic deterioration, which may or may not have coincided with the decline of certain civilisations. The identification of these events, their causes and the question of their historical impact is considered in Chapter 9, but as a general observation, historical records provide relatively little insight. There are plenty of examples of custom and practice (e.g. the production of olive oil and wine, the location of bridges and the design of buildings), which can be used to infer that climate change had a part to play in events, but their interpretation is the subject of dispute. Specific references to the weather are available here and there. So for instance a compilation of the examples referring to Italy between around 300 BC and AD 1300 suggest warmer, drier conditions in the third to fifth centuries AD, and again in the eighth and tenth centuries AD, with wetter conditions at other times. But, apart from the occasional tantalising clues such as the weather diary of Ptolemy which indicates that between AD 127 and 151 the climate of Alexandria was much wetter than now, the historical records provide little clear evidence as to just how much the climate varied between 3000 BC and AD 1000.

Table 8.4 Changes during the Holocene (8000 BC–AD 1000)

Period*	Central England	Europe	Sahara	North America
Sub-Atlantic 900 BC onwards.	15.1/4.7/9.3** (900–450 BC)	Cooler, wetter 900–450 BC, then warmer and drier until ~400 AD then colder and more variable until ~1000 AD.	Generally arid but some evidence of more rain in North Africa	Northern Great Plains drier than now until 1200 AD, with extreme droughts 200–370, 700–850, and 1000–1200 AD. White Mountains coldest ~900 AD.
Sub-boreal 1000–3000 BC	16.8/3.7/9.7	Generally dry and warm until ~1500 BC, much cooler and wetter until ~1200 BC, then warmer.	In the western desert the moist period ended ~2350 BC thereafter much drier	Cooling in White Mountains ~1200 BC. Northern Great Plains warmer (~1–2°C) and drier until ~2000 BC.
Atlantic 3000–5500 BC.	17.8/5.2/10.7 wetter than present	Rapid peat build-up in N. Europe, but cold spell in Alps (5500–4500 BC).	Lake Chad still 30–40 m higher than present, but to the east arid conditions set in around 3500 BC.	Sharp cooling in White Mountains 3300–2800 BC.
Boreal 5500–6900 BC	16.3/3.2/9.3	~1–2°C warmer	Wetter in Egypt	Laurentide ice sheet expanded (Chochrane Readvance) 4700–6500 BC.
Pre-boreal 6900–8300 BC			Lake Chad 52 m higher (area 400 000 km²)	

*The dates quoted for these periods, whose titles are widely and flexibly used to attach a coherent time-scale to evidence of climatic change, vary considerably. Here a consensus view is taken on the timing of these events.

**The three figures are quoted for central England temperature – high summer (July and August), winter (December to February) and the annual value – based on biological evidence. The values for the twentieth century are 15.8/4.2/9.4.

8.8   The medieval climatic optimum

The amount of historical evidence on the weather mounts in the ninth and tenth centuries. This shows that there are more and more pointers to the fact that the climate of northern Europe and around the North Atlantic became warmer during the ninth and tenth centuries. In part, this is inferred from the expansion of economic and agricultural activity throughout the region. Grain was grown farther north in Norway than is now possible. Similarly crops were grown at levels in upland Britain, which has proved uneconomic in recent centuries. The Norse colonisation of Iceland in the ninth century and of Greenland at the end of the tenth century are also seen as clear evidence of a more benign climate.

There is little doubt the warmer conditions over much of northern Europe extended into the eleventh and twelfth centuries, but the evidence from other sources provides a complicated picture, which has led to a vigorous debate (see Box 8.1) as to the scale of what is known as the *Medieval Climatic Optimum*. The principal evidence comes from tree rings. For example, work done by Keith Briffa and colleagues at the Climatic Research Unit, University of East Anglia, (see Section 4.5.1) shows that northern Fennoscandia had a warm period between 870 and 1100. These data also show a warm period around 1360 to 1570, but similar measurements made on trees growing in the northern Urals show no evidence of the earlier warm period. Instead the warmest periods are in the thirteenth and fourteenth centuries together with the late fifteenth century.

Recent speleothem data from a cave high in the Austrian Alps provide a more clear-cut picture (Fig. 8.20). It is evident that for this part of Europe the warm period extended from around 800 to 1350. More striking is the periods of particular warmth that occurred in the eleventh century and around 1250, which match or exceed levels reached in recent decades. What is more, these data are a remarkably close match with temperature figures obtained from an ocean sediment core taken on the Bermuda Ridge, where high sedimentation rates produce good resolution of decadal changes. Even more intriguing is that both sets data correlate closely with figure for the level of carbon 14 in the atmosphere at the time: a measure of solar activity at the time (see Section 6.5).

These temperature variations are found in other records. The Greenland ice-core data (see Section 4.5.2) shows that warmer conditions started earlier, around AD 600, and reached a peak at the beginning of the twelfth century before showing a marked decline in the next 200 years. Other records, however, show little evidence of conditions that reflect this warm period. So the overall picture lacks coherence,

Box 8.1 The 'hockey stick'

The scale of climate change over the two millennia or so, including the 'Medieval Climatic Optimum' and the 'Little Ice Age', leading up to the warming in the twentieth century, has been the subject of a more contentious debate. The essential issue is how much of the recent warming is due to human activities and how much is attributable to natural climate change (see Chapter 10). Recent syntheses of proxy records, notably from tree rings, have been used to calculate temperature trends over the last 1800 years. When combined with instrumental records since the mid-nineteenth century one particular analysis presented a striking picture (see Fig. B8.1). The shape of this time series, which has become known as the 'hockey stick', suggests that the recent warming trend dwarfs earlier natural variations over the last two millennia or so.

There is a lively debate as to whether the hockey stick is an accurate assessment of climatic variations during the last two millennia. Because the recent warming is so much more marked that earlier natural fluctuations, it implies that the gentle long-term downward trend, which was the most obvious natural variation, was reversed dramatically around the end of the nineteenth century as a result of human activities. If true, it would appear that current efforts to moderate our impact on the climate are, to say the least, a bit late.

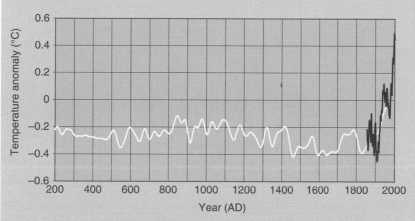

Fig. B8.1 The hockey stick: a reconstruction of northern hemisphere mean surface temperature over the past two millennia based on high-resolution proxy temperature data (white curve), which retain millennial-scale variability, together with the instrumental temperature record for the northern hemisphere since 1856 (black curve). This combination indicates that late twentieth century warmth is unprecedented for at least the past two millennia. (Based on Mann & Jones, 2003, data archived at the World Data Center for Paleoclimatology, Boulder, Colorado, USA.)

Box 8.1 (cont.)

The debate centres on the methods used to generate the hockey stick. It is argued that it seriously underestimates past natural climatic variability, notably the Medieval Climatic Optimum and the Little Ice Age (see Sections 8.8 and 8.9). This may well have been an example of the *segment length curse* (see Section 4.5.1).

Recently work has been carried out to examine more closely longer-term aspects of millennial-scale climate variability in order to understand patterns of natural climate variability, on decade to century timescales. These have used mainly tree-ring data of annual to decadal resolution, but have combined them with lake and ocean sediments data, which have a lower time resolution, but provide climate information at multicentennial timescales that may not be captured by tree-ring data. One example, which uses *wavelet analysis* (see Section 5.6) that handles longer-term fluctuations better, shows larger variability (Fig. B8.2) than the 'hockey stick', with temperatures – similar to those observed in the twentieth century before 1990-occurred around AD 1000 to 1100, and minimum temperatures that are about 0.7 °C below the average of 1961–90 occurred around AD 1600. This large natural

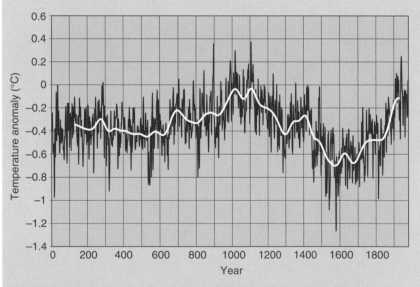

Fig. B8.2 A estimate of northern hemisphere temperature anomalies during the last two millennia using proxy records that aims to give adequate weight to longer-term cycles. (Moberg *et al.* 2005, data archived at the World Data Center for Paleoclimatology, Boulder, Colorado, USA.)

variability in the past, which is consistent with the results in Figure 8.20, suggests an important role of natural multidecadal and centennial variability that is likely to continue.

This debate as to the scale of natural variations prior to the recent warming is set to run and run. In the meantime, the IPCC AR4 has concluded that the weight of current multiproxy evidence suggests greater twentieth-century warmth than anything during most, or all, of the last two millennia, and that the last century was the warmest in at least the past 1300 years. This reinforces the conclusion that we have to continue to respond to observed changes within the framework of existing international agreements. If, however, the hockey stick proves to be correct, then the real issue will be how we do to adapt to the profound change already in full flow while, at the same time, not making matters worse.

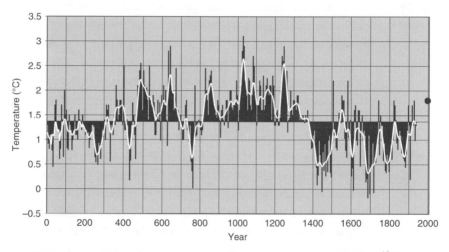

Fig. 8.20 The temperature record derived the proportion of oxygen-18 ($\delta^{18}O$) per mille (‰) in a speleothem in Spannagel Cave in Central Austria at an altitude of 2347 m. The black bars are the data average values for every 5 year period, the white line is the 11-term binomial smoothing of this data and the black dot at 2000 is the current annual temperature of the cave. (Mangini, Spotl & Verdes, 2005, data archived at the World Data Center for Paleoclimatology, Boulder, Colorado, USA.)

which may reflect the limited geographical coverage of the records and the fact they relate only to some seasons. For this reason the IPCC concluded in AR4 (see Box 8.1, and Section 10.2) that on a global scale the Medieval Climatic Optimum did not equal the warmth of the late twentieth century, which was reckoned to be the warmest period in at least the last two millennia.

8.9   The Little Ice Age

By comparison with the uncertainties of preceding centuries, the evidence of the cooler period between the mid-sixteenth and mid-nineteenth centuries appears to be built on firmer foundations. It is the best-known example of climate variability in recorded history. The popular image is of frequent cold winters with the Frost Fairs on the Thames in London. Elsewhere in Europe the same image of bitter winters prevails together with periods when cold wet summers destroyed harvests. Widely known as the *Little Ice Age*, the period has been closely studied by climatologists for many years. This growing body of work shows that, as with all aspects of climatic change, the real situation is more complicated than the simple stereotype suggests.

The first question is when did the Little Ice Age start? While the standard answer is that began in the middle of the sixteenth century, examination of Figure 8.20 suggests that in northern Europe the cold set in at the beginning of the fifteenth century. Following a warmer period during the first half of the sixteenth century, the cold returned later in the century and then occurred in pulses every 90 years or so, until the end of the nineteenth century. These broad features are confirmed by work by Jurg Luterbacher and colleagues at the University of Berne in Switzerland, which has produced multiproxy reconstructions of monthly and seasonal surface temperature fields for Europe back to 1500. These show that winter temperatures were from 1500 to 1900 were for much of the time about a degree Celsius below values during the twentieth century. In contrast, summer temperatures did not experience systematic century-scale cooling relative to present conditions (Fig. 8.21).

Climatic observations during the fifteenth century are relatively sparse. This may reflect the fact that, after the calamitous events of the fourteenth century with its population collapse due to plague (the Black Death), famine and almost ceaseless warfare, the climatic events of the fifteenth century seemed benign by comparison. There is, however, no doubt that the climate in Europe deteriorated in the second half of the sixteenth century. The glaciers in the Alps expanded dramatically and reached advanced stages at the end of the 1590s. Studies of a wide range of historical records of both the weather and related behaviour of crops, and flora and fauna by Christian Pfister at the University of Berne clearly show that in Switzerland the period

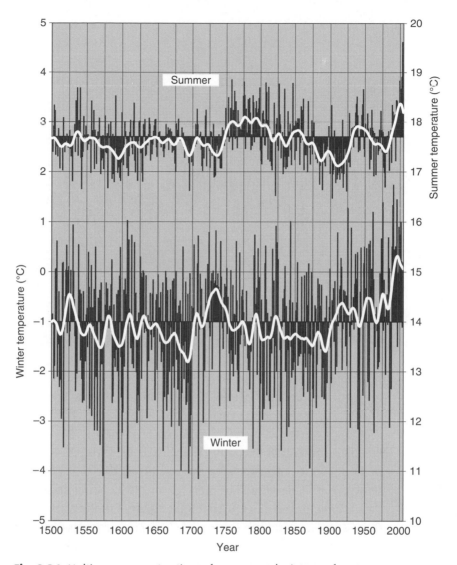

Fig. 8.21 Multiproxy reconstructions of summer and winter surface temperature fields for Europe back to 1500. (Luterbacher *et al.*, 2005, data archived at the World Data Center for Paleoclimatology, Boulder, Colorado, USA.)

1570 to 1600 featured an exceptional number of cool wet summers. They also confirm that the winters of the 1590s were particularly severe, while the Ladurie wine harvest dates show the poor summers of the late sixteenth century. The striking feature of the Burgundy wine harvest series is the relative stability of summer temperatures (Fig. 8.22).

The interpretation of the Little Ice Age as a period of sustained cold does not stand up to inspection in the subsequent decades. The growing seasons

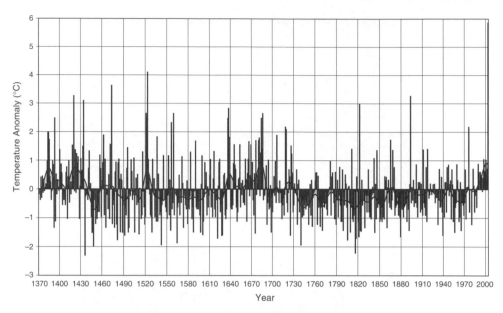

Fig. 8.22 Wine harvest dates for Burgundy, which provide an accurate indication of the temperature for the growing season, extending back to 1370. (From Chuine *et al.*, 2004, data archived at the World Data Center for Paleoclimatology, Boulder, Colorado, USA.)

returned to more normal conditions after the 1590s, although the incidence of cold winters remained high. Records kept by Dutch merchants of when the canals were frozen and trade interrupted show that from 1634 until the end of the seventeenth century the winters were roughly 0.5 °C colder than in subsequent centuries, with the winters of the 1690s being particularly severe. The 1690s stand out more generally as being exceptionaly cold in both winter and summer, with frequent harvest failures, notably in Finland, France and Scotland (see Section 9.5). This cold period is frequently linked to the lack of sunspots at the time: the Maunder Minimum (see Section 6.5).

Instrumental records for central England temperature (CET) started in 1659 and the records for de Bilt in the Netherlands in 1705 (see Section 4.1) provide more detail. The CET series confirms the exceptionally low temperatures of the 1690s and, in particular, the cold late springs of this decade. Equally striking is the sudden warming from the 1690s to the 1730s (Fig. 8.23), which stands out in the winter record for Europe (see Fig. 8.21). In less than 40 years the conditions went from the depths of the Little Ice Age to something comparable to the warmest decades of the twentieth century. This balmy period came to a sudden halt with the extreme cold of 1740 and a return to colder conditions, especially in the winter half of the year. Thereafter the next 150 years or so do not show a pronounced trend.

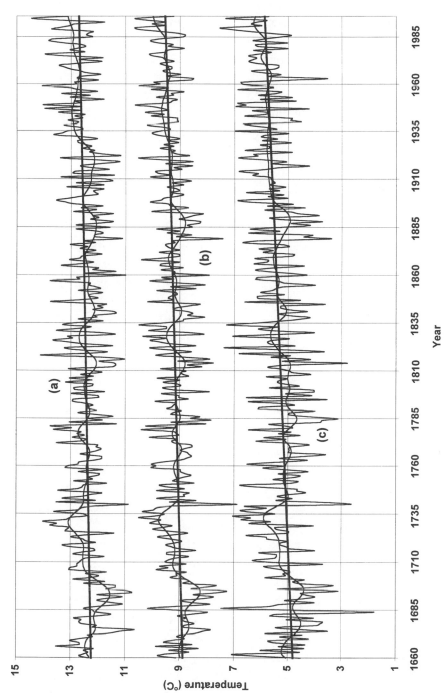

Fig. 8.23 Temperature records for Central England since 1659 showing how most of the increase in the annual temperature (a) has been the result of warming in the winter half (October to March) of year (b), as to opposed relatively little change in the summer half (April to September) of the year (c). (From Burroughs, 2001, Fig. 4.21.)

Various other series for other European cities from the mid-eighteenth century onwards confirm this conclusion. Although the winters begin to become noticeably warmer from around 1850, the annual figures did not show any appreciable rise until well into the twentieth century. Furthermore the temperatures for the growing season (April to September) remained virtually steady until the present day (see Fig. 8.23). The lower temperatures in the Little Ice Age, in Europe, were concentrated in the winter half of the year.

A more striking feature is the general evidence of interdecadal variability. So, the poor summers of the 1810s are in contrast to the hot ones of the late 1770s and early 1780s, and around 1800. The same interdecadal variability shows up in more recent data. The 1880s and 1890s were marked by more frequent cold winters, while both the 1880s and 1910s had more than their fair share of cool wet summers.

A similar variable story emerges from other parts of the world. In eastern Asia the seventeenth century was the coldest period, with another set of cold decades around 1800, but the rest of the nineteenth century does not show the frequent periods with low temperatures seen in Europe. Conversely, the North American records show the coldest conditions were in the nineteenth century. Tree-ring data suggests that the seventeenth century was also cold in northern regions, but in the western United States this period appears to have been warmer than the twentieth century. Limited records for the southern hemisphere indicate that the most marked cool episodes occurred earlier, principally in the sixteenth and seventeenth centuries. Evidence from the tropics is even sparser, but measurements of the ice cores from the Peruvian Andes suggest that there was a sustained cooler period from around 1500 to 1800, but ice core data from other parts of the world show a more varied picture.

So the most obvious conclusion is that there was not a sustained cold period in the sixteenth to nineteenth centuries. Certain intervals were, however, colder than others. Only a few short cool episodes appear to have been synchronous on the hemisphere and global scale. These are the decades 1590s to 1610s, the 1690s to 1710s and 1800s and the 1880s to 1900s. Synchronous warm periods are less evident with the 1650s, 1730s and 1820s being the most striking. The geographical extent of climatic anomalies lacks synchronicity, the coldest episodes in one region often not being coincident with those in other regions. Roger Bradley and Phil Jones, who edited a comprehensive study of the climate since 1500 (see Further Reading), that reaches the overall conclusion that the term 'Little Ice Age' should be used with caution.

8.10   The twentieth-century warming

The evidence of global warming since the late nineteenth century draws principally on instrumental records. The reliability of these, and what it means for interpreting the causes of the warming are analysed in Chapters 4 and 6. The best estimate of the warming that between 1901 and 2005 that annual global surface temperatures have risen by $0.65 \pm 0.2 \,°C$, with the rise being slightly greater in the southern hemisphere than in the northern hemisphere (Fig. 8.24a, b and c). Particularly striking is that 1998 and 2005 were the warmest years in the record and that five of the six warmest years in the record occurred in the period 2001 to 2005. Most of warming occurred between 1910 and 1945, and 1979–2005. Over these periods global temperatures rose at a rate of $0.14 \,°C$ per decade and $0.17 \,°C$ per decade respectively. In between there was a slight cooling, which was more marked in the northern hemisphere (Fig. 8.24b). In the southern hemisphere the warming trend has been more gradual and continuous (Fig. 8.24c).

In both hemispheres, air temperatures over the land have risen at about double the rate of those over the ocean since 1979 ($0.25 \,°C$ per decade versus $0.13 \,°C$ per decade). This warming has been accompanied by a decrease in those areas affected by exceptionally cool temperatures and, to a lesser extent, increases affected by exceptionally warm temperatures. In recent decades night minimum temperatures have increased more rapidly than day maximum temperatures. Over the period 1950 to 1993 the diurnal temperature range has decreased by $0.08 \,°C$ per decade.

Another feature of the warming this century is the regional patterns. In recent decades, the most pronounced warming has been over much of the northern continents and a marked cooling in the north-west Atlantic, with a lesser cooling across the central northern Pacific. These changes are most marked in winter – December to February. On a seasonal basis the warming over the mid-latitude northern hemisphere continents has been restricted largely to winter and spring while the cooling of the northwest Atlantic and mid-latitudes of the North Pacific has partially compensated the overall warming trend. This recent warming is in contrast to the warming period around the 1940s that was concentrated at higher latitudes of the northern hemisphere. In addition, the Southern Hemisphere warmed less during this earlier warm period.

Efforts to establish equivalent trends in global annual land precipitation have made less progress. A variety of analyses of available data have produced equivocal results. Much depends of the handling of the data. Some

(a)

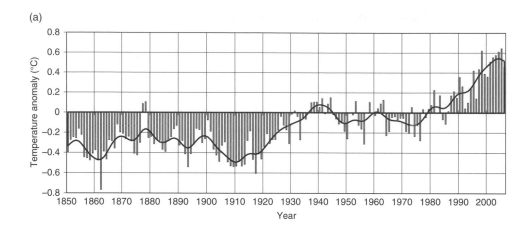

(b)

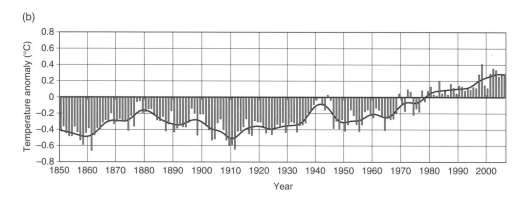

(c)

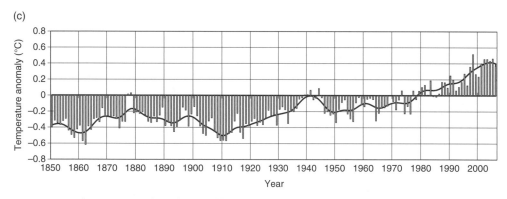

Fig. 8.24 Combined annual land-surface air and sea-surface temperature anomalies (°C) 1850 to 2006, relative to 1961 to 1990 (the bars are the annual figures and solid smoothed curves are 21-term binomial smoothing of the annual data): (a) northern hemisphere; (b) southern hemisphere; (c) globe. (Data published by the UK Meteorological Office as the HadCRUT3 series on the Hadley Centre website: http://www.hadobs.org.)

analyses approach the problems of gaps in the coverage by spatially infilling by either interpolation or, with more recent data, the use of satellite estimates of precipitation. Furthermore, even with these efforts, the somewhat surprising result that there is no clear evidence that the warming during the twentieth century has led to a comparable trend in precipitation. This conclusion is, however, of questionable value. It does not address the real issues of current climate change concerning the decadal changes that have occurred around the world, such as the drought in the Sahel. In practice the global figure is made up of much larger regional anomalies, which counterbalance one another, and what really matters concerning these sustained changes in the climate in any particular part of the world.

There is, however, a form of consensus on how things have changed latitudinally in the northern hemisphere this century. At high latitudes (55–85° N) there had been a rising trend, although part of this may be due to improved instrument design. In mid-latitudes in both hemispheres there is no appreciable trend. The most striking conclusion is the drying of the northern subtropics (10–30° N). This trend is dominated by the desiccation in the Sahel region of Africa, which experienced a 25% decline in successive 30-year periods (1931–60 and 1961–90) (see Fig. 7.5). Elsewhere in the tropics and subtropics there is little evidence of long-term changes. For instance, the trend in the all-India summer monsoon index, if anything, shows a hint of a wetter trend, and less variability in recent decades compared with the period before 1920.

The recent rise in global temperature and shifts in regional precipitation are only part of the current concern about our changing climate. In terms of the economic impact the most damaging consequence would be a marked increase in extreme weather events (see Section 9.8). This is not an easy thing to define, as it depends on the choice of extremes and reliable measurements of such events over a sufficiently wide area to be representative of global trends.

A major international effort has recently produced a suite of climate-change indices derived from daily temperature and precipitation data from around the world. The primary focus of this work has been to compute and analyse the incidence of extreme events. Probability distributions of indices derived from approximately 200 temperature and 600 precipitation stations, with near-complete data for 1901–2003 and covering much of the northern hemisphere midlatitudes (and parts of Australia for precipitation) were analysed for the periods 1901–1950, 1951–1978 and 1979–2003.

The results analyse the consequences of the significant warming throughout the twentieth century. Differences in temperature-indices distributions are particularly pronounced between the most recent two periods and for those indices related to minimum temperature. The global figures shows that the most significant temperature changes have been the decline in frost days (days with a minimum temperature less than 0 °C), the lengthening growing season (see Section 4.2), the intra-annual temperature range has declined, and the incidence of warmer night time temperatures has increased. The annual occurrence of cold spells significantly decreased while the annual occurrence of warm spells significantly increased. The trend in warm spells is, however, greater in magnitude and is related to a dramatic rise in this index since the early 1990s. All of these observations suggest that the variability of temperature has declined. As far as rainfall is concerned, there has been a significant increase in particularly wet days and periods of wet weather, and a decline in dry periods.

In terms of the points made in Box 8.2, recognising the non-Gaussian nature of the distributions, these results, for the periods 1951–1978 and 1979–2003, can be described as follows:

(a) The mean of the distribution of cold nights and cold days (i.e. the proportion of minimum and maximum temperature were below the 10-th percentile) has shifted to higher temperatures. In the case of cold nights this has involved an increase in the mean and a reduction in the variance (effectively the reverse of Fig. B8.3c). In the case of cold days, there has been a smaller shift in the mean to higher temperatures and little change in the variance (cf. Fig. B8.3a);

(b) The distribution of warm nights and warm days (i.e. the proportion of minimum and maximum temperature were above the 90-th percentile) has broadened, so the mean has shifted to higher temperatures and the variance has increased, with the shift in warm nights being greater (cf. Fig. B8.3c).

These changes have been consistent for all seasons. So, overall, on the basis of the last 50 years, a warmer world has featured an increase in high maxima, which, if anything has been more than offset by low minima. The consequences of the compensating changes in extremes are considered in Chapter 9.

More generally, climatologists have concluded that there is no consistent trend in interannual temperature variability in recent decades, and no consistent pattern for rainfall variability. The same story emerges in respect of intense rainfall. Because of the non-Gaussian nature of precipitation statistics, it is not possible to identify clear shifts in the distribution of various precipitation indices. Precipitation indices show a tendency towards wetter conditions throughout the twentieth century, but it is difficult to establish a spatially coherent description of these changes. The one

Box 8.2 Climate change and variability revisited

In considering extremes we need to return to the distinction between climate change and climatic variability (see Section 1.2). What we now need to do is quantify the two concepts in terms of the statistical analysis presented in Section 5.3. For temperature statistics, if the climate is

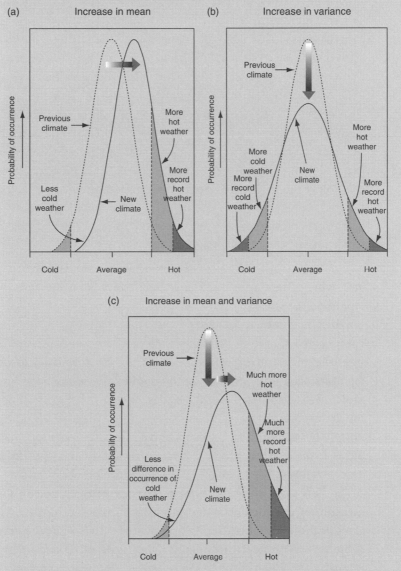

Fig. B8.3 How climate change and shifts in climatic variability affect the incidence of extremes.

Box 8.2 (cont.)

stationary over the period of the record, these parameters can be presented as a bell curve; the peak of the curve is the mean and the width of the curve is a measure of the variance (Fig. B8.3a). If the climate undergoes a warming without any change in the variance then the whole bell curve moves sideways. The consequence of this shift is that there are fewer cold days and more hot days, and a high probability that previous record high temperatures will be exceeded. If, however, there is an increase in variance but no change in the mean the bell curve becomes fatter and lower (Fig. B8.3b). The consequence is that there are more cold and more hot days and a high probability that previous records for both the coldest and hottest days will be broken. If both the mean and the variance increase then the bell curve both shifts sideways and becomes lower and fatter (Fig. B8.3c). The effect of this change is relatively little change in cold weather, but a big increase in hot weather, with previous record high temperatures being exceeded. So, in principle, we can calculate how the incidence of extreme temperatures will change for a predicted rise in mean temperatures and a predicted change in variance. In practice, the statistics for extremes do not show the normal bell curve; they are highly non-Gaussian (see Section 5.3).

area where there appears to be a clear trend is an increase in heavy precipitation days.

When it comes to changes in weather patterns, again the picture is complicated. There is little evidence of clear trends in the number and intensity of extratropical depressions. As for tropical storms there is a vigorous debate in progress as to whether there has been an increase in more intense storms in recent decades. As with other areas of uncertainty so much depends on the quality of observations. Those who argue that there is no reliable trend claim that subjective measurements and variable procedures make existing tropical cyclone databases insufficiently reliable to detect trends in the frequency of extreme cyclones.

The first thing to do is define a measure of what constitutes a reliable measure of the incidence and intensity of tropical storms. The standard classification of hurricanes is known as the *Saffir–Simpson Scale* (see Table 8.5). To reflect the steeply rising scale of damage with increasing intensity, we will use the NOAA *Accumulated Cyclone Energy (ACE) index* here. This index is the sum of the square of the 6-hourly sustained wind speed, measured in knots, for all systems in a given tropical basin for a defined

Table 8.5 Properties of tropical depressions, storms and hurricanes

Type	Category[1] Pressure	Central wind speed (mb)	Maximum sustained storm surge (knots)	Height of (feet)	Damage[2]
Tropical Depression			<35		
Tropical Storm			35–64	<4	
Hurricane	1	≥980	65–83	4–5	Minimal
Hurricane	2	965–979	84–95	6–8	Moderate
Hurricane	3	945–964	96–113	9–12	Extensive
Hurricane	4	920–944	114–135	13–18	Extreme
Hurricane	5	<920	>134	>18	
Catastrophic					

[1] The categorisation of hurricanes is defined as the Saffir–Simpson scale after the meterologists who developed it. Tropical depressions and storms are not the subject of categorisation and classified solely on windspeed as this is the only distinguishing measure of their behaviour – they may or may not graduate to hurricane status or simply fizzle out.
[2] The damage categories relate almost entirely to the impact of storms on the shoreline and further inland, but can include losses at sea.

period, while they are of at least tropical-storm strength. The important feature of this index is that it provides a measure of the destructive potential of the storms, which gives greater weight to the intense storms. The ACE index for the Atlantic Basin since 1951 has shown a marked variation over the last 50 years or so. This may be related to the Atlantic Multidecadal Oscillation (see Section 3.7).

Similar substantial interdecadal variations in the ACE index occur in different parts of the tropics. Recent analysis of worldwide tropical cyclone frequency and intensity to determine trends in activity over the past 20 years has sought to identify the consequences of a 0.2°–0.4 °C warming of SSTs. The data indicate that the increasing trend in tropical cyclone intensity and longevity in the North Atlantic basin is offset by a decreasing trend for the northeast Pacific. All other basins showed small trends, and there has been no significant change in global net tropical-cyclone activity. There has been a small increase in global Category 4–5 hurricanes from the period 1986–1995 to the period 1996–2005. Most of this increase is likely due to improved observational technology. These findings indicate that other important factors govern intensity and frequency of tropical cyclones besides SSTs.

8.11   Concluding observations

The clear message is there is no shortage of evidence of climate changes on every timescale. The scale of these changes has clearly had major impacts on the world around us, but although there is a lot of evidence, it falls far short of providing a complete picture of the size of changes and when and where they took place. This limitation applies to even recent instrumental observations and becomes increasingly the case as we go back in time, as the uncertainties about the Little Ice Age and the Medieval Climatic Optimum demonstrate. The principal features of the recent ice ages and longer-term fluctuations over the geological timescale stand out, but almost in every period there remain more questions than answers.

These gaps assume greater importance in exploring the consequences of climate variability and climate change. In many instances, from mass extinctions in the distant past to assessing the implications of recent weather-related economic and social upheavals, it is not certain what part fluctuations in the climate have played in events. So, whenever climatic factors are cited as the cause of other changes, these claims need to be examined closely. The first stage must be to obtain independent evidence of the climate changing in the manner proposed. Often the available information is either too sparse or ambiguous to draw categorical conclusions, or there is a degree of circularity in the analysis. Where these objections are justified, the only approach has to be to accept that until new evidence based on improved measurements becomes available, no conclusion can be reached. Where the evidence is more convincing the second stage is to demonstrate that the inferred changes could have had the impact proposed. To do this we must explore the examples of where the evidence of climatic change can be linked with other significant consequences, to establish which aspects of the climate really matter.

QUESTIONS

1. It has been suggested that one explanation for the possible occurrence of the 'Snowball Earth' (see Section 4.1) was that the tilt of the Earth's axis of rotation was much greater at the time. Why would this alter the temperature distribution between the equator and the poles and how could the axis be shifted into its current position?

2. If the area of the world's oceans is 360 million km^2, calculate the amount the sea level should have dropped if the amount of additional ice locked up in the ice sheets during to last glacial maximum was 60 million km^3. To the extent that the figure you calculate is greater than that given Section 4.4, explain what other physical effects could explain this difference.

3. What proportion of the current ice sheets over Antarctica and Greenland would have to melt to produce an average sea level rise of 10 m? Is this a useful figure, or do other factors have to be taken into account when considering the impact of this change on different parts of the world?

FURTHER READING

A complete reference list is available at the end of the book but the following is a selection of the best books or articles to follow up particular topics within this chapter. Full details of each reference are to be found in the Bibliography.

Alexander *et al.* (2006): A thorough description of the work done to examine the changes in the incidence of extreme events around the world.

Benton (1995): An excellent review article on current thinking about how life has diversified on Earth and the role played by mass extinctions in this process.

Bradley and Jones (1995): A series of papers by many of the leading lights in the study of climate change over the last 1000 years, which provides many valuable insights into the challenges of building up a reliable of past climates.

Dawson (1992): This textbook on the geology and climate of the last ice age provides an accessible and comprehensive description of what is known about the global environment between 125 000 and 10 000 years ago. Its only limitation is that it does not include the latest work on the evidence of sudden changes in the climate, which would make it even more valuable; a new edition is eagerly awaited.

Frakes *et al.* (1992): A good presentation of many aspects of climatic change on geological timescales, together some interesting theories on the periodic nature of these long-term fluctuations.

Grove (1988): A scholarly analysis of the evidence of the Little Ice Age, which concentrates principally on the fluctuations in the extent of glaciers in the mountain ranges around the world.

Lamb (1972) and (1977): This two-volume classic work on all aspects of climate change includes comprehensive information on the early studies of the evidence of past climatic ups and downs.

Lamb (1995): A fascinating analysis of the historical impact of climatic fluctuations throughout recorded history.

9   Consequences of climate change

In nature there are neither rewards norpunishments – there are consequences.   **Robert G. Ingersoll, 1833–1899**

Identifiable consequences of climate variability and climate change provide a measure of the significance of climatic events in our lives and the world around us. In many disciplines failure to understand how a changing climate may have influenced outcomes can lead to partial or inaccurate interpretation of past events. So identifying the most important consequences provides a checklist of the issues, which require more research and are central to predicting what future changes may occur and what their impact could be. Equally important is the ability to draw on the evidence of past consequences of climate change to help our thinking about the implication of future changes in the climate.

The analysis of the consequences of past climate change falls naturally into the two areas identified in Chapter 8. First, there is the long term and often dramatic fluctuations that occurred before the start of the Holocene, some 10 kya. Then there is the Holocene with its relatively stable climate. This stability means the consequences of climatic shifts are intertwined with other events and so the central issue is whether they have played a significant part in human economic and social development. This separation does not mean there are not periods of great climatic stability before the last ice age. Indeed, as implied in Section 8.1 there may well have been vast periods of geological time when the climate was far more benign than in recent millennia. Here, however, it helps to distinguish between interpreting the impact of long-term climate change on many earth sciences, and the more immediate issues of understanding how current climatic fluctuations may now influence our lives.

9.1   Geological consequences

In examining the geological record, there is the fundamental issue of whether it is possible to unravel the climatic factors from other causes of geological change. Clearly, climatic events like the waxing and waning of ice sheets, the desiccation of continental interiors or the drying up of oceans had major impacts on the geology of the large parts of the Earth. The real question is: were they 'cause' or 'effect'? If they were the result of more basic processes in the Earth's geological history linked principally to plate tectonics and changed levels of volcanism then their import in understanding climate change is less profound than if they played a significant part in driving the pace and direction of the underlying geological changes. This boils down to two issues. First, do shifts in the climate exert a significant feedback on tectonic activity (e.g. volcanism) by, say, changing the load on the Earth's crust due to the build-up and collapse of polar ice sheets, or changing sea levels? Secondly, if the climate can exert influence on tectonic activity, is it affected by extraterrestrial effects (e.g. the Earth's orbital parameters, fluctuations in the Sun's output, or even the motion of the solar system through the galaxy)?

As yet, there is no agreement on whether climate change exerts a significant effect on tectonic activity. There is, however, sufficient evidence to suggest that there are various ways in which the climate might create, or help relieve, the stresses in the Earth's crust. For instance, during the last ice age the eruption of volcanoes in the Mediterranean region appear to have occurred more often at times of rapid sea-level change. This supports the proposition that changes in the loading of the Earth's crust by the build-up and collapse of ice sheets over northern Europe did influence the level of volcanic activity in the region. How this fed back into climate change at the time is not yet clear.

A more subtle global effect of climate change occurs in the form of variations in the length of the day. If the atmosphere circulates more rapidly, the principle of the conservation of angular momentum requires that the Earth should slow down an ever so tiny amount to compensate for this atmospheric acceleration. Measurements in recent decades have shown that when climate warms up, say, as a result of a major ENSO warming event (see Section 3.7), the equatorial winds speed up and the Earth slows down. In the case of the 1997–98 event the length of the day increased by about 0.4 ms. This suggests that during sudden changes of the climate the Earth could speed up or slow down by significant amounts, which in turn could set up stresses and strains in the crust which might lead to increased volcanic activity, and hence, to an additional perturbation of the climate.

As for the extraterrestrial influences on the climate, these are subjects of intense speculation (see Chapter 6). The most obvious is the variation of Earth's orbital parameters, which is a major factor in ice-age dynamics. Then, if there is a link between ice-sheet dynamics and volcanic activity, this is another factor that needs to be included in a comprehensive theory of climate change. Extending this thinking further, if one of the consequences of shifts in the climate is to alter tectonic activity, then the possibility of extraterrestrial influences playing a more central role in the process has to be considered.

At the practical level, it is clear from the geological record discussed in Chapter 8 that climatic factors are part of immense changes. Reserves of coal, natural gas and oil are the product of a series of events that include major changes in the climate. For example, the sequence of a highly productive warm shallow sea laying down large quantities of organic matter followed by a period of desiccation which dried out the sea and capped the organic matter with an impermeable layer of evaporites is a classic formula for creating oil and gas fields. The fact that other forces, such as tectonic uplift, which isolated the sea and accelerated the rate of desiccation, may also have driven these processes does not alter the fact that the climate played an important part in the outcome. This means that understanding the geological record requires a knowledge of how both the global and regional climate behaved at the time the strata were laid down. So it is not surprising that oil companies are among the organisations that are funding research into how the oceans of long ago deposited organic material to inform their search for undiscovered oil fields.

9.2   Flora and fauna

The impact on flora and fauna has already been touched upon in presenting the evidence of past climatic fluctuations. There are, however, certain features of using evidence of the distribution of flora and fauna in the past that need to take account of how plants and animals evolved to meet the challenge of a changing climate. So it may not be safe to assume that animals which appear to have required a warm climate (e.g. dinosaurs) were not capable of living in cooler parts of the world which exhibited a marked seasonal cycle. For instance skeletal remains indicate that at least eight different species of dinosaur lived on the North Slope of Alaska around 75 Mya, when the region was above the Arctic Circle and experienced a mild climate (see Section 8.1). The dinosaurs lived in a deltaic environment among forests of deciduous conifers and broad-leaved trees, and probably remained there all year round, coping with long winter darkness and low temperatures.

How dinosaurs survived the darkness of an albeit mild Arctic winter raises some difficult questions. One answer could be that they migrated, as do

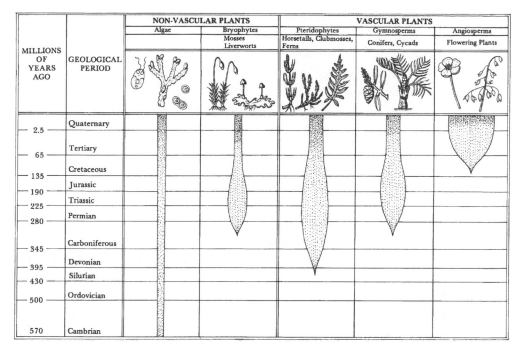

MILLIONS OF YEARS AGO	GEOLOGICAL PERIOD	NON-VASCULAR PLANTS		VASCULAR PLANTS		
		Algae	Bryophytes	Pteridophytes	Gymnosperms	Angiosperms
			Mosses Liverworts	Horsetails, Clubmosses, Ferns	Conifers, Cycads	Flowering Plants
2.5	Quaternary					
65	Tertiary					
135	Cretaceous					
190	Jurassic					
225	Triassic					
280	Permian					
345	Carboniferous					
395	Devonian					
430	Silurian					
500	Ordovician					
570	Cambrian					

Fig. 9.1 How the plant kingdom has developed over the geological timescale. (From *100 Families of Flowering Plants.* Cambridge University Press, Hickey & King, 1988, Fig. 1.)

modern caribou. It is argued, however, that the baby dinosaurs, with their short stumpy legs, would have been unable to migrate far. Compared with their parents they were proportionately much smaller than rangy young caribou with their long spindly legs. There would have been sufficient food for survive the winter in the Arctic, but as cold-blooded reptiles, how did they remain active and see to forage. At least one of the species had unusually large eyes, which may be part of the answer.

Similarly, the assumption that fossilised relatives of current warmth-loving plants (e.g. cycads – see below) could only have lived in year-round temperate or warm climates may underestimate the adaptation of plants long ago. The development of the plant kingdom (Fig. 9.1), and in particular vascular plants (plants having a vascular system for conducting water and food solutions) provides a good example of the interaction with climate change. These plants fall into three groups:

(a) pteridophytes (ferns, horsetails, clubmosses);
(b) gymnosperms (conifers, cycads); and
(c) angiosperms (flowering plants).

The pteridophytes (spore-bearing plants) were the first to appear, emerging from the protective environment of the oceans some 400 Mya. Then came the gymnosperms, woody plants bearing cones with naked seeds (i.e. not enclosed in an ovary). In this group pollen grains from the male cones are carried by the wind on to ovules (unfertilised seeds) produced on the scales of female cones. The most recent group are the angiosperm or flowering plants, which have their ovules enclosed in a protective structure – the ovary. This usually extends upwards to form the style and stigma. Pollen grains landing on the stigma can only reach the ovules by forming a tube that grows down through the style and into the ovary.

Since their emergence in the early Cretaceous (around 120 Mya) flowering plants have overtaken conifers and cycads to become the dominant form of land plants and now number some 250 000 species. As such they provide the best example of how slow variations of the climate have exerted powerful controls over the development and distribution of species. For much of the last 100 Myr, as the continents drifted into their current positions, the warm climate of the Late Cretaceous and early Tertiary may have been a crucial factor in allowing flowering plants to diversify and increasingly monopolise terrestrial vegetation – a dominance they continue to hold to this day. By the Eocene, some 50 Mya, tropical and sub-tropical plants had extended their range as far north as western Europe, and the Pacific Northwest of the United States. The subsequent cooling trend pushed this range back somewhat.

The dramatic changes in the extent of flowering plants is more easily discerned with the onset of the Pleistocene and the periodic expansion and contraction of the ice sheets in the northern hemisphere. In North America the plants were able to migrate north and south with the shifting climatic zones; so many of the consequences for plants were limited, and can only be detected in pollen analysis (Fig. 9.2), which show the waxing and waning of many species with rising and falling temperatures. In Europe, the barrier of the Alps posed a greater challenge to migration north and south. As a result, species that are common in China and the United States, such as *Actinidia* (the Kiwi fruit family), *Liriodendron* (Tulip Tree) and *Liquidamber* (sweet gum) which were present in Europe, but could not escape to the south of the Alps, were driven to extinction by the successive waves of ice. This process continues to this day with the extent of various species adapting rapidly to shorter fluctuations. In Denmark the population of holly (*Ilex aquifolium*) was largely wiped out by three successive severe winters in the early 1940s, while the range of the nine-banded armadillo (*Dasypus novemcinctus*) spread north through

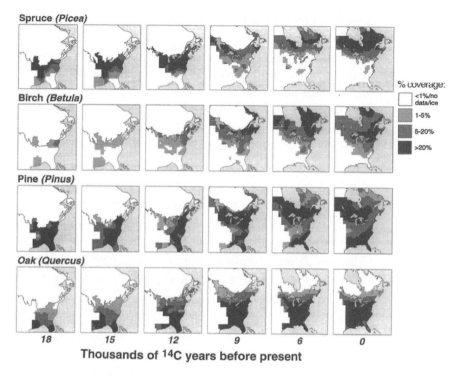

Spruce (Picea)

Birch (Betula)

Pine (Pinus)

Oak (Quercus)

% coverage:
<1%/no data/ice
1-5%
5-20%
>20%

18   15   12   9   6   0

Thousands of ^{14}C years before present

Fig. 9.2 Postglacial changes in the distribution and abundance of some major tree types in eastern North America based on pollen-analysis data. (From IPCC, 1995, Fig. 9.3.)

Texas and Oklahoma in the 1940s and 1950s only to retreat from the colder winters of the 1960s. Changes like these provide evidence of both past climate change and the possible consequences of future global warming (see Chapter 11).

Even more striking is the theory that climatic change played a particularly important part in the evolution of humans. The deterioration of the climate since the mid-Pliocene (see Fig. 8.9), and, the switch to bigger swings in the climate around 2.5 Mya, as the 41-kyr cycle kicked in, may have provided the stimulus for evolution of our species. The increasing aridity of this period, notably during the cold parts of the ice-age cycle, amounting to over 60 events in the last 2.8 Myr, favoured cold and drought-resistant taxa. It also and provided the conditions for savannah-dwelling hominids in Africa to evolve more rapidly. The drier, more open landscape with more pronounced seasons led to the need to forage further and longer and provided the spur for early hominid anatomy and physiology to be exposed to novel selective pressures. In particular, the development of bipedalism and relative hairlessness appear to have been the response to these challenges. So our very existence may be a consequence of the closing of the Isthmus of Panama (see Section 8.1), and the gradual onset of the ice age cycle around 2.8 Mya.

The emergence of modern humans (*Homo sapiens sapiens*) during the penultimate ice age between 200 and 150 kya, and the disappearance of Neanderthals *(Homo neanderthalensis)* between 35 and 25 kya, may also be the consequence of the dramatic climatic fluctuations which occurred during the ice ages (see Section 8.4). Like so many aspects of evolutionary theory, this is, however, an intellectual snake pit. Put at its simplest level this is an area where there are many more theories than pieces of evidence. From what little we know it is apparent that intelligent tool-making hominids have inhabited Europe for about half a million years. But modern humans emerged from Africa, appearing in the Middle East around 100 kya and may have lived alongside Neanderthals for many thousands of years.

This was, however, only a temporary movement. The decisive migration out of Africa did not occur until around 80 to 70 kya. Modern humans entered Europe around 45 kya and appear to have cohabited with Neanderthals for the next 10 to 20 kyr, although the DNA evidence suggests there was no interbreeding. The latter were physically well endowed to survive the capricious climate of the last ice age, having an immensely powerful physique adapted to cold climates and for endurance and prolonged locomotion over irregular terrain. Nevertheless, by the time the LGM they had disappeared, and it was modern humans, with their more advanced tool-making skills, who were able to exploit the more benign postglacial conditions.

9.3   Mass extinctions

So far we have focussed principally on the gradual changes that have occurred over geological time. The massive changes in fauna over these timescales are, however, only part of the story. In addition, there were sudden huge changes that appear in the fossil record as *mass extinctions*. This is another fertile area of scientific debate. The arguments centre on not only the part played by climate but also on what caused these catastrophes, how quickly they developed, and even just how many of them there were. Much depends on the quality of the fossil record and its ability to reveal the pattern of the history of life. While some palaeontologists take the view that the record is sufficient to draw conclusions, others argue that before around 300 Mya the record is inadequate because fewer fossils have survived and less work has been done to find those that remain.

Over the last 600 Myr a broadly consistent picture emerges. This shows that the diversity of all types of marine and continental life, including microbes, algae, fungi, protists, plants and animals, increased exponentially from the end of the Precambrian, and now totals between 5 and 50 million species. This diversification was interrupted by mass extinctions; the largest of these included the *Big Five* (see Section 8.1), at 440, 365, 255, 210 and 65 Mya. Pride of place goes to the Late Permian extinction (255 Mya) in which just over 60% of all families of species were wiped out. As many as 96% of all species may have disappeared; the mortality being higher on land than in the oceans. A further mass extinction in the early Cambrian at 520 Mya could be added on to this list, although the number of species involved and uncertainties about how rapidly it occurred mean it has to be treated with caution. Furthermore, a number of more minor, but still significant extinctions, have been identified in the fossil record, the last of which was in the late Eocene around 38 Mya. The identification of these less significant events depends in part on the improved fossil record, as we get closer to the present. Nevertheless, the picture is of a continuum of events with increasingly frequent minor extinctions being responsible for the elimination of many species. What is not clear is whether the Big Five fit in to this continuum, or are in some way fundamentally different. But, overall, species are at low risk of extinction most of the time, and this condition of relative stability is punctuated at rare intervals by a vastly higher risk of extinction.

A general explanation for major mass extinctions is made in the work of Vincent Courtillot, at the University of Paris (see Further reading). This provides a well-argued presentation of the close links between volcanic traps (see Sections 6.4 and 6.10) and mass extinctions. While not denying that an asteroid impact appears to have played a part in the late-Cretaceous extinction, in the case of the other extinctions there is no evidence of impact craters coinciding with these catastrophic events. On the contrary, there is plenty of evidence to suggest that massive volcanic flows are more likely to have been to blame. In particular, the greatest mass extinction – at the end of the Permian – appears to have coincided with the eruption of vast lava flows in eastern Siberia. Even the end-Cretaceous event coincided with the formation of the Deccan traps, which may have played a substantial part in the climate changes at the time.

Attempts have been made to establish a pattern in the timing of the major extinctions. One proposal that has attracted considerable attention is that a hitherto unknown celestial body, Nemesis (the *Death Star*) – small unseen binary companion to the Sun – was capable of disturbing the paths of comets every 28 Myr. These bombard the solar system and occasionally hit the Earth. There has been a vigorous debate about the evidence of the regularity of impacts or of extinctions (see van Andel; Further Reading). Courtillot identifies seven identifiable mass extinctions in the last 250 Mya – spaced between 20 and 60 Myr apart but decides there is no periodicity in their occurrence. The debate has, however, been rekindled by a recent study by Richard Stothers, at the NASA Goddard Institute for Space Studies in New York. This provides much more convincing arguments of a periodicity in major impact craters on the Earth dated since 250 Mya. The 11 craters more than 35 km in diameter have a spacing close to 30 Myr. The 20 smaller craters with a diameter greater than 5 km are clustered either side of the major crater peak at periodicities of 29 and 43 Myr.

All this suggests that the mass-extinction debate will continue to run and run. Improved remote sensing techniques will probably identify more craters, while improved dating will shed more light on whether or not their spacing is periodic. So the balance of argument between sudden huge bouts of volcanic activity, catastrophes resulting from collisions with asteroids or comets, or simply the capacity of non-linear responses (see Section 6.1) of the global ecosystem to produce cataclysmic events [e.g. the Late Palaeocene Thermal Maximum (see Section 6.10)], together with their climatic consequences and consequent evolutionary crises, will swing back and forth.

Whatever the role of the climate in mass extinctions the essential point is that throughout geological time the diversity of life has increased. While extinctions and other environmental changes have resulted in major setbacks, the upward trend is clear. This means that, in adapting to these changes, many life forms evolved in conditions that were changing continually. The long cooling from greenhouse to icehouse (Section 8.2) punctuated, as it was, by colder episodes exerted powerful selective pressures. As a consequence species adapted rapidly to survive.

All this reinforces the warning about trying to drawing conclusions about past climates on the basis of geographical distribution of recognisable types of flora that may have evolved appreciably to survive. How this adaptability will come into play in responding to future climate change is a big unknown. As many species have developed genetic defences to past changes, some of them will be equipped to meet future challenges. What we do not know is which species will survive, nor which genetic defences will provide the best protection against both natural variations and the consequences of human activities.

9.4   Sea levels, ice sheets and glaciers

Having considered the changes in sea level in geological terms in Section 8.3, the more recent changes provide a stepping-stone to the challenges of living in the Holocene. The changing volume of freshwater locked up as land-ice around the world has various consequences. It affects worldwide changes in sea level (termed *eustasy*). As such, it was an integral part of the dramatic changes in the climate at the end of the last ice age (see Section 8.5).

In the depths of the LGM global sea levels stood some 130 m below current levels. The collapse of the North American and Fennoscandian ice sheets following the last ice age produced a rapid rise in sea level (see Fig. 8.15). By the middle of the Holocene rising sea levels had flooded some 25 million square kilometres of the continental shelves around the world; redefining the map to the form we now know. Vast areas from the North Sea and the Bay of Biscay, much of the land between Indo-China and Indonesia (*Sundaland*) and the land bridge between Papua New Guinea and Australia, plus the great land bridge between Siberia and Alaska (*Beringia*) had disappeared below the waves. This inundation came to a virtual halt about 2 kya, and the level did not change appreciably until around 100 years ago.

Instrumentally based estimates of modern sea-level change show evidence for onset of rising levels at the end of the nineteenth century. Recent estimates for the last half of the twentieth century (1950–2000) give ~2 mm year^{-1} global mean sea-level rise. Satellite observations now provide precise sea-level data with nearly global coverage. This data set shows that between 1993 and 2005 sea level has been rising at a rate of 3.1 mm year^{-1} (Fig. 9.3), a rate significantly higher than during the previous decades. It is, however, unclear whether the higher rate of sea-level rise in the 1990s is due to anthropogenic global warming, or a result of natural climate variability, or a combination of both effects. The data also confirm that sea level is not rising uniformly over the world. While in some regions (e.g., western Pacific) sea-level rise since 1993 is up to five times the global mean, in other regions (e.g., eastern Pacific) sea level is falling.

On the basis of ocean-temperature data for the past 50 years it is estimated that thermal expansion has contributed ~0.4 mm year^{-1} to sea-level rise. For the recent years (1993–2003), thermal expansion accounts for ~1.5 mm year^{-1} and is the largest part of sea-level rise. On average over the past four decades, loss of mass by glaciers and ice caps has accounted for ~0.5 mm year^{-1} sea-level rise, and more in recent years. Observations beginning in the 1990s indicate that the Greenland and Antarctic ice sheets

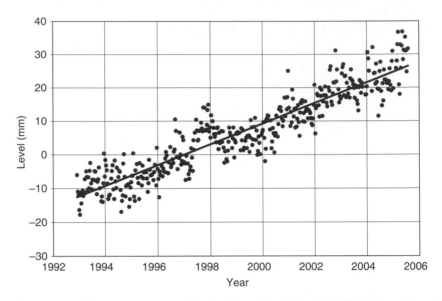

Fig. 9.3 Satellite measurements of the rise in sea level since 1993. (Data from University of Colorado website: http://sealevel.colorado.edu/current/sl_ib_global.txt.)

are also contributing substantially. In particular, satellite measurements suggest a sharp increase in the melting of the Greenland ice sheet since 2004. Overall, for the years since 1993 there is fair agreement between the observed rate of sea-level rise and the sum of known contributions.

Locally, changes in the extent of glaciers and ice caps affect the lives of those who live in their shadow. Written reports of glacier-length changes go as far back as 1600 in a few cases, and are directly related to low-frequency climate change. The general picture is of decline in mean length variations of glaciers (Fig. 9.4). The retreat of glacier termini started after 1800, with considerable mean retreat rates in all regions after 1850 lasting throughout the twentieth century. There was a slowdown between about 1970 and 1990 and a rapid retreat in the 1990s. The only exceptions were where increases in precipitation led to advances of glaciers in Western Scandinavia and New Zealand.

On the larger scale the waxing and waning of the ice sheets during the ice ages have left their mark on the landscape (see Sections 8.5 and 9.1) and are still affecting our lives. This is a consequence of the continuing impact of the deformation they caused to the Earth's crust. The term *isostasy* is given to the theory, which proposes that where the crust is thickest it extends into the mantle beneath it. Because crustal material is less dense than the mantle, it effectively forms a 'root' under mountain ranges that holds them up. The same is true where an ice sheet builds up.

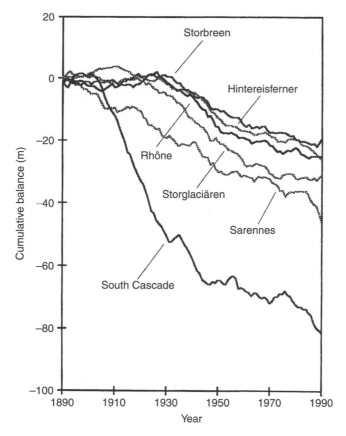

Fig. 9.4 Cumulative mass balances, in metres of water equivalent, for the glaciers Hintereisferner (Austria), Rhône (Switzerland), Sarennes (France), South Cascade (United States), Storbreen (Norway), and Storglaciären (Sweden). These are among the few glaciers with long observational time series that have been extended using well-calibrated hydrometeorological models. All values are relative to 1890. (From IPCC, 1995, Fig. 7.3.)

When the ice sheet melts away, the less dense crustal 'root' is no longer needed to hold it up, and rebound occurs. During the height of the last ice age around 18 kya the additional load on areas like the Canadian Shield and the Baltic Shield caused the crust to dip downwards by as much as 700 m. This was accommodated by the creep of mantle material away from the region of application of the extra load. Melting of the northern-hemisphere ice sheets caused a reverse flow of mantle material and slow uplift (*glacial rebound*) which continues today at rates up to 10 mm per year (Fig. 9.5). This means that in some coastal regions in North America and northern Europe this process is of as much importance as more general sea-level rises.

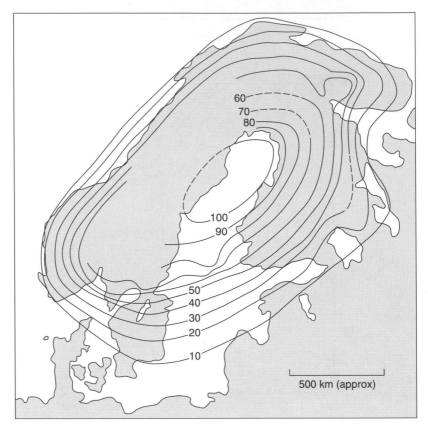

Fig. 9.5 A map of the present elevation in metres of the shoreline of 5000 years ago in Scandinavia to illustrate isostatic 'rebound' following deglaciation. (From Smith, 1982, Fig. 15.5.)

9.5   Agriculture

The transition from the ice age to the Holocene had a dramatic impact on humankind. Moreover, in moving to the relative calm of the Holocene we need to return to the concepts explored in Section 1.2. This effectively means moving from issues of climate change to climate variability. There have been fluctuations during the last 10 kyr (e.g. the desiccation of the Sahara – section 8.7), which may be the consequence of specific physical causes. By comparison with the cataclysmic changes during and after the last ice age, apart from the change in the Sahara, they amount to little more than the flutter (see Section 8.5). Prior to the stability of the Holocene the much greater climatic variability that prevailed until around the end of the Younger Dryas would have made organised agriculture impossible.

We now need to look at this claim more closely and consider the impact of subsequent fluctuations on agriculture and the consequent implication for human activities.

The emergence of agriculture is a fiercely debated subject. Clearly, the stability of the climate in the Holocene made agriculture a viable proposition. What is far less clear is when humans started to exploit plants in an organised manner. Many hunter–gatherer groups exploited seeds and fruits from wild stands of plants. In so doing they appear to have developed the means of storing what they had gathered to last through periods of scarcity. Recent evidence from an ancient settlement, Ohalo II, in northern Israel, suggests that this survival strategy was adopted as early as 23 kya. There is also evidence of grinding grain to make bread, which would have requires storage to sustain the practice.

It was not until after the LGM that evidence of the formal management of wild plants began to emerge. After 15 kya the warming in the Bølling period plus some evidence of reduced variability during this amelioration provided opportunities for a more settled lifestyle. Most notably this occurred in the *Fertile Crescent* of south-west Asia, where people could exploit the abundance that came with warmer temperatures and greater rainfall. With settlement, the opportunities for *proto-agriculture* grew. This may well have involved maintaining stands of native grasses, notably wild barley, and einkorn and emmer wheat, that grew in the mountains of eastern Anatolia and the Zagros Mountains, and removing weeds to raise yields. Although the Older Dryas interrupted this climatic improvement, populations in favoured areas probably rose until the savage cold and aridity of the Younger Dryas plus the dramatic increase in variability brought this period of advance to a grinding halt.

Recent archaeological studies have provided evidence supporting the Malthusian interpretation of the origins of agriculture. Information has come from the site at Abu Hureyra, on the Euphrates, in what is now northern Syria, suggests that systematic cultivation of cereals started at least as early as 13 kya, some 400 years before the Younger Dryas. Cultivation of crops started in response to a steep decline in wild plants that had served as staple foods for at least the preceding four centuries. Work by Gordon Hillman of University College London, UK and his colleagues found that the wild seed varieties gathered as food gradually vanished, before the cultivated varieties appeared. Those wild seeds most dependent on water were the first to die out, then one by one the more hardy ones followed. So, the hunter-gatherers turned to cultivating some of the foods they had previously collected from the wild. In an unstable environment, the first farmers started simply by transferring the hardiest wild plants to more suitable habitats and cultivating them there.

The establishment of agriculture as the dominant source of human food took most of the first half of the Holocene. The spread of agriculture across Europe during this period, together with its independent development in the Far East and Mesoamerica, means that by 5 kya the vast majority of humankind was dependent on its success. The fact that agriculture is so sensitive to climatic fluctuations meant that weather became vital to our forebears' survival. This remained the case until the industrial revolution. For instance, in medieval Europe the purchase of food represented some 80% of the expenditure of working people. So fluctuations in food supply and hence prices had an immediate and costly impact on the majority of the population. When famine occurred the death rate rose. Examination of mortality statistics in England in the seventeenth and eighteenth centuries concluded, however, that the cumulative effects of weather-induced price variations over five years was essentially zero, as all they did was alter the timing of deaths which would in any case soon have occurred.

The socially disruptive effects of high prices and food shortages are more difficult to quantify. Many historians have proposed links between such fluctuations and more widespread disorder. For instance, the French Revolution in 1789 is sometimes linked with damaging summer storms that compounded the already vulnerable food supplies and spread social unrest. But the scale of the changes in the climate at this time were not that exceptional and so placing too much emphasis on specific extreme weather events in the preceding year or two seems to overlook wider social and economic factors. To be convinced of the connection we need to find reasons why many other equally unpleasant bouts of weather did not produce similar outbursts of civil unrest. The explanation is probably that, while weather-induced price rises may have a catalytic effect in stimulating unrest, the other social and political factors, which contribute most to revolt, were not present.

This cocktail of factors makes it equally hard to identify the overall impact of periods of climatic deterioration on agricultural societies. As noted in Section 8.9, the Little Ice Age, rather than being a sustained period of lower temperatures, was, for much of the time, not appreciably colder than the twentieth century but was punctuated by markedly colder decades that hit different parts of the world at different times. In Europe the greatest agricultural disasters were in the 1310s, the 1590s and the 1690s. Each, in its own way, provides useful insights into the major short-term disruption.

Although the extent of the Medieval Climatic Optimum (see Section 8.8) is uncertain, there is considerable evidence that by the late thirteenth century the climate in northwest Europe had become stormier and an analysis of historical records of winter weather in Europe during the fourteenth century suggests the first quarter of the century featured an exceptional number of cold winters. But it is the run of extraordinary cold wet summers from 1314 to 1317 and their associated harvest failures that come ringing down through history as the greatest weather-related disasters ever to hit Europe. From Scotland to northern Italy, from the Pyrenees to Russia there are an unparalleled number of reports of failed harvests, starvation and pestilence. Grain prices rose to levels unsurpassed in the subsequent 150 years. Where detailed records exist, mortality rose dramatically with over 10% of the population of the rich Flemish town of Ypres dying in the summer of 1316. But, in spite of the awful impact at the time the lasting consequence of this series of bad summers was small. What they demonstrated was the vulnerability of a population whose numbers were at the limit of what could be provided for by the agriculture of the time. The arrival of the Black Death in 1348 and the almost constant warfare throughout the continent for much of the century reduced the population to a level where the vagaries of the weather had less impact on social history.

The link between the climate, agriculture and the ability to feed a growing population re-emerged to haunt Europe at the end of the sixteenth century. This was a period of more frequent cold winters and cool wet summers north of the Alps. In England poor harvests and rising grain prices meant the standard of living of the working people fell to the lowest level recorded in the last seven centuries (Fig. 9.6). The growing social unrest led to panic legislation. Parliament passed a Great Act codifying a mass of sectored legislation and local experiments in poverty relief. It also restored many of the restrictions on enclosures of common land and the conversion of arable land to pasture that had been repealed only four years before.

Although agriculture was always stretched to meet the needs of the population in Europe during the seventeenth century, the exceptional cold of the 1690s had serious consequences. The scale of impact varied from place to place, and this provides insights into the vulnerability of different communities. France, with its high population levels, was hit first, when two poor harvests precipitated the worst famine since the early Middle Ages, in 1693. England, by comparison, escaped relatively unscathed, but the continued cold hit Scandinavia particularly hard. In Finland the famine of 1697 is estimated to have killed a third of the population. But, possibly the most lasting effect was in Scotland. Here, between 1693 and

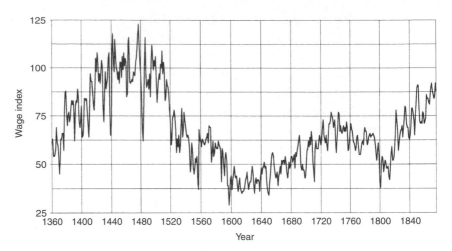

Fig. 9.6 The index of the purchasing power of builders' wages in England over six centuries. (From Burroughs, 1997, Fig. 2.5.)

1700, the harvests, principally oats, failed in seven years out of eight in upland parishes. The death rates rose to between one- and two-thirds of the population in many of these parishes, exceeding the figures recorded during the Black Death. The economic consequences of these awful years probably, more than anything, made the union with England in 1707 inevitable.

In other parts of the world, the agricultural consequences of climate fluctuations have more often been the result of drought. There are many examples of how sustained drought has destroyed apparently thriving farming communities, but it is only where this leads to a radical change in managing subsequent activities that it can be said to have had real consequences. A good example of this response is how the United States Government rose to the challenge of the Dust Bowl years of the 1930s. Ever since settlers had introduced arable farming to large areas of the Great Plains, periodic series of hot, dry summers had destroyed crops and driven many farmers from the land. Both in the 1890s and the 1910s many areas experienced widespread depopulation as crops failed.

In 1934 and 1936 the average wheat yields across the Great Plains fell by about 29% compared with the trend. In the hardest hit regions of Kansas and Oklahoma there was massive outward migration with more than half the population leaving. But the Democratic Government recognised that without central action the problems of recurrent drought would continue. It concluded that much of the agriculture on the Great Plains was not appropriate to such an arid region and so marginal land was purchased and retired from cultivation and seeded with grass. This was combined with educational programmes for farmers to plant trees for shelter-belts, grow crops better suited to drought conditions and introduce conservation methods (e.g. contour ploughing, water conservation and strip ploughing to allow part of the land to lie fallow). Although the pressures of the Second World War led to some marginal land being pressed back into service in the wetter 1940s, subsequent droughts have built up the acceptance of the need for state and federal laws to protect the land from over-exploitation.

Analysis of the events in the Sahel, together with lengthy studies of the variation of the Indian monsoon, has led to the conclusion that tropical SSTs are a major factor in rainfall variations in the sub-tropics (see Further reading). In particular, the ENSO is a good predictor of wet and dry years in various parts of the Tropics. In the 1990s it appeared that the links with such vital rainfall patterns in the Sahel, southern Africa and across South America, together with the Indian Monsoon, could be accurately predicted months in advance (see Section 3.7). So, understanding the quasi-cyclic fluctuations of the climate due to the ENSO had widespread agricultural consequences. In recent years more extensive measurements in other ocean basins, plus the erratic input of intraseasonal oscillations (the MJO, see Section 6.2) have shown that the prospect of reliable long-range seasonal forecasts requires a lot more work before farmers will be able to make decisions about planting the right crops for dry or wet seasons.

Although the prospect of improved forecasting may lead to some advances in the ability of agriculture to respond to climate variability, the general message is that we remain as vulnerable as ever to extreme weather events. This means that if the climate should become more variable or undergo some more marked change the implications for food production around the world are bleak. In essence, any change will be a bad thing, especially as the growing population in the developing world makes ever greater demands for increased productivity.

This gloomy conclusion, together with recent painful lessons of the consequences of prolonged extreme weather disrupting agriculture and rural communities, exerts a major influence on current farming policy in many countries. It is recognised that without some government involvement extreme weather events exaggerate other fluctuations in supply and demand and do great damage to both farming interests and the community at large. So, possibly the most important consequence of future changes in the climate for the agricultural sector is to identify the right level of government intervention in managing the markets. The ability to extract the right message from past agricultural experience in planning for future climate variability (see Chapter 10) will be an important part of this process.

9.6   The historical implications of climatic variability

The rise and fall of ancient civilisations is a source of enduring historical fascination. While many instances can be defined in terms of armed conflict, there have been instances where more widespread decline occurred. The possibility that these periods of decline (*Dark Ages*) might be related to climatic events has only emerged slowly. While there is some archaeological evidence that the climatic events around 5.2 kya (see Section 8.7) had a significant effect on the development of early civilisations, it is not until the Bronze Age that the historical impact of the Dark Ages becomes apparent. Across the Middle East there is widespread archaeological evidence of periods of major destruction. Originally thought to have been caused by earthquakes, four successive destruction levels were present in all sites, the most prominent of which were detected at the end of the Early Bronze Age (~4.3 kya), at the end of the Middle Bronze Age (~3.65 kya), and at the end of the Late Bronze Age (~3.2 kya). The first and the last of these three events coincide with the periods of rapid climate change (see Section 8.7).

The geographical extent of these events was far too great to be attributed to tectonic activity. A more likely explanation is that the climate change around 4.2 kya caused agricultural disasters and prolonged droughts that consequently led to a breakdown of social order in population centres in western Asia, North Africa and Eastern Europe. Furthermore, similar ecological and social upheavals appeared to have occurred at around the same time in China and the Americas. Explanations of these periods of climatic deterioration tend to concentrate on either natural climatic variability or more specifically to large volcanic eruptions.

The most frequently cited of these upheavals is the rapid demise of Akkadian civilization (in what is today Syria) around 4.2 kya is associated with the onset of an unprecedented dry episode. Only a hundred years before the collapse, Sargon of Akkad had conquered the plains of Mesopotamia, and taken control of the Sumerian city-states, to establish a domain from the Persian Gulf to the headwaters of the Euphrates River. This was the first example of one state conquering other independent societies to form a single entity. The empire's breadbasket was in northern Mesopotamia. A string of fortresses was built to control imperial wheat production. To the south, irrigation canals were extended, a new bureaucracy established and palaces and temples built from imperial taxes.

This empire lasted less than a hundred years. Evidence from Tell Leilan, in northern Mesopotamia, shows that the site was abandoned suddenly only decades after the city's massive walls had been constructed, its religious quarter renovated and its grain production reorganized. At Tell Leilan, analysis of the various layers of debris shows that after the Akkadian occupation there is an interval devoid of signs of human activity, containing only the clay of deteriorating bricks. The abandonment began about 4.2 kya. Soil samples from that time showed abundant fine, windblown dust and few signs of earthworm activity and much reduced rainfall. All this suggested that the people of Tell Leilan abandoned the site in the face of a sudden shift to a much drier and windier climate. This precipitated the collapse of the Akkadian empire's northern provinces. Only when wetter conditions returned, some 300 years later, was Tell Leilan reoccupied.

At about the same time drier conditions brought famine to Egypt with the end of the Old Kingdom, which had had bountiful floods, and the onset of what is now known as the *First Intermediate Period*. The Nile floods, which normally protect Egypt from the worst climatic calamities, were catastrophically low. The records talk of hot winds from the south bringing dust storms that obscured the sun so that men could not see. This disruption lasted over a century. It led to a breakdown in the economic system in Egypt with local governors taking responsibility for their own provinces, by conserving water supplies, and reducing the number of hungry mouths by repelling famine-stricken outsiders.

The next Dark Age occurred in conjunction with the climatic deterioration of the thirteenth century BC (see Section 8.7). The abandonment of lakeside settlements as water levels rose in northern Europe about 1250 BC can be attributed to sustained shifts in the climate at the time. In the eastern Mediterranean these changes appear to have triggered large-scale demographic movements that are usually associated with the name 'Sea Peoples'. They are best known for waging two campaigns against Egypt, which ended in battles in 1232 BC and 1183 BC. Although they were repulsed on both occasions, after these attacks Egypt ceased to be an imperial power. Other civilisations were, however, even worse hit. The Hittite Kingdom in central Anatolia, which had represented a serious challenge to Egypt during the preceding century, was totally extinguished in its Anatolian heartland, although there was some continuation of the culture in Syria. In Greece the might of Mycenae was snuffed out.

The Sea People were probably part of a great migration of displaced people, possibly as a result of climate-change-induced crop failures in their homelands at the time. Their successful progress appears to have focused on attacking capitals and cities important to administration. In these cities they destroyed government buildings, palaces and temples, while leaving residential areas and the surrounding countryside untouched. By doing this, they destroyed the local leadership, and could win fairly easy victories. They appear to have destroyed Mycenae, and then moved on to Troy, which they destroyed in 1250 BC. They then moved into the Levant and on to Egypt where they met their match in the two battles mentioned above.

Although these Dark Ages have been linked to climate change, how this combines with other archaeological evidence, such as warfare, invasion and social instability, requires careful handling. So, in arguing the climatic case, we must not fall into the trap of selective use of the evidence. In particular, the problem with dating events (see Section 4.5) makes it all too easy to attach too much significance to what seem like striking coincidences. Although extensive work has produce a lengthy tree-ring series for the Aegean and Anatolian region, which provides an accurately dated series from around 1800 to 800 BC, it has yet to pinpoint clear evidence of climatic change. The one event that does stick out is a spurt in growth at around 1650 BC, which may be a result of the massive volcanic eruption of Thera on the island of Santorini in the Aegean Sea. This date matches the eruption date recorded in Greenland ice cores but is nearly 200 years earlier than the date based on archaeological evidence associated with the decline of the Minoan Civilisation on Crete. An accepted reconciliation of these arguments has yet be reached.

The decline of the classical Mayan culture in the ninth century AD is, however, a good example of how the climatic case can be constructed. This civilisation reached a pinnacle in the eighth century when the population density in the Mayan lowlands, which extend over modern day Guatemala, Belize, Honduras and Mexico, was far higher than current levels. Because of the accuracy of the Mayan calendar and their propensity to erect large monuments containing detailed records of events, it is possible to date the cataclysmic decline accurately. Furthermore, recent measurements of sediments in Lake Chichancanab, in what is now Mexico, show that the period around 750 to 900 AD was the driest in the last 8000 years. This suggests that a period of drought may have reduced the capacity of society to support an overbearing theocracy and building major monuments. This would have increased the susceptibility of many Mayan cities to revolt against these demands. But it would depend on each city's circumstances and this could explain why the cessation of records in different cities occurred at different times, with those cities on riverside location maintaining records longest.

Even more clear-cut is the disappearance of the Norse colony in Greenland. The availability of records of contacts with the colony from its establishment in the late tenth century until the fourteenth century, together with data from ice cores drilled in the Greenland ice sheet (see Section 4.4.2.), provide a pretty clear picture of climatic deterioration. But, while the colder weather is not in dispute, the extent to which it destroyed the colony is still the subject of debate. An alternative explanation is that the rigidity of the social structures and the reluctance to adopt Inuit technology, more appropriate to the harsh climate, may have had as much to do with the collapse as the cooling trend.

Turning to more recent history, the shifts in the climate identified over the last millennium (see Sections 8.8 to 8.10), while significant, have not exerted an overwhelming influence on the course of history. Indeed the argument among most historians is whether these changes have had any appreciable impact whatsoever. So, rather than seeking to establish the shadowy consequences of climatic shifts on history at large, it is better to examine those social and economic activities where changes are most likely to have had an impact and then ask the question as to whether they could have had wider implications. This approach also prepares the ground for considering questions of how future changes in the climate may impact on the most vulnerable sectors of our modern society.

9.7   Spread of diseases

The incidence of epidemics of disease is sometimes linked to climatic factors. Any analysis must, however, take account of variations in food supplies and population levels. So a period of benign climate (e.g. the Medieval Climatic Optimum in Europe – Section 8.8) can lead to sustained population growth, which then magnifies the damaging consequences of subsequent bad weather. Reduced food supplies at times of historically high population levels increases vulnerability to diseases. Hence any interpretation of the role of climate variability in famine and social decline has to take account of demographic trends.

There are additional intrinsic reasons for the coincidence between major pandemics and periods of serious social disruption. These are part of the very nature of many of the diseases that afflict the human race. The symbiotic relationship between a disease (e.g. bubonic plague, influenza or typhus), and its animal and human hosts means potentially more virulent mutations in the disease organism may remain quiescent because the local animal or human populations have some degree of immunity. When, however, some major upheaval occurs due to, say, drought, earthquakes or floods, populations are dispersed, mixing with others who have less immunity, and a new plague may emerge with horrifying rapidity.

Examples of this chain of events, which may have had a climatic component, include the Justinian Plague of the sixth century, the Black Death in the fourteenth century and the global cholera pandemic of the 1830s. Each of these cases is worth examining as it provides insights into the complexity of the processes involved. The epidemic that struck Constantinople in 542, in the reign of Emperor Justinian, was the first reliably identified instance of bubonic plague. While historical sources claim it came from Ethiopia, its origin may be linked to the sudden cooling associated with the 'mystery cloud' of 536 (see Section 6.9). The consequent crop failures from Rome to China led to widespread famine. Against the background of growing demographic pressures, this provided a combination of social pressures ideal for the emergence of a pandemic. There is no way of quantifying the contribution of climatic disturbance to these events. Nevertheless, over the next century or so the population of Europe halved and the more heavily populated Mediterranean region may have suffered more grievously. Effectively the region relapsed into rural convalescence and centuries of a Dark Age.

The analysis of the origins of the Black Death involves the same complex mixture. The combination of demographic pressures, bad weather and crop failures was already a feature of northern Europe in the early fourteenth century (see Section 9.6). But, as noted in Section 8.8 there is no clear evidence of a coherent climatic trend across the northern hemisphere at the time. So, given the plague appears to have originated in China, a more local explanation may need to be invoked. China was devastated by horrendous floods in 1332, which reportedly killed several million people, disrupted large parts of the country and caused substantial movements of wildlife, including rats, in which bubonic plague is endemic. This disruption may have triggered the emergence of the Black Death, and, given that China is often wracked by devastating summer floods, once this virulent new version of bubonic plague had come into being, it was only a matter of time before it made its presence felt. So, while a climatic event may have set the ball rolling, it was only a tiny but essential component of the subsequent cataclysmic pandemic.

The cholera pandemic is an even more complicated story. Cholera appears to have been unknown outside Bengal before 1800. In 1815 the massive eruption of Tambora (see Section 6.4) appears to have disrupted the global climate, and may have been the cause of harvest failures in Bengal. The resulting famine triggered the first cholera pandemic, which eventually reached Europe and the eastern United States in 1832. In Russia the mortality was extremely high and in New York the death rate exceeded 100 per day in the summer of 1832.

These historical examples of how climatic shifts may have triggered global pandemics are of relevance to the analysis of the potential impact of global warming (see Section 10.3). The possible spread of tropical diseases and their insect vectors to higher latitudes is widely regarded as one of the more worrying feature of current climatic change. The experience of the spread of various tropical diseases during recent ENSO events provides insight into the potential for these changes to occur, but such concern must be viewed in the context of the advances that have been made in public health in the last century.

Of all the climate-related health hazards in tropical and sub-tropical developing countries malaria is by far the most damaging. It is, in essence, a disease of the tropics and the sub-tropics, because a key part of the life cycle of the parasite (*sporogony*) depends on a high ambient temperature. Malaria also depends on adequate conditions for mosquito breeding, which is relatively clean water, usually due to rainfall forming in still pools. So, malaria is endemic in the humid tropics where there is heavy rain throughout the year. But, in the sub-humid tropics, where wet and dry seasons alternate, it has a distinct seasonality as mosquitoes breed during the rainy season. Additionally, the intensity of malaria transmission depends on the specific mosquito vectors that are present. Mosquitoes of the genus *Anopheles* transmit all malaria. Some anopheles species, especially those in sub-Saharan Africa, show a high preference for taking their blood meals from humans (*anthropophagy*) as opposed to animals such as cattle. This is a cruel example of where evolution has worked to the distinct disadvantage of humans, as these human-biting vectors lead to more intensive transmission of the disease.

Malaria causes between 1.5 and 2.7 million deaths around the world annually; 90% of the victims are children under five years of age. The impact of malaria is greatest in Africa, and, in particular, in the sub-Saharan region and southern Africa. Regional climate influences both the development of the malaria parasite and behaviour of the carrier mosquito. Climatic variability from year to year is particularly important. The incidence of the disease is affected by periods of drought or heavy rainfall. During a prolonged drought period, a population's immunity to malaria declines. When the rains return, epidemics often occur; so an understanding of how seasonal rainfall throughout the tropics is affected by ENSO and the SSTs in other tropical areas (see Section 3.7) is an essential in responding to the scourge of malaria.

Human health has formed an important part of the general debate on geography and economic development. Jeffery Sachs and colleagues at the Center of International Development at Harvard University, have explored the impact of malaria on the economics of development in tropical countries, notably Africa. Their conclusion concerning the economic impact of malaria is that the location and severity of malaria are mostly determined by climate and ecology, not poverty as such. Areas with intensive malaria are almost all poor and continue to have low economic growth. The geographically favoured regions that have been able to reduce malaria have grown substantially faster since effectively eradicating the disease. As places like Hong Kong and Singapore clearly show, once a state achieves both a high level of public health and sufficient wealth effectively to control its environment for its principal economic activities, it can become largely independent of its climatic circumstances.

9.8   The economic impact of extreme weather events

The economic impact of more frequent extreme weather (e.g. droughts, heat waves, hurricanes and winter storms) is the source of much comment in the context of global warming. This is a difficult issue to deal with in terms of the definitions we have been working with, as it can lead to the selective interpretation of perceived changes in climate variability. Nevertheless, despite the fact that there is, as yet, little evidence of a significant shift in the incidence of extreme events (see Section 8.10), it is widely assumed that global warming will lead to greater climate variability; so an examination of the impact of past extremes is part of the process of both interpreting current trends and also predicting the future impact of climate change.

The most damaging forms of weather extreme during the twentieth century have varied from one part of the world to another. In many parts of the tropics and subtropics, failure of seasonal rainfall causes the greatest hardship. The industrialised countries in the mid-latitudes experience a wider variety of damaging events. For example, in North America the combination of blizzards, floods, hurricanes, tornadoes, summer droughts and heat waves, or winter freezes can all cause immense economic damage. Europe experiences a similar mixture, and although hurricanes and tornadoes are not a comparable threat, intense extratropical depressions are a major issue.

The evidence of changes in the incidence of extremes during the twentieth century has been examined in Section 8.10. In terms of health consequences, the two most important forms of temperature extremes are cold spells and heat waves. Since 1951, the annual occurrence of cold spells significantly decreased while the annual occurrence of warm spells significantly increased. The trend in warm spells is, however, greater in magnitude and is related to a dramatic rise in this index since the early 1990s. The significant decreases in cold spells occurred predominantly in central and northern Russia, parts of China and northern Canada; however, a large part of the USA showed a significant increase in this index. Significant increases in warm spells were seen over central and eastern USA, Canada and parts of Europe and Russia.

In recent years public concern has focused on heatwaves, and, in particular, the severe event in Europe in the summer of 2003. Many places had record-breaking temperatures that were about 10–15 °C above average. In Rome the temperature maxima exceeded 35 °C for 42 days during June to August. Night-time minima were also exceptionally high. Perhaps the most evocative measure of the exceptional nature of the summer emerged from an analysis of the Burgundy Pinot Noir wine harvest extending back to the fourteenth century (see Fig. 4.5). This concluded that the summer of 2003 was an unprecedented event. It was nearly 6 °C warmer than the average for the period 1960–89, whereas the next highest anomaly during the whole period was just over 4 °C in 1523.

This heatwave was deadly. Over 30 000 people perished in Western Europe in about 20 days in August, close to 1500 a day, and more than in any European weather-related disaster in more than 50 years. In France alone the death toll was around 1000 a day for nearly 15 days. Mostly they were elderly in nursing homes, without air conditioning, and the fact that their sons and daughters were vacationing in early August, away from the big cities, leaving the emergency personnel understaffed or the older people unattended in their homes.

These figures need to be put in the context of wider weather-related mortality statistics. The striking feature of the figures for countries in the temperate latitudes, especially in the northern hemisphere, is the dramatic rise in mortality every winter. There is also a rise in unusually hot weather. The largest rises in cold-related mortality are not in the coldest countries, but in those that are relatively mild but have occasional cold spells. For example, in Europe the highest proportion of winter deaths are in Ireland and Portugal, while the lowest figures are in colder countries such as Norway and Finland. Indeed, the city with the lowest average winter temperature in the world, Yakutsk, in Siberia, has no excess winter mortality; the people there have learnt to adapt fully to the extraordinarily cold weather.

Approximately half of excess winter deaths are due to coronary thrombosis. These peak about two days after the peak of a cold spell. Approximately half the remaining winter deaths are caused by respiratory disease, which peak about 12 days after peak cold. The rapid coronary deaths are due mainly to thickening of the blood resulting from fluid shifts during cold exposure; some later coronary deaths are secondary to respiratory disease. Heat-related deaths often result from thickening of the blood resulting from loss of salt and water in sweat. In the UK, it is estimated that excess cold spells account for between 20 000 and 50 000 deaths a year. By comparison, hot spells in summer account for around 1000 deaths a year.

With the possible exception of some tropical countries, global warming can be expected to reduce cold-related deaths more than it increases the heat-related deaths, but statistics on populations in different climates suggest that, given time, people will adjust to global warming with little change in either mortality. Some measures may be needed to control insect-borne diseases during global warming, but current indications are that cold will remain the main environmental cause of illness and death.

Turning to other destructive weather events, there is an additional underlying question of whether the changes recorded are real or the product of growing public awareness and increasing population in areas where the extremes occur. A good example of this phenomenon is the rise in the number of tornadoes recorded in the US since the 1950s (Fig. 9.7). Clearly the number of observations has increased, but what is less certain is how much is the result of climatic developments and how much is due to more efficient reporting of smaller twisters. The fact that there is no measurable trend in the incidence of strong to violent tornadoes (F3 to F5) in the records since 1950 supports this conclusion. Other measures of the impact of tornadoes show that the number of deaths each year has declined since 1950, but this could be attributed more effective warnings leading to people taking shelter. Perhaps more significant, the number of days each year when tornadoes have been observed has shown little increase in recent decades and the number which have caused fatalities each year has shown no rise.

The second factor, which must be addressed, is whether, in focusing on events related to global warming, the index overstates the economic impact of any trend. If the observed rise in the events used in the index is matched by, say, a decline in severe winter weather and killing spring frosts then the economic consequences are greatly reduced.

A more direct measure of economic impact is trends in insurance losses. These have been rising rapidly in recent decades, but even this increase

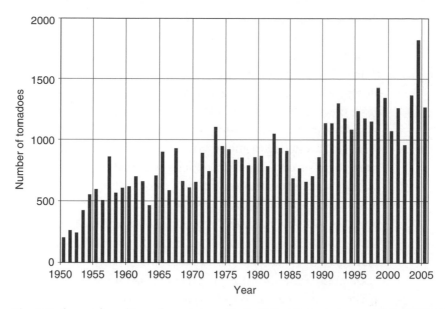

Fig. 9.7 The number of tornadoes observed in the United States between 1953 and 2005 showing a strong upward trend. What is not clear is whether this trend is the result of climate change or more comprehensive observations. (From Burroughs, 2001, Fig. 5.8.)

must be treated with caution. Any analysis must take full account of changing levels of insurance cover, population densities in vulnerable areas (e.g. coastal sites), inflation, and even fraudulent claims. These corrections must be combined with the fact that there is little evidence of an increase in extreme events, and a few major disasters can have an excessive influence on thinking (e.g. Hurricane Andrew hitting Miami in 1992 or the storms of October 1987 and January 1990 in the UK). Only by taking full account of all the relevant factors and also a sufficiently lengthy perspective is it possible to form a balanced view about whether there is a significant climatic component in current economic developments.

Hurricane losses in the United States provide a good example of how this type of correction needs to be applied. The Gulf and east coasts of the United States suffer on average nearly US $5 billion (1995 US $) hurricane damage each year. But over 83% of this damage is caused by severe hurricanes (Category 3 and above: see Table 8.3), which make up barely a fifth of the tropical cyclones that make landfall along these coasts. Moreover, a single massive storm can have a disproportionate impact on thinking. For instance, Hurricane Andrew was estimated, at the time, to be the most costly storm in US history causing some US $30 billion of damage. Following close on the heels of Hurricane Hugo in 1989, which cost some US $6 billion, there was a widespread assumption that these huge losses were the product of a dramatic worsening in the climate.

What was missing in this instant reaction to such major events was a thorough analysis of the trends in the costs of hurricane-caused damage along the United States coast. This needs to address two principal issues. First, there is the question of the capacity of hurricanes to do damage. Here NOAA has produced estimates of the potential of every storm since 1950 to do damage. This statistic combines the square of the sustained wind speed of the storm over successive 6-hour periods summed throughout its lifetime. The use of a square-law relationship reflected that higher wind speeds generate much greater damage. Known as the *Accumulated Cyclone Energy (ACE) index*, it provides a useful measure of the vigour of hurricane seasons in the Atlantic Basin. The annual values of the ACE index since 1950 (Fig. 9.8) clearly show that there was a marked lull in hurricane activity between 1970 and 1994. Prior to 1970 and since 1995 hurricane seasons were more vigorous.

The second factor to consider is to include economic indices in the analysis. This requires accounting not only for inflation, but also changes in coastal population and wealth, both of which have risen dramatically in recent decades in areas where hurricanes strike. When this normalisation process was conducted in the late 1990s, it demonstrated that there has not been a rising trend in damage in recent decades. This approach produces a similar set of figures to Figure 9.8 with little clear trend overall, but a clear lull in the

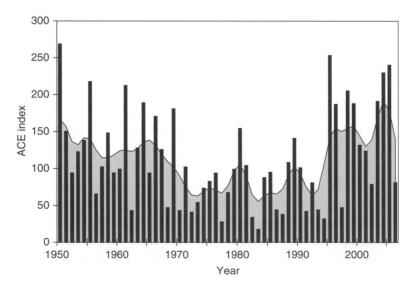

Fig. 9.8 A measure of the annual damage potential of hurricanes in the Atlantic basin together with a seven-term binomial smoothing of the annual figures.

1970s and 1980s. Furthermore, to put Hurricane Andrew into perspective, it was estimated in 1995 that if the hurricane, which devastated Miami in 1926 were to occur in 1995 the damage would exceed US $70 billion.

A further aspect of these calculations is the nature of the planning processes. These must involve not only adequate provision of defences and emergency services but also decisions about where developments can take place and who should take responsibility for the consequent risks involved. This boils down to the fundamental question is how much can government invest in defences to enable people to live in attractive, but inherently risky places. Recent estimates of the long-term effects of changes in society dwarf the effects of any projected changes in tropical cyclones. By 2050, for every additional dollar in damage that is expected to result from the effects of global warming on tropical cyclones, there is likely to be between US $22 and US $60 of increase in damage due to population growth and wealth in vulnerable areas.

All these numbers have been brought into sharp focus by recent events. These have shown that it is not just the scale of the meteorological event that matters, but also the level of preparation for, and management of emergencies. The overwhelming experience of Katrina in 2005 tragically demonstrated the consequences of inadequate defences. It killed over 1800 people and the estimated economic losses are expected to exceed US $200 billion. As far as the weather forecast as concerned, the prediction of the course and intensity of Hurricane Katrina were remarkably accurate. Although Katrina reached Category 5 in the Gulf of Mexico it had just subsided to a category 3 storm by the time it hit New Orleans. In principle, the levees defending the city were designed and engineered to withstand a storm of this intensity. The fact that the huge damage has been attributed principally to failures in providing adequate storm defences plus deficiencies in constructing and maintaining these defences has brought the political nature of climate change into even sharper focus.

In terms of the preparedness of the emergency services, it was estimated that evacuating New Orleans would take at least 72 hours. This figure was reasonable for those who owned cars and were prepared to heed the warnings. Unfortunately, it did not take account of the fact that about a one-third of the populace lacked the mobility to move to safety. Furthermore, a disproportionate number of the people stranded in the city were elderly, infirm and/or poor. Then, when the city was flooded the emergency services, which were supposed to deliver the response phase of the management system, proved wholly incapable of rising to the challenge; quite simply they were overwhelmed.

A similar set of issues arises from other major climatic disasters in recent years in the developed world. For instance the consequences of extensive flooding, whether down the Mississippi in 1993, or along various major European rivers in recent years, has raised a series of planning issues. These relate to flood-control management, the provision of adequate insurance for people living in the flood plain, the extent to which people should settle in the flood plains, and the management of emergency services. So the economic consequences of extreme weather events have to be combined with how our social and economic systems respond to events to provide a balanced picture. Furthermore, there may be occasions when large-scale events have a combination of consequences: some good, some bad.

A good example of the balancing nature of some events occurred with the 1997 El Niño. Despite the huge amount of adverse publicity in the US, the overall economic impact there was surprisingly favourable. This is for two reasons. First, El Niño years are almost always quieter than normal hurricane seasons in the Atlantic, and this was, most certainly the case in 1997 (see Fig. 9.8). Given the scale of damage caused by hurricane events in the USA, this represents a substantial benefit. Second, as widely forecast in 1997, El Niño years usually lead to milder winters in Canada and the northern USA. This was the case in 1997/8 and the reduction in winter mortality and in heating bills represents a huge saving to the country.

On the downside, the Pacific North West coast was battered by storms throughout the autumn and winter. Much of California had record or near record rainfall between December and March with floods and mudslides, while the southeast, from Louisiana to Florida, also had near record amounts of rainfall in this period. Even so, California had the benefit that the Sierra Nevada ski resorts had exceptional snow conditions and a long and lucrative season. Offshore, abnormally warm waters had a major impact on marine life and fishing industries. Nevertheless, while there are difficulties in treating all these impacts in a balanced manner, the overall estimate is that the benefits outweighed the costs to North America.

The implications of these comments is that, while variations in the incidence of extreme weather events represent the most immediate threat of climate change, interpretation of claims of increases in specific types of extreme require careful analysis. So we need to develop reliable up-to-date estimates of the potential damage caused by various types of extreme, and accurate measures of how these extremes will increase or decline if the climate changes in a given manner. Combined with growing confidence in our capacity to forecast future climate change, we will then be able to make more accurate predictions of the economic risks of going ahead with any particular development. Increasingly, these arguments will come to dominate the decision-making process.

9.9   Summary

The consequences of changes in the climate are substantial, but they must not be exaggerated by focusing only on the downside. It is essential that any analysis of the impact of climatic events accurately reflects all aspects of the argument. This need for balance works both ways. In examining the past, only a thorough appreciation of the other explanations of what may have caused changes can enable climatologists to demonstrate that climatic factors offer the key to better understanding. This means demonstrating that without the climatic component other explanations of, say, historic events are insufficient. Conversely, in considering the future an accurate assessment must be made of how the consequences of changes in the climate compare with other threats confronting society.

The need for balance and a sense of proportion applies not only to interpreting past events but also highlights the importance of better measurements of past climates. In many instances, the weakness in the climatic explanation of past changes lies in the inability to establish precisely when and by how much the weather shifted around the time in question. This applies not only to the rise and fall of civilisations, the occurrence of famines, the changing patterns of flora and fauna, and the spread of diseases, but also to more remote questions such as the evolution of the human species and the occurrence of mass extinctions. In using this analysis of the past to inform our thinking about the potential consequences of future climate change, we must be sure that there is not real bias in our measures of the past. What is clear is that our knowledge of past changes in the climate, together with our understanding of how the climate functions, have improved in recent years. So now we need to move on to the next stage, which is to consider whether we can successfully model the climate and predict how it will behave in the future.

QUESTIONS

1. During past changes in the climate many forms of flora and fauna were able to migrate to areas where the new climate suited them. Why could this option not be available to many species as the climate changes in the future?

2. Why could the cessation of written records by past civilisations give an exaggerated impression of social decline?

3. List the various impacts of the ENSO around the world in terms of whether they have a negative or positive impact on social and economic affairs. To what extent do these impacts tend to cancel one another out, and, if they do, are there ways in which those who gain can compensate those who lose out?

4. What action could insurance companies take to reduce their exposure to loss from various forms of extreme weather event and who would suffer from such action?

FURTHER READING

A complete reference list is available at the end of the book but the following is a selection of the best books or articles to follow up particular topics within this chapter. Full details of each reference are to be found in the Bibliography.

Brown *et al.* (1992): This contains a wide variety of geological information including useful discussions of palaeoclimatology and volcanoes.

Burroughs (2005a): A detailed analysis of past examples of the economic and political consequences of climatic change and assessment of their implications for the future.

Burroughs (2005b). An analysis of the debate on climate change and the part human activities play in current changes, combining a historical perspective, economic and political analysis together with climatological explanations of the impact of extreme weather events on aspects of society.

Fagan (1999). A lively and eclectic analysis of the nature and impact of ENSO on the world with a sweeping presentation of the historical consequences of climate change.

Fagan (2004). A different perspective on the implications of the climatic amelioration after the LGM and how climate change moulded many aspects of human history, which provides a respected archaeologist's observations on how vulnerable human societies have been to shifts in the climate.

Grove (1988). An immensely thorough review of the evidence and consequences of the Little Ice Age, which extends its comprehensive analysis to cover the contraction and expansion of glaciers around the world throughout the Holocene.

Lamb (1995): A fascinating review of the influence of climate on human history by the leading authority in the field.

van Andel (1994): An intensely illuminating set of discussions of many aspects of the geological consequences of climate change.

10   Modelling the climate

Vaulting ambition, which o'erleaps itself   **Shakespeare (Macbeth)**

Producing computer models that accurately reflect the complexity of the Earth's climate is an immense challenge. Even when focusing on the priority areas of natural variability and the impact of human activities, the task is daunting. Extending it to cover all the potential causes of climate change may prove overwhelming. So the approach must be to develop systems that inspire confidence in our ability to address the priority areas, and then go on to explore the sensitivity of the climate to other factors. The only way to do this is to create detailed computer models of the global climate, to evaluate the various hypotheses. The first stage is to establish whether these models are capable of producing a realistic representation of the climate and can respond to the most obvious quantifiable perturbations. The next stage is to check the relative importance of other possible causes of climate change, both to ensure we have correctly identified the priority areas and also to gain a better understanding of past events and decide how much confidence to attach to predictions of future changes.

Given these broad aims, the objective of this chapter is to describe the essential features of computer models of the climate and then to assess their performance in terms of reproducing known features of the climate. After this, the ability of the models to handle the natural variability of the climate and the possible impact of human activities will be considered. In the light of this analysis, the challenges facing modellers will be discussed in terms of whether reliable predictions of climate change are a realistic proposition within the foreseeable future.

The pace of climatic opinion is governed by the mighty labours of the IPCC. With its publication of assessment reviews every five years or so, it draws a new line in the sand with each round of activity. Since the first report was published in 1990, the scale of the activity has grown and the results assessed as part of each round show a steady growth in our understanding of the nature of climate change. The fourth assessment review (known as AR4), which was published early in 2007 represents the increasingly firm consensus that most of the warming in recent decades is the consequence of human activities. How this consensus has been formed and what it means for the future constitute the central features of this chapter and the next.

10.1   Global circulation models

In considering climate models, the obvious starting point is the numerical weather predictions that provide our daily forecasts. Because we are so familiar with these products of computer models, it is easy to relate to their performance. Weather forecasting is a problem in mathematical physics. The computer models used for forecasting contain a very large and complex array of equations based on the physical and dynamical laws that govern the birth, growth, decay and movement of weather systems. They incorporate the principles of conservation of momentum, mass, energy and water in all its phases; the Newtonian equations of motion applied to air masses, the laws of thermodynamics and radiation for incoming solar energy and outgoing heat radiation, and equations of state of atmospheric gases. Parameters that are specified in advance include the sizes, rotation, geography and topography of the Earth, the incoming solar radiation and its diurnal and seasonal variations, the radiative and heat-conductive properties of the land surface according to the nature of the soil, vegetation and snow and ice cover, and the surface temperature of the oceans.

The physical state of the atmosphere is updated continually drawing on observations from around the world using surface land stations, ships, buoys, and in the upper atmosphere using instruments on aircraft, balloons and satellites. The model atmosphere is divided into 70 layers and, in the most advanced models, each level is divided up into a network of points about 40 km apart. Within this global model regional models can be nested that provide forecasts on a grid scale of 12 km and on local level down to a grid scale of 4 km. Each of these points is assigned new values of temperature, pressure, wind and humidity with each run of the model and the governing differential equations are integrated in 15-min steps at each point to provide forecast values up to 10 days ahead. The output is hundreds of forecast charts of pressure, temperature, wind, humidity, vertical motion and rainfall, which are used to provide a variety of forecasts.

So the scale of weather forecasting provides a good indication of the resources that can be brought to bear on the issues of climatic change. But, as we all know, weather forecasts have obvious limitations. They have shown significant improvements, with three-day forecasts now being as good as those forecasting a day ahead 20 years ago. This slow progress highlights the unpredictable nature of the atmosphere. Of particular interest in terms of the longer-term behaviour of the climate is the handling of regimes (see Sections 3.2, 3.6 and 7.2). Standard weather forecasts do not predict sudden switches between stable circulation patterns well. At best they get some warning by using statistical methods to check whether or not the atmosphere is in an unpredictable mood. This is done by running the models with slightly different starting conditions and seeing whether the forecasts stick together or diverge rapidly. This *ensemble* approach provides a useful indication of what modellers are up against when they seek to analyse the response of the global climate to various perturbations and to predict the course it will follow in the future.

To start with, the general circulation models (GCMs) used by climate modellers cannot represent the global climate in the same detail as the numerical weather predictions because they must run their simulations for decades and even centuries ahead in order to consider possible changes, and not just ten days. Because of the high cost of running the most advanced computers and the simple limitations of time taken to complete the calculations, the models have to use a lower spatial resolution. Typically, the most advanced models now have a horizontal resolution of between 125 and 400 km, but retain much of the detailed vertical resolution, having around 20 levels in the atmosphere. So while they use the same set of mathematics and physics as weather-forecasting models, they cannot analyse atmospheric patterns to the same degree.

Coupling the ocean processes to atmospheric GCMs is a major challenge. The thermal capacity of the oceans is massive compared to the atmosphere and can provide to, or drain from, the atmosphere massive amounts of heat energy. Representing their heat storage and the absorption of greenhouse gases by the oceans, long-term simulations of climate require a full three-dimensional ocean model. This must simulate the formation of deep water and the detail of the western boundary currents. Changes in the intensity and location of deep-water formation can have profound effects on the atmosphere. In the past, changes in the thermohaline circulation of the oceans have resulted in major atmospheric responses. The models must also be able to handle shorter-term fluctuation such as ENSO.

Recent developments in climate modelling, which take into account not only surface processes at the ocean–atmosphere interface but also those at work deep in the oceans, have produced considerable improvements. An oceanic GCM requires much higher spatial resolution to capture eddy processes associated with the narrow major currents, bottom topography and basin geometry. High-resolution ocean models are therefore at least as costly in computer time as are atmospheric GCMs. Further integration of other climate system component models, especially the cryosphere and the biosphere, are also necessary in order to obtain more realistic simulations of climate on decadal and longer time scales.

The performance of the best atmosphere–ocean global circulation models (AOGCMs) has progressed appreciably in recent years. This rapid increase in computing power has permitted not only improved spatial resolution, but also improvements to numerical schemes and parametrisations (e.g. sea ice, atmospheric boundary layer, ocean mixing). More processes have been included in many models, including a number of key processes important for forcing (e.g. the formation of aerosols), which affects the properties of clouds, are now modelled interactively in many models.

The simplest measure of performance is the global values of temperature and precipitation that they predict. The notable feature of the temperature results is that, while on average there is reasonable agreement with observed values, the scatter between different models is several degrees Celsius, which is large compared with both past climate change and predicted future changes. The latitudinal variation of simulated temperature from actual values is also considerable. The biggest discrepancies are at high latitudes and over the continents. By contrast, differences over the ocean are smaller.

With few exceptions, the absolute error (outside polar regions and other data-poor regions) is less than 2 K. Individual models typically have larger errors, but in most cases still less than 3 K, except at high latitudes. Some of the larger errors occur in regions of sharp elevation changes and may result simply from mismatches between the model topography (typically smoothed) and the actual topography. There is also a tendency for a systematic cold bias over land and warm bias over oceans. Outside the polar regions, relatively large errors are evident in the eastern parts of the tropical ocean basins, a likely symptom of problems in the simulation of low clouds.

The simulation of precipitation produces a substantial scatter. The broad features of peaks associated with tropical rainfall and the mid-latitude storm tracks appear in all the models. The variations are related to the intensity of rainfall and are largest in the tropics. These discrepancies are important as the precipitation rate is a measure of the intensity of the hydrological cycle and also influences the ocean thermohaline circulation. More generally the models with higher average temperature have larger precipitation rates, as might be expected. So, a warmer world is likely to have a more vigorous hydrological cycle. In addition, the seasonal average mean sea-level pressure is represented reasonably well.

When it comes to simulation of snow and sea-ice cover the models continue, however, to have major unresolved difficulties. In TAR it was observed that some models had an excess of winter snow cover, which persisted into summer, while others lost all their sea ice in the Arctic and almost all their sea ice in the Southern Hemisphere throughout the year. Clearly, the failure to simulate accurately these essential features of the climate is an area that required more work. Another feature of the latest models is that substantial uncertainty remains in the magnitude of cryospheric feedbacks within AOGCMs. This contributes to a spread of modelled climate response, particularly in high latitudes. On the global scale, the surface albedo feedback of snow and ice is positive in all the models, with a spread among current models that is much smaller than that of cloud feedbacks. Understanding and evaluating sea-ice feedbacks is complicated by the strong coupling to polar cloud processes and ocean heat and freshwater transport.

Until recently the representations of the circulation of the oceans had fundamental limitations. The latest high-resolution models have, however, produced more realistic simulations of ocean water mass structure, thermohaline circulation (Great Ocean Conveyor: Section 3.8) and ocean heat-transport figures (see Section 2.1.4). Most models, however, show biases in their simulation of the Southern Ocean, leading to some uncertainty in modelled ocean heat uptake when climate changes. So further progress is needed to establish confidence in the models and to examine the possible impact of global warming on the future circulation patterns of the oceans.

Modelling clouds remains a substantial source of uncertainty. This is a critical test because clouds exert such a strong influence on both incoming solar radiation and outgoing terrestrial radiation (see Section 3.3). So, any inadequacies in their simulation could have major implications for the reliability of the models. The most fundamental issue is whether climatic change will lead to coherent shifts in cloudiness, which reinforce any change (*a positive feedback mechanism*) or damp down the change (*a negative feedback mechanism*) (see Box 1.1). The formulation of climate models has developed through improved spatial resolution, and improvements to numerical schemes and parametrisations (e.g. sea ice, atmospheric boundary layer, ocean mixing). More processes have been included in many models, including a number of key processes important for forcing (e.g., aerosols are now modelled interactively in many models).

Although the latest models represent clouds better, these processes remain a significant part of the uncertainty in predicting human-induced climate change. There has been some improvement in the simulation of certain cloud regimes, notably in the case of marine stratocumulus, which plays an important role in cooling over the tropical oceans. At best, however, the models do a modest job of portraying the latitudinal and seasonal distribution of global cloudiness. They systematically underestimated the cloudiness in low and middle latitudes in both winter and summer. In higher latitudes they overestimated cloudiness, especially over Antarctica, even allowing for the uncertainty of satellite measurements in discriminating between clouds and underlying snow and ice in these parts of the world.

More generally, the differences between temperature values models arise primarily from inter-model differences in cloud feedbacks. The impact of changes in boundary-layer clouds on incoming shortwave radiation, and to a lesser extent mid-level clouds, constitutes the largest contributor to inter-model differences in global cloud feedbacks. The relatively poor simulation of these clouds in the present climate is a reason for some concern. The response to global warming of deep convective clouds is also a significant source of uncertainty in projections since current models predict different responses of these clouds. Cloud feedbacks have been evaluated using observations. They indicate that climate models exhibit different strengths and weaknesses, and it is not yet possible to determine which estimates of the climate-change cloud feedbacks are the most reliable.

There is yet one more problem with modelling clouds. This is the question about whether global warming will alter the precipitation properties of clouds. There is a heated argument among atmospheric scientists as to what will be the overall impact of warmer oceans pumping more water vapour into the atmosphere. If it leads only to more moisture in the lower atmosphere and heavier rainfall from clouds, its impact will be limited. If, however, it results in an increase in humidity throughout the troposphere, it will lead to a positive feedback producing greater warming because water vapour is the most important greenhouse gas (see Section 2.1.3). Although the strength of this feedback varies among models, its overall impact on the spread of model climate sensitivities is reduced by lapse-rate feedback. Several new studies indicate that modelled tropospheric relative humidity is consistent with observations. Taken together, these results strongly favour a combined water vapour lapse rate feedback of around the strength found in AOGCMs.

10.2   Simulation of climatic variability

A fundamental test of the utility of AOGCMs in predicting climate change is their ability to replicate the natural variability of the current climate. Until recently this was done by running GCMs for the equivalent of many centuries of the global climate without any external perturbations to see how much they go up and down of their own accord. Increasingly, however, the real test of the models is the successful simulation and prediction over a wide range of the modes of variability exhibited by the climate that are described in Chapters 3 and 6 (Sections 3.6 to 3.9, and 6.2).

On the shortest timescales the models do a reasonable job of reflecting the synoptic variability of weather systems in the climate. They do, however, underestimate the variability of the regions of maximum storminess. The models have tended also to underestimate the amount of blocking (see Section 3.2) and, hence, the contribution this important phenomenon makes to climate variability. Recent studies have found that GCMs now simulate the location of northern hemisphere blocking more accurately, but simulated events are generally shorter and less frequent than observed events. This analysis is, however, complicated by the fact that no commonly accepted definition of blocking exists. It is the synoptic equivalent of an elephant: it is difficult to define, but you know one when you see it. So, it is not easy to compare the different blocking studies. Nevertheless, given the importance of blocking in the observed decadal climatic variations during the twentieth century in the northern hemisphere (see Section 3.7), it is essential to get a better handle on what drives these changes.

The challenge that has to be faced is to define a blocking index that does not effectively prejudice the outcome of the results of GCM blocking simulations, which seek to explore the changing incidence as a consequence of altering, say, radiative forcing. Otherwise the results will simply reflect any bias that is built in to the initial conditions of the models; a conclusion that might be likened to a redefinition of the old computer rule as *bias in equals bias out*. Furthermore, both GCM simulations and analyses of long datasets suggest the existence of considerable interannual to interdecadal variability in blocking frequency. This highlights the need for caution when assessing blocking climatologies derived from short records, whether observed or simulated. Blocking events exercise less climatic influence in the southern hemisphere, which may explain why no systematic intercomparison of observed and simulated blocking climatologies has been carried out.

Comparison of the output of AOGCMs averaged over timescales of a few months to several years suggests that the standard deviation of the near-surface temperature was in reasonable accord with observed values. In the tropics, simulation of the MJO (see Section 6.2) remains unsatisfactory. How and where energy is fed into the tropical atmosphere is critical for the development of the MJO, so models that simulate some broad features of the MJO, that on closer examination do not past muster. Similarly, variability with MJO characteristics is simulated in some models, but this does not occur often enough or with sufficient strength for the oscillation to stand out clearly above the background variability. As a consequence, the underestimation of the strength and coherence of convection and wind variability of the MJO, with the result that its influence on, say, rainfall variability in the monsoons, the genesis of tropical cyclones and the development of ENSO events, are poorly simulated by the models.

Interannual variability is an equally complex matter. Some of the AOGCMs display changes that resemble certain aspects of ENSO behaviour (see Section 3.7). Despite some progress, serious systematic errors persist in both the simulated mean climate and the natural variability. These include problems with handling the behaviour of the *Intertropical Convergence Zone (ITCZ)* in simulating the annual cycle in the tropics. Along the equator in the Pacific most AOGCMs fail to adequately capture the zonal SST gradient; the equatorial cold tongue is confined too close to the equator. Furthermore, they tend to produce anomalies that extend too far into the western tropical Pacific and fail to capture the meridional extent of the anomalies in the eastern Pacific. Moreover, typically the thermocline is far too diffuse. Most, but not all, models produce too rapid ENSO variability – peaking in the range of two to three years – and have difficulty capturing the correct phase locking between the annual cycle and ENSO. In addition, some models fail to represent the spatial and temporal structure of El Niño–La Niña asymmetry.

Models show dominant modes of extratropical climate variability that reflect the various oscillations that were discussed in Section 3.7 [e.g., the North Atlantic Oscillation (NAO) or Arctic Oscillation, the Antarctic Oscillation (AAO), the Pacific Decadal Oscillation (PDO) or the Interdecadal Pacific Oscillation (IPO)] still cause problems for the models. In the case of the NAO most models produce too much sea-level-pressure variability. The year-to-year variance of the NAO is correctly simulated by some coupled GCMs, while others are significantly too variable; for the models that simulate stronger variability, the persistence of anomalous states is also greater than observed. The magnitude of multi-decadal variability (relative to subdecadal variability) is too low.

Although the spatial structure of the AAO is well simulated by the latest models, other features such as the amplitude, the detailed zonal structure, and the temporal spectra do not always compare well with the NCEP-analysis data. These features vary considerably among different simulations of multiple-member ensembles, and the temporal variability of the AAO in the NCEP re-analysis is problematic when compared to station data.

Coupled models do not seem to have difficulty in simulating IPO-like variability, even in models that are too coarse to properly resolve equatorially trapped waves important for ENSO dynamics. Finally, there has been little work evaluating the amplitude of Pacific decadal variability in coupled models. One model showed that the variability has roughly the right magnitude but a more detailed investigation using recent models with a specific focus on IPO-like variability would be useful.

In the case of the Atlantic Multidecadal Oscillation (AMO), coupled models simulate this 65- to 80-year fluctuation. This multidecadal variability appears to be driven by variations in the surface heat flux, and hence by atmospheric variability, it appears to be coherent with parallel fluctuations in the strength of the thermohaline circulation of the North Atlantic. The mechanisms that control these variations are, however, quite different across the ensemble of coupled models. In particular, coupled interactions between the ocean and the atmosphere differ appreciably, with oceanic variability being dominant in some and atmosphere playing a bigger part in others. The relative roles of high- and low-latitude processes differ also from model to model. These uncertainties are important in understanding current climate change as the presence of strong Atlantic multidecadal variability may mask any anthropogenic weakening of the THC for several decades.

An alternative means of checking how well GCMs deal with potential climate change is how they handle specific forms of natural perturbation. The best example of this process is the impact of large volcanic eruptions. Because these produce a reduction in the amount of solar radiation reaching the Earth's surface (see Section 8.4), which is of the same scale as the warming effect equivalent to a doubling of CO_2 in the atmosphere, the performance of models in computing the impact of major volcanoes is a good test of their ability to predict the consequences of human activities. The predicted changes due to the eruption of the volcano, Mount Pinatubo, in the Philippines in June 1991 (see Fig. 10.1) show models can do well in predicting the global consequences of specific perturbations. This success suggests that, while GCMs still have many limitations, their ability to predict the incremental consequences of human activities (see Section 11.2) makes them more valuable than their absolute performance might suggest.

Another popular test of model performance is to explore how they simulate past climate. The most frequently examined are conditions during the Holocene optimum around 6 kya (see Section 8.6) and the extreme opposite of the coldest period at the end of the last Ice Age at about 21 kya (see Section 8.4). While the models produce what look like reasonable

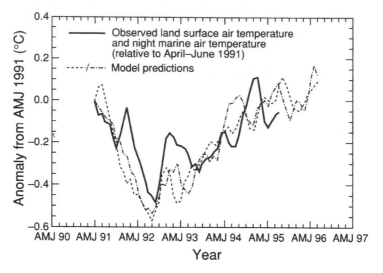

Fig. 10.1 The predicted and observed changes in global land and ocean-surface air temperature after the eruption of Mount Pinatubo, in terms of three-month running means from April to June 1991 to March to May 1995. (From IPCC, 1995. Fig. 5.20.)

representations, the limitations in our knowledge of the climatic conditions over time make it difficult to establish whether such work is a real test of the models. Indeed, the benefit of these studies may be more to do with testing the physical implications of possible mechanisms of climate change against available evidence than to validating the models.

10.3   The challenges facing modellers

The potential to improve GCMs depends on making progress in a range of areas. There are obvious gains to be had in using bigger and faster computers. The most interesting developments are likely to take place, however, in better physical representations of the processes identified in Section 10.1. These include improved handling of clouds, the changing conditions of the land surface, such as soil moisture, deforestation and agriculture, plus snow cover and sea ice. Progress in these areas will depend on improved experimental studies of what is actually happening.

10.3.1   Clouds

Almost every aspect of the representation of clouds in GCMs requires improvement (see Section 10.1). At the most basic level the understanding of the physics of how much sunlight is absorbed and reflected by different types of cloud, and then how much heat radiation is absorbed and emitted by them, must be quantified more accurately. Then the spatial and temporal distribution of different clouds must both be measured and then reproduced by GCMs. In addition the processes controlling precipitation must be defined more accurately. Without progress on these fronts it will not be possible to provide more reliable estimates of how cloud cover varies over longer timescales and hence what part these variations play in climatic change.

The essence of the challenges facing modellers is found in the need to move away from the parametrisation of clouds, to a more detailed treatment of the physical processes involved in the formation and behaviour of clouds. These improvements require better measurements of the radiative properties of water droplets and ice crystals both in respect of sunlight and terrestrial radiation for different types of clouds. An improved knowledge of the shape and size distribution of cloud particles is part of this understanding. In addition, measurements from the ground, and by aircraft and satellites of how single clouds and the combination of a number of clouds alter how radiation is absorbed, emitted or scattered by the atmosphere will be needed. Only when these are available will GCMs be able to produce more climatically realistic representations of current cloudiness.

Building up an improved climatology of cloud types using satellite measurements to identify how the amounts of various clouds vary on every timescale from diurnal to decadal is the next step. Once the geographical distribution of cloud types has been established accurately it can be used as a test for improving the performance of GCMs. Such measurements may also be used to check another potentially important aspect of human activities. This is the formation of particulates, notable from emissions of sulphur dioxide from the combustion of fossil fuels (see Section 4.8). What are needed are accurate figures on how much sunlight is absorbed and reflected by these particulates and how their net effect is modified by the albedo of the underlying surface.

In addition to the direct consequence of particulates, there is the indirect impact on cloud formation. This influences how much impact they have on the extent and duration of cloud cover. At present modellers do not even know for certain whether these will lead to an increase or decrease in cloudiness. Satellite measurements of ship trails show that particulates increase cloud cover. So it is estimated the net effect of particulates generated by human activities on the properties of clouds will probably be some additional cooling, but the amount cannot yet be quantified with any precision.

10.3.2 Tropical storms

The consequences of the limited resolution of AOGCMs extend into the contentious area of tropical storms, including cyclones and hurricanes (see Section 8.10). While the simulation of extratropical cyclones has improved, the models cannot resolve tropical cyclones and simulate their intensity. Some models can, however, simulate the large-scale conditions necessary to infer their frequency and distribution. They can also handle how the tracks of tropical cyclones are affected by the structure of the tropical SST in any given year (e.g. the switch from El Niño to La Niña conditions), and models are able to simulate those differences. This is of importance in predicting changes in the tracks of destructive cyclones. Nevertheless, the lack of resolution leaves a big gap in our understanding of the part played by tropical storms in global energy transfer and how they will contribute to the predicted warming during the twenty-first century.

One way round this challenge has come from work by Kevin Trenberth and colleagues at NCAR. They have explored various aspects of global energy fluxes including using a high-resolution regional model, with a resolution of 4 km covering a radius of 400 km, embedded in the lower resolution AOGCM. Using examples of recent major hurricanes in the Gulf of Mexico (Ivan in 2004 and Katrina in 2005), these studies have shown that there is a strong relationship between intensity and potential destructiveness of such storms with SSTs. This suggests an increased likelihood of heavier rainfalls and probably more intense storms. The work concludes that since 1970 there has been increase in more intense hurricanes with the amount of heavy precipitation rising by about 8%.

This analysis then used the surface latent heat and sensible heat fluxes plus precipitation as a function of maximum wind speed in the model simulations, to draw some wider conclusions. Relying on the global best track data on frequency of hurricanes at various strengths for 1990–2005, it calculated the total heat loss by the tropical ocean in hurricanes category 1 to 5 within 400 km of the centre of the storms. This was estimated to be equivalent to $1.3 \, \mathrm{W \, m^{-2}}$ heat loss spread over the tropical ocean area from 20 °N to 20 °S: sufficient to cool this area by $1.0 \, \mathrm{°C \, year^{-1}}$ over a 10 m thick layer. Given that this calculation considered only the effect of the central region of major hurricanes, the loss of total potential energy (*enthalpy*) due to hurricanes, computed on the basis of the precipitation, is about a factor of 3.4 greater than these numbers, owing to the addition of the surface fluxes outside 400 km radius and moisture convergence into the storms typically as far from the eye as 1600 km.

Changes over time reflect basin differences and a prominent role for El Niño, and the most active period globally was 1989 to 1997. Strong positive trends from 1970 to 2005 occur in these inferred surface fluxes, in precipitation arising from increases in intensity of storms, and also higher SSTs. This highlights the importance of surface energy exchanges in global energetics of the climate system and indicates that climate models are markedly deficient by not adequately representing tropical cyclones.

10.3.3   Land-surface processes

Many features of the land surface vary on short timescales. They do, however, exert a powerful influence on the climatic conditions where we all live. So, while it may be possible to parametrise many longer-term aspects of the processes, it is essential that GCMs accurately represent their climatic consequences. As with clouds, the current evidence is that land-surface processes are not handled well by the models and the way forward is to obtain better measurements of how the properties of the land surface affect the behaviour of the atmosphere. These measurements will need to include changes in albedo (see Section 2.1.4.) with the seasons and between wet and dry conditions, the balance between how much precipitation is either stored in the soil, or evaporates, or runs off in rivers, the impact of soil moisture on heat and moisture flux at the surface, and the consequences of winter snow cover.

The consequences of soil-moisture variations are a particularly good example of how the challenges facing modellers are interlinked. Weather forecasting work (see Section 10.3.5) has shown that more accurate treatment of soil moisture can produce significant improvements in predicting rainfall. But in climate models, unless the general treatment of rainfall is realistic, weaknesses in the modelling of precipitation will overwhelm any improvements in the representation of land-surface processes. This basic challenge of identifying what are the most important parameters affecting a given climatic variable may be the biggest obstacle to progress. Nevertheless, only by exploring how the models respond to more realistic physical representations of land-surface processes may answers be found to some of the problems of simulating precipitation more accurately.

The impact of snow cover may produce more direct progress. The models do not handle global snow cover well and the scale of the effects of changes in the extent of snow at the end of the northern winter across the boreal forests is potentially large (see Section 9.8). These changes need to address the coupled issues of both how fluctuations in the climate may be amplified by parallel change in snow cover, and also how deforestation might alter these effects more radically, especially at high latitudes (see Table 2.1).

The related issue of desertification must also be examined in more depth. The possibility that the high albedo of deserts may reinforce the dry conditions needs further examination. Some theoretical analyses have suggested that the albedo increase due to expanding deserts may be balanced out by a parallel decline in local cloudiness. More striking is the satellite observations that seem to show that the expansion and contraction of the deserts is dominated by larger-scale weather patterns (see Section 9.8). So when rainy seasons return the vegetation rapidly becomes re-established and reduces the albedo. If, however, longer-term climatic change led to a more permanent shift in rainfall patterns then the albedo effects could be important. So the models need to be able to handle these changes.

10.3.4   Winds, waves and currents

The importance of atmosphere–ocean interactions in climatic change means an improved understanding of how energy is exchanged between the atmosphere and the sea surface is central to better modelling. The physics of how the winds stir up waves and are in turn influenced by the roughness of the sea requires improved analysis. Satellite measurements are now providing improved information about wave heights and wind speeds. These will contribute, together with oceanographic studies, to establish better figures for the amount of heat, momentum and moisture between the sea surface and the atmosphere as a function of the temperature contrast between the two and wind speed. This information will inform judgements about the handling of 'flux adjustments' in GCMs, especially as higher-resolution models become available. Some of the most recent modelling work suggests that progress is being made to produce models that can make lengthy realistic simulations of the global climate without recourse to 'flux adjustments'.

On the larger scale, how the atmosphere and oceans combine to drive the ENSO and the ocean currents must be refined in the models. As current GCMs cannot reproduce the broad effects of the ENSO (see Section 10.2), this has to be a major priority in producing more realistic computer simulations of the climate. Beyond this, the processes which maintain the Great Ocean Conveyor have to be modelled so that we can form a more informed view about the stability of this system. Some preliminary model results suggest that increasing CO_2 levels may produce conditions that reduce the strength of thermohaline circulation in the North Atlantic, especially if the levels rise rapidly. The fact that there is some evidence that the pattern of circulation in the North Atlantic has shown some signs of changing in this way confirms the importance of producing improved models of ocean circulation.

10.3.5   Other greenhouse gases

The contribution to the greenhouse effect of other trace constituents of the atmosphere, apart from CO_2, has been recognised throughout this book. In addressing the impact of human activities it is important, however, to consider changes in all of the radiatively active gases. This analysis can be divided into two areas. The first group is the various anthropogenic emissions that will effectively accumulate in the atmosphere [e.g. methane, sulphur dioxide and chlorofluorocarbons (CFCs)]. The second group is those constituents that respond in a complex manner to changing climatic conditions, and so become an integral part of the modelling process, and water vapour and ozone are the most important. Although the former do participate in various atmospheric photochemical processes, their principal contribution to climate change is the amount they add to the radiative forcing of the atmosphere. This increase can be combined with the effect of CO_2 to provide an overall figure for human activities. By comparison, water vapour and ozone variations are much more challenging.

Because water vapour is the dominant greenhouse gas (see Section 2.1.3.), its accurate representation is central to realistic modelling. This analysis must also recognise how sensitive the concentration of water vapour is to the temperature of the atmosphere, especially in the vertical. But the radiative properties of water vapour vary with atmospheric concentration and so the vertical distribution will be altered in a warmer world.

Similar issues arise with ozone. Here the vertical distribution is the product of photochemical processes (see Box 2.2.). Human activities are altering this distribution. Near ground level, rising pollution levels in urban areas are producing a widespread increase in O_3 levels especially in the summer. In the stratosphere, where the majority of O_3 is found, the build-up of CFCs has led to destruction of O_3, notably over Antarctica each austral spring. These changes must be adequately represented in GCMs to provide a realistic representation of how human activities will affect the climate.

10.3.6 Exploitation of numerical weather prediction

The scale of numerical weather prediction (NWP) (see Section 10.1) underlines how, in spite of developing massive computer to produce incredibly detailed models of the global weather system, the forecasts have limitations. Progress in improving day-to-day weather forecasts also has direct relevance to the challenges facing climate modellers. This can provide rapid insights into which refinements to the physical representation of the climate system have the most impact on performance. By using the latest experimental observations of how atmosphere interacts with the oceans and the land surface, weather forecasting models can improve the representation of various physical processes. The impact of these refinements can be checked in terms of improvements in forecasting performance. This process offers climate modellers the prospect of being able to identify which factors are likely to have the most impact on their longer-term predictions and how best to represent them in GCMs.

The other area where NWP output can assist in both producing better GCMs, and analysing their output, is re-analysis work (see Section 4.3). Because this work collates all the data collected over the years, it enables the initial conditions used by the models to be defined more realistically. Equally important the output is used to check the performance of the models in terms of their ability to simulate more reliable measurements of the behaviour of the global climate over the annual cycle and from year to year. These measurements include not only temperature and pressure measurements but also improved observations of the transport of energy and water vapour around the globe.

10.4 Summary

Climate modelling has many features of the meeting of an irresistible force and an immovable object. The computational apparatus available to modellers is massive, and when compared with many aspects of managing complexity in our lives, the analysis it provides is far more comprehensive. Furthermore our understanding of the physical processes involved has come on with leaps and bounds. Nevertheless, the realisation of just how enormous a challenge the production of a reliable representation of the global climate is has grown as fast, if not faster.

Confronted with what looks like an intractable problem, it would be all too easy to throw our hands up in the air and declare it is all too difficult, and that nothing can be concluded until we have better measurements, bigger computers and more sophisticated GCMs. But this extreme 'wait and see' approach is not realistic. We have to make decisions now. Even taking no action to respond to the threat of global warming amounts to a decision and has to be justified in terms of current knowledge. The only way forward is to exploit the work that has been done while recognising its limitations when deciding what are the most worthwhile things to do. So, having identified the strengths and weaknesses of modelling the climate we can now consider how this work can be used to plan the future.

QUESTIONS

1. The ocean–atmosphere models show a variation in poleward water transport in the North Atlantic of 2 to 26 Sv ($1 Sv = 10^6 m^3 s^{-1}$). If this water has an average temperature of 11 °C when it arrives at high latitudes whereas the deep water returning southwards has a temperature of 3 °C, estimate the range in the amount of heat being transported northwards in these models. How does this variation in energy compare with the amount of solar energy falling on the region (50–60° N and 20–60° W) in winter and summer?

2. Estimate how much of the existing ice sheets on Antarctica and Greenland would have melt to raise the global sea level by 10 m. Is this a useful measure of the impact in any given part of the world? If not, what other local factors matter?

3. Is an estimate of the average rise in sea level a useful measure of the future impact of global warming in terms of damage to property and loss of life? If not, identify which measures of climate change might represent the most important aspects of a warming, placing particular emphasis on the consequences of the early stages of any rise.

FURTHER READING

A complete reference list is available at the end of the book but the following is a selection of the best books or articles to follow up particular topics within this chapter. Full details of each reference are to be found in the Bibliography.

Burroughs (1991): This provides descriptions of how satellite technology is being used to obtain better measurements of how various physical processes are contributing to climate change.

Gurney *et al.* (1993): A thorough presentation of many of the results that have been obtained from weather satellites which are of relevance to climatic studies.

Harries (1990): A more fundamental analysis of how weather satellites work, and a guide to the contribution they can make to studying climate change.

IPCC (1995): The definitive statement on where the meteorological and climatological community have got to by the end of 1995 in measuring climate change and modelling the climate. The analysis of the performance of GCMs provides the ideal start for finding out more about where this rapidly advancing subject has got to and the challenges it now faces.

Trenberth (1992): A thorough introduction to the physical and computational principles that form the basis for modelling the global climate.

11   Predicting climate change

Some people ask, "What if the sky were to fall?"
Terence (Publius Terentius Afer), c190–150 BC

The obvious starting point in this chapter is to consider what we mean by predictability. A helpful approach is to say that it is the extent to which it is possible to predict the climate with a theoretically complete knowledge of the physical laws governing it. Using this definition of predictability, it follows from the discussion of computer models of the climate in Chapter 10 that predicting how the climate may change over the next century or so faces four basic challenges. The first is model error and how much we can improve the mathematical representation of the workings of the climate. The second is the closely related question of reducing the initial-condition error to ensure that the model starts with as accurate a representation of the state of the climate system as possible. The third is how do we measure the results of our studies. Finally, there is the issue of the potential for generalising the concepts and results of the model in framing policy.

The final concept of the generalisation of the results of modelling studies leads us into the economic and social consequences of any future climate change. Here the prediction is even more difficult. Because economic models lack the well-defined physical laws that underpin climate models the uncertainties are much greater. In particular, these uncertainties hinge on public behaviour in the light of any emerging evidence of climate change. When combined with the risk of perverse responses to efforts by governments to curb the emissions of greenhouse gases, this may lead to wholly unexpected changes in energy use and hence human impact on climate change.

In addressing these challenges we have to decide how much is known about natural fluctuations in the climate and what this means for the future. Second, is the whole question of monitoring the current state of the climate and whether we can make substantial improvements of the input to the model? Then there is the issue of how human activities will develop and how these will interact on the climate and respond to any changes in the climate. These issues are best considered separately and then, in the light of any conclusions reached, lumped together. In doing so, the objective will be to draw on all the material in this book to sum up our knowledge on climate variability and climate change can be used to plan for the future. Always recognising that, if events in the coming years show we have moved into a different world, then we will have to revise our view of the future and hence our policies.

11.1 Natural variability

The relative stability of the climate throughout the Holocene poses a problem when considering how the climate may change over the next hundred years. It is all a matter of whether we can regard this stability as being the current natural order of things. If so, then an accurate measure of fluctuations over this period defines the current natural variability of the climate. If not, then we may have to include some elements of climate change that occurred prior to the Holocene in the analysis. In short, we must decide which version of natural fluctuations is the best choice before we can decide whether human activities are having a significant impact.

The obvious approach is to consider the forms of climate change that were presented in Section 1.2. The absence of a well-defined trend over the last few millennia means that we can start with periodic change. Although there is a strong case for orbital variations being the pacemaker for the ice ages, these will only exert an influence on longer timescales. One estimate is that we are headed for another ice age in about 23 000 years time. More immediately, there is very little evidence of regular changes on shorter timescales (i.e. from few years to a few millennia). Apart from the ubiquitous nature of the '20-year cycle' which has variously been attributed to solar, lunar, or atmosphere–ocean autovariance, there is less substantial evidence of cycles around 100 and 200 years which may also be linked to solar activity (see Section 6.5). There is also growing evidence that there may be some quasi-cyclic variation of between one and two millennia in duration.

Much depends on the reconciliation of the current debate of the rectitude of the hockey stick (see Box 8.1). If the conclusion is that the temperature rise in the twentieth century dwarfs any fluctuations of the underlying cooling trend during the last two millennia then the case for natural variability playing a significant part in the warming during the twenty-first century is small. But then we are confronted with the daunting prospect that we already have a century during which anthropogenic warming has been the dominant climatic effect.

The alternative is that re-examination of the proxy records, plus new observations may lead to the conclusion that the hockey stick underestimates the scale of natural variability during the last two millennia. If this were to show that the Medieval Climatic Optimum and the Little Ice Age were of comparable magnitude to the twentieth-century warming then natural variability could yet have a significant part to play in the climatic development of the coming decades.

Beyond these immediate questions is the issue of more sudden and sizeable shifts. Here the issue is whether using our knowledge of climate change during the Holocene underestimates the capacity of the global climate, in its current form, to change more radically. Recent interglacials have been short-lived, last around 10 kyr. The last (the Eemian) reached temperature levels some 2 °C above current levels (see Section 8.4). We have to go back to around 420 kya for a long-lasting interglacial – a period when the orbital motions of the Earth bore a marked resemblance to current patterns. So it may be that we are not living on borrowed time (see Fig. 4.13). It is in the nature of the non-linear behaviour of the global climate, however, that we have no way of telling whether it will remain this stable, or when it might flip into a new state. All we can say is that, at a time when human activities are predicted to have an appreciable impact, the chances of the climate becoming more unstable will increase. At the same time there is always the possibility of a sudden natural change occurring which has nothing to do with human activities.

The implications of these conclusions are to undermine confidence in any predictions of future natural climatic variability. To the extent that there are any 'cycles', their impact is likely to be small, and difficult to quantify until an accepted physical cause is identified. This places particular emphasis on finding out whether solar activity plays a part in these changes, given the cyclic variation in sunspot numbers and the close parallel between their longer-term behaviour and global temperature trends. The success of predicting the amplitude and period of successive sunspot cycles does not have an impressive track record. Moreover, if we include possible magnetic field variations then things become even more difficult.

The question of whether the climate could shift to a more variable pattern as a consequence of natural changes, rather than as a result of human activities, remains in the realms of speculation. Progress depends upon coupled atmosphere–ocean models reaching a degree of sophistication where we have confidence in the changes they may predict in, say, oceanic thermohaline circulation. This will require modellers to overcome fully the challenges set out in Section 10.3. In the meantime, the priority must be to use all the available sources of palaeoclimatic information to generate a better understanding of natural variability and what triggers more dramatic shifts in the climate. Without this, it will take far longer to decide whether future changes are part of the predicted impact of human activities, and hence in theory avoidable, or simply part of the normal ups and downs of the climate. If, however, the climate does undergo a sudden shift, it will be entirely academic as to whether it was a result of natural variability or human activities, as it will be too late to do anything about it.

11.2   Predicting global warming

In spite of the limitations in the capacity of general circulation models (GCMs) to reproduce all the details of the current global climate described in Chapter 10, they are the only physically realistic way to predict the impact of human activities on the climate. Furthermore, although their predictions cover a considerable range of climatic conditions, they are consistent in their broad prediction. This is that the radiative forcing due to the build-up of greenhouse gases in the atmosphere (see Section 2.1.4.) will exert a dominant influence on the future global climate change. The consensus view of the IPCC in AR4 is that based on the available observations and the strength of known feedbacks simulated in GCMs the equilibrium warming for doubling CO_2, or *climate sensitivity*, is likely to lie in the range 2 to 4.5 °C, with a most likely value about 3 °C. This climate sensitivity is most unlikely to be less than 1 °C. For fundamental physical reasons, as well as data limitations, values substantially higher than 4.5 °C still cannot be excluded, but agreement with observations and proxy data is generally worse for those high values than for values in the 2 to 4.5 °C range.

Because the various GCMs adopt such a wide range of approaches to handling the many challenges in simulating the global climate, it is not practical to present a review of the different predictions they produce. Instead, it is more illuminating to consider how the models on average handle the change in global temperature during the twentieth century. There are now a considerable number of coupled model simulations covering this period. They use a variety of forcings in different combinations. These simulations use models with different climate sensitivities, rates of ocean heat uptake and magnitudes and types of forcings. The results of simulations that incorporate anthropogenic forcings, including increasing greenhouse-gas concentrations and the effects of aerosols, plus natural external forcings provide a consistent explanation of the observed temperature record

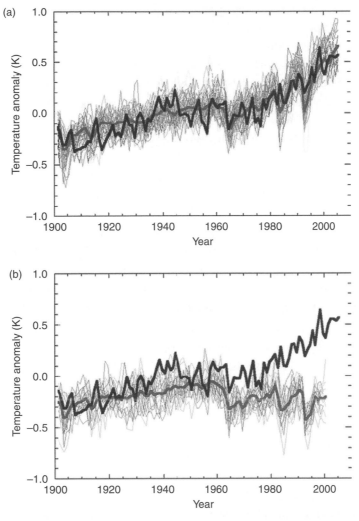

Fig. 11.1 Global mean temperature anomalies, as observed (black line) and as modeled when the simulation include (a) both anthropogenic and natural forcings and (b) natural forcings only. The multimodel (13 models) ensemble mean is shown in grey and individual simulations in thin black lines.

(Fig. 11.1a). Simulations that include only natural forcings (Fig. 11.1b) do not simulate the warming observed over the last three decades.

These results are pretty convincing. They use a variety of different forcings. For example, some anthropogenically forced simulations include both the direct and indirect effects of sulphate aerosols whereas others include just the direct effects. Similarly, the effects of tropospheric and stratospheric ozone changes are included in some simulations but not others, whilst the naturally forced simulations include a variety of different representations of changing solar and volcanic forcing. Despite the inclusion of this additional uncertainty there is a clear separation between the simulations with anthropogenic forcings and the simulations without. In addition, the models generally simulate large-scale natural internal variability quite well, and also capture the cooling associated with volcanic eruptions on shorter timescales.

Predicting future warming requires us to make assumptions about future emissions of greenhouse gases. In AR4 this involves considering a wide variety of scenarios. These are coherent, internally consistent and plausible descriptions of possible future states of the world. We cannot explore all these alternatives here. Instead, we will take a specific scenario and consider what this means in terms of future climate changes. Here we will take what can be described as an economically and politically optimistic but climatically risky combination (defined as the medium emissions scenario A1B). This view of the future involves rapid economic growth, global population that peaks at around 8.7 billion mid-century and declines thereafter to around 7 billion in 2100, and speedy introduction of new and more efficient technologies. Major underlying themes are economic and cultural convergence and capacity building, with a substantial reduction in regional per capita income. Technological change in the energy system will be balanced across all sources.

In this view of the world the level of CO_2 in the atmosphere rises to just over 700 ppm by 2100. Global mean temperature projections by AOGCMs for this scenario suggest a warming of 2.2 to 4.3 °C above present levels (1980–2000) towards the end of this century (2090–2100). The warming is predicted to be greater at high latitudes, especially in the northern hemisphere, and somewhat less in the tropics. This is consistent with the standard view of past climate change, notably the ice ages (see Section 8.5).

This consensus does not get away from the uncertainties discussed in Section 11.1. Because much of the warming over the last century may have been the product of the natural variability of the climate, it could be largely coincidental that the models produce a reasonable fit. In spite of the recognised limitations of the models, this conclusion is probably uncharitable for at least one good reason. This is the fact that the models are not 'tuned' to produce a good fit with observed temperature trends. Instead, they are built on sound physical principles. This means that once constructed and kicked off with a realistic set of starting conditions they run on their own accord to simulate how the world should have responded with and without the atmosphere being modified by human activities. The fact that the output of 13 different models of the world looks rather like the world we live in, and the changes induced by human activities look similar to events in recent decades, is good reason for taking these forecasts seriously.

11.3   The predicted consequences of global warming

If confidence in models is justified, the next issue is what are the implications for our lives if the global temperature rises by 2.2 to 4.3 °C by the end of the twenty-first century? This depends on the answer to three principal questions. The first is, will global warming alter regional weather patterns? Second, will the weather become more extreme? How much will sea level rise?

The analysis of the performance of GCMs in Section 10.1 notes the scatter between the outputs of different models. It is hardly surprising, therefore, to discover that various simulations of future regional weather patterns in a warmer world show a wide range of possible developments. These differences reflect not only the physical components of the model, but also how human activities are considered, especially those that will vary regionally in their impact (e.g. particulates) and whether all their physical effects are handled (e.g. the direct and indirect impacts of particulates). For instance, the UKMO model suggests that the incorporation of sulphate particulates significantly alters the simulation of the impact of global warming on the Indian monsoon. Without particulates the model, along with most GCMs, predicts an increase in rainfall. This behaviour is consistent with a general conclusion that increased temperatures in the tropics will mean more water vapour in the atmosphere and hence a stronger hydrological cycle. When particulates are included the UKMO model produces drier conditions over India.

The same uncertainties apply when considering how warming will affect higher latitudes. The cosy notion that warming will simply produce a gradual displacement of climatic zones to higher latitudes, so England would eventually have a climate like southern France, is probably a gross oversimplification. What will matter more is whether, instead of a general warming, there is a significant shift in the incidence of weather regimes (see Section 9.3). There is some uncertainty in the handling of the changing behaviour of phenomena such as blocking. Nevertheless, there is a growing risk of high temperature extremes with heat waves projected to be more intense, more frequent and longer lasting in a future warmer climate. Cold episodes are projected to decrease in a warmer world with a decrease in diurnal temperature range almost everywhere. The decreases in frost days, observed during the twentieth century almost everywhere in the mid and high latitudes, is likely to continue with the length of growing season extending.

The global hydrological cycle will probably become stronger. Current models show that precipitation generally increases in the wet parts of the tropics (e.g. where monsoon regimes prevail) and over the tropical Pacific. There is a general decrease in the subtropics, and increases at high latitudes. Globally averaged mean water vapour, evaporation and precipitation increase. Intensity of precipitation events increases (i.e., proportionately more precipitation falls for a given precipitation event), particularly in tropical and high latitude areas that experience increases in mean precipitation. Conversely, even in areas where mean precipitation decreases (most subtropical and mid-latitude regions), precipitation intensity increases mainly because there are longer periods between rainfall events. There is a tendency for drying of the mid-continental areas during summer, and hence a greater risk of droughts in those regions.

As for the other extremes, which are part of the popular perception of global warming (e.g. stronger and more frequent hurricanes and mid-latitude storms), the results from models are equivocal. Tropical cyclones play a key role in redistributing the heat from the summer sun over the oceans. As SSTs increase, the amount of water vapour in the lower troposphere will, in general, increase. This will provide more energy to fuel tropical storms, which generally require SSTs to exceed about 26 °C to form. As SSTs increase the potential area over which such storms can form will probably increase. But, as noted in Section 8.10, the evidence of an increase in intense storms in recent decades is not convincing.

Among the other factors that will influence the generation and tracks of tropical storms are ENSO and variations in monsoons. The maximum wind speed depends critically on SSTs and atmospheric structure. Increases in greenhouse gases decrease radiative cooling aloft, thus potentially stabilising the atmosphere. Furthermore, it is not simply a matter of frequency or number of storms; the intensity and duration are likely to matter more. The energy in a storm, which accounts for the intensity and duration of tropical storms, is proportional to the wind velocity squared. So, one big storm may have much greater impacts than several smaller storms.

In the case of extratropical depressions, the models tend to predict a reduction in the intensity of these storms and of the incidence of gales over Europe. This reflects the fact that if the polar regions warm more than the tropics the temperature gradient driving these storms will be reduced and the storm tracks will move to higher latitudes. If the strength of mid-latitude circulation declines, especially in winter, then it could lead to more meandering weather patterns and a greater incidence of blocking. One consequence of this change would be more frequent cold winters in some parts of the northern hemisphere, especially in the early stages of the warming. This change might be matched by hotter summers. The impact of these shifts on rainfall patterns is not yet clear, and the inconclusive figures on current trends, apart from the drier conditions in the Sahel, provide few clues to future patterns.

The consequences of shifting regional patterns and greater climate variability are likely to impact most on agriculture. While seasonal forecasting may alleviate the worst effects of these fluctuations on some annual crops, the role of government in enabling farmers to survive from year to year will play a major part in responding to these challenges. The other economic consequences will depend crucially on the nature of the increased variability. Here the scale of the impact will depend on what action public and private institutions take to influence the economic decisions we all make. For instance, the extent to which coastal areas vulnerable to increased hurricane activity are developed will depend on whether local authorities tighten up planning regulations and whether insurance companies are willing to provide adequate cover at economically acceptable rates.

The same processes govern the related issue of rising sea levels. In climatological terms the rate of rise is linked to precipitation patterns at high latitudes. If warming leads to increased precipitation over Antarctica then any greater peripheral melting of the ice shelves will be more than balanced by the accumulation of snowfall at higher levels. Over Greenland the situation is more finely balanced, and recent figures suggest a rapid increase in melting that is unlikely to have been matched by increased precipitation at higher levels.

At lower latitudes, where glaciers and ice caps have been receding since the late nineteenth century, warming has clearly outweighed any increase in precipitation. So, for much of the coming century the rise in sea level will be dominated by the thermal expansion of the oceans plus the melting of ice caps and mountain glaciers at lower latitudes. The melting of the ice sheets of Greenland and Antarctica could only become a major problem in the longer term (see below). The latest estimate of the median rise in sea level is 29 cm by 2100 with a range from 14 to 43 cm using medium emissions scenario A1B. This compares with the figure of 10 to 25 cm over the past 100 years (see Section 9.4).

This relatively modest figure does, however, contain a more powerful message. This is that once the process of thermal expansion, which is the cause of most of this sea-level rise, is set in motion it is likely to continue for centuries. So even the lowest predictions imply that in due course the sea level will rise appreciably. This rise is bound to increase the damaging effects of both tropical and extratropical storms.

How a rise in sea level will affect different parts of the world also depends on how the Earth's crust will adjust to past and future changes in the load of the major ice sheets. This glacial isostatic adjustment means that the impact will vary from one coastal site to another. Superimposed on this post-glacial rebound will be regional isostatic and tectonic effects. In addition, local factors such as groundwater extraction and land reclamation further complicate matters. So analysis of the specific impact of a rise in sea level will depend on a combination of improved global models of post-glacial rebound and local geological knowledge.

In spite of all these uncertainties, the actual progress of sea-level rises is likely to be an inexorable process that can be monitored closely. In the short term its impact will be dominated by the increasing vulnerability of low-lying coastal areas to severe storms. So the overall social response to these events will be an extension of the institutional reactions identified above. Local and regional governments will have to reach long-term decisions about how much is invested in coastal defences as opposed to allowing some land to be lost to the sea. At the same time, individuals will have to make decisions about the risks they are prepared to take in respect of themselves and their property. The extent to which they can get adequate insurance cover is bound to influence these decisions. In the longer term, of course, mounting costs will provide ever-greater incentives for central governments to make more substantial and lengthy provisions to prevent damage rising to politically unacceptable levels.

The more dramatic scenarios of the melting or collapse of the Greenland and Antarctic ice sheets appear to be distant prospects because of the likely effects of increased precipitation over the ice sheets in the next century or so. But uncertainty surrounds the stability of the West Antarctic ice sheet, which rests on a rock bed well below sea level. Depending on the physical assumptions made about the conditions under the ice, theoretical models can produce equivocal predictions about whether this ice sheet could collapse catastrophically if global warming and the sea level rose to some threshold value. This prospect has been the subject of a large number of studies and the overall judgement is that possible scenarios range from the

Ross Ice Shelf disintegrating over the next 200 years and the West Antarctic Ice Sheet collapsing rapidly over 50 to 200 years raising sea level 60 to 120 cm per century, to snowfall increasing and the Antarctic ice sheet making a negative contribution to the sea level until global temperature has risen at least 8 °C, something that is not predicted to happen until 2200 at the earliest. In the case of Greenland the picture is less rosy. Some of the models suggest that only a 2 °C temperature rise is needed to set off an irreversible melting of the Greenland ice sheet.

On a wider scale, it is clear that the predicted levels of global warming in the next 100 years will affect all forms of life on the planet. In many cases the impact will be non-linear, in that species may be able to adapt to small rises in temperature but above a certain level the damage will escalate rapidly. An estimate of the consequences of specific levels of warming appears in a report (*The Economics of Climate Change*) prepared by the former chief economist to the World Bank, Sir Nicholas Stern, for the UK Treasury published in March 2007. This is distilled in the Table 11.1.

It could be said that these conclusions may be unduly pessimistic, unless they are interpreted in wider terms. For example, the imminent demise of corals may prove to be correct, but it is unlikely to be a result of temperature rise alone. Often called the *rainforests of the oceans*, corals first emerged more than 200 Mya. A number of living coral reefs are believed to be more than 2 Myr old, though most began growing within the last 10 kyr. There is plenty of evidence that they survived the higher temperature of the Eemian interglacial (see Section 8.4) and the warmth of the mid-Holocene. What they now face is the combined threats of local assault from destructive fishing methods and coastal development, and climate change, which may combine a number of threats. For example, it has been proposed that the *coral bleaching* events in the Caribbean associated with the high temperatures following the record-breaking El Nino in 1998, may instead have been caused by pathogens transported on Saharan dust; climate change is rarely simple.

More generally the estimates of extinction levels have been the subject of spirited debate. The focus of the discussion centres on how to extrapolate figures for species that are identified as being at risk of extinction within three generations or 10 years (the *Red List*) over longer periods and include species with shorter lifetimes. So the figures in Table 11.1 may be unduly pessimistic. Nonetheless, even if the environmental figures are on the high side, the overall picture in Table 11.1 is a wake-up call for policy-makers around the world. The available policy options will be considered in Section 11.5.

Table 11.1 Predicted outcomes of global temperature rise

T rise	Water	Food	Health	Land	Environment
1 °C	Glaciers in Andes melt, threatening water supplies to 50 million people	Cereal yields rise in temperate zones	At least 300 000 a year die from climate-related diseases. Winter mortality rates drop in Europe and US.	Permafrost in Arctic regions thaws threatening buildings and roads.	10% of land species may face extinction; coral reefs suffer 80% bleaching
2 °C	20 to 30% less water available in Southern Africa and the Mediterranean	Crop yields drop in tropical region (by up to 10% in Africa)	Up to 60 million more Africans would be exposed to malaria.	Coastal flooding affects 10 million more people each year	15–40% of species may face extinction, including polar bears and caribou
3 °C	Serious droughts in southern Europe; up to 4 billion suffer water shortages	Up to 550 million people are likely to go hungry.	Up to 3 million people would die of malnutrition (if carbon fertilisation weak).	Coastal flooding affects up to 170 million people each year.	20–50% of species face extinction; the Amazon rain-forest begins to die.
4 °C	Up to 50% less water available in the Mediterranean and Southern Africa	Agriculture stops in parts of Australia and crop output in Africa drops 35%.	Up to 80 million more Africans would be exposed to malaria	Coastal flooding affect up 300 million more people each year	Half the Arctic tundra disappears
5 °C	Himalayan glaciers could disappear affecting hundreds of millions in China and India.	Ocean acidity disrupts marine ecosystems and possibly fish stocks.		Cities such as London, New York and New York and low-lying coastal areas threatened by rising sea levels	

11.4   Doubts about the scale of global warming

Those involved in assessing the contribution of human activities to global warming fall into two principal camps. The majority regard its impact is already a proven fact. The minority consider the evidence as inadequate and its reality will only be established when a far more sustained warming has been observed. The median view is the consensus reached by the IPCC in AR4, published in 2007 that 'It is very likely that greenhouse gas forcing has been the dominant cause of the observed warming of globally averaged temperatures in the last 50 years.' It then goes on to note that 'An increasing body of evidence suggests a discernible human influence on other aspects of climate including sea ice, heat waves and other extremes, circulation, storm tracks, and precipitation'. On regional changes it makes a more cautious observation that 'Projected future warming shows scenario-independent geographical patterns similar to those over the past 50 years. Warming is expected to be greatest at high northern latitudes and over land (roughly twice the global average), and least over the Southern Ocean and the North Atlantic'. Perhaps the most worrying feature of this conclusion is that it can be interpreted, in terms of precipitation, that wet regions will get wetter and the dry areas will get drier: so more floods and more droughts.

The differences in opinion about this centrist view reflect the uncertainties in defining the natural variability of the climate and modelling its behaviour. The vociferous minority argue that the IPCC conclusions are not justified by the arguments presented in AR4. The doubts of a more guarded but extensive minority revolve around both the reliability of the forecasts based on both the physical uncertainties in the models and the greater doubts about the economic costs of some of the proposed strategies to reduce the emissions of greenhouse gases are excessive. Needless to say, those industries that have a major interest in maintaining the status quo are in the vanguard of this opposition.

These tensions become much greater when the question of how the impact of a warmer world will affect different parts of the world; at the national level there will be winners and losers. If this is the case then with a sufficiently accurate model it will be possible to define how, say, the spatial distribution of global atmospheric temperatures and precipitation will change as a result of human activities. One approach is to concentrate on the three-dimensional patterns that might reflect human rather than natural factors. In the case of temperature patterns, two particular features have attracted most attention. First, the regional distribution of temperature change in recent decades (see Fig. 8.23) has been more accurately reproduced by models that include the effects of sulphate aerosols (see Section 11.2). The second is how the stratosphere has cooled while the lower troposphere has warmed (see Section 4.2). Here again the observed changes are consistent with the model predictions of the consequences of the build-up of greenhouse gases and with the basic physics of the greenhouse effect (see Box 2.1). Both these developments support the thesis that the observed trends cannot be explained away without inclusion of a significant contribution from human activities.

On the other hand, much of the variance in global surface temperature observed in recent decades is attributable to the ENSO and the North Atlantic Oscillation. The models continue to improve, but still have difficulty representing aspects of these and other oscillations. This has implications for handling both the most obvious features of the variation in the incidence of these broad climatic fluctuations, and the various attendant weather regimes (see Section 3.2) that are a major part of the process of observed climate change in recent decades. In particular, the problems with simulating the MJO highlight the complications of modelling the link between the chaotic behaviour of the weather and the apparently more ordered world of oscillations. So, there is the sneaking suspicion that a lot of what is being observed, while difficult to explain without including human activities, is due to natural variability.

The emerging consensus on the causes of the current warming underlines the strengths and weaknesses of using GCMs to detect the impact of human activities. The models have concentrated on the most obvious physical effects and, as discussed in Section 10.3, there are a variety of other processes that could play a significant part in future changes. It is not just natural variability we have to be worried about. At the very least, the indirect effects of particulates on cloudiness, the effects of changes in other greenhouse gases, changes in the albedo of land surfaces, the role of the biosphere through the carbon cycle (see Section 3.5), and solar variability need to be included in the models. The challenge is to obtain an adequate understanding of the impact on radiative forcing of these processes, which is set out in Fig. 7.4. Depending on the assumptions made about the impact of particulates, volcanoes and solar activity, the quality of the fit can be improved, but at the cost of introducing ever-greater uncertainty about just how important is the contribution of rising CO_2 levels.

Nevertheless, as Figure 11.1 shows, the central figure for the sensitivity to increasing greenhouse gases equivalent to doubling CO_2 levels, together with reasonable estimates of aerosol, volcanic and solar effects produces an excellent fit of past trends. A more sceptical view is that increasingly we run the risk of falling into the trap of 'tuning' our models to come up with a better fit. So while the physical credentials of the models remain sound, the temptation to select values for the human perturbations, from within an acceptable range of uncertainty, which fit current observations, could lead to a distortion of our judgement.

In the case of stratospheric cooling, other factors intervene in a different manner. Because of the decline of ozone (O_3) levels in recent years, notably at high latitudes, the amount of solar radiation absorbed in the stratosphere has been reduced (see Box 2.2). This has led directly to a cooling at these levels. This change, although almost certainly due to human activities, is physically separate from the analysis of the build-up of greenhouse gases in the troposphere. So the models needs to consider both how changes in O_3 alter both solar absorption and also affect outgoing terrestrial radiation given that O_3 is a powerful greenhouse gas. It is also part of the solar conundrum, as the greater variability of UV emissions from the sun alter the distribution of ozone in the stratosphere and hence can influence global weather patterns (see Section 6.5).

In spite of these doubts, the fact of the matter is that nine of the ten warmest years on record globally has occurred in the last decade. So, if this record-breaking run of warm years continues, it suggests that the models are on the right track. Nevertheless, the climate may have a few tricks up its sleeve. The latest unexpected developments are a cooling of the ocean down to a depth of 750 m between 2003 and 2005. A similar cooling occurred between 1980 and 1983. The product of the Argo floating buoy system (see Section 4.2), these latest measurements have identified detailed features of substantial heat losses that are exceedingly difficult to accommodate within the models. The other intriguing development is the scale of the changes in the measurements of the total global cloudiness over the last two decades (see Section 3.3). Again these measurements are not yet an integral part of current simulations. Both of these developments may well turn out to be nothing more than temporary fluctuations that will not alter the long-term rise in global temperature.

This is all well and good, but what if the upward trend in surface temperatures experiences a significant interruption that cannot be attributed to a known cause (e.g. a major volcano)? The modellers will have much greater difficulty in convincing those who have to make decisions about global warming that their actions should be based on the output of GCMs. It will not be realistic to assume that producing an additional factor, like a rabbit from the hat, just as particulates may have provided an explanation for the interruption in the warming trend between around 1940 and 1970, will satisfy potential users.

The power of the arguments for taking action about global warming is that the principal cause is identified as the emission of greenhouse gases, and that these emissions can be reduced without politically unacceptable social and economic costs. If, contrary to current trends, reality turns out to be a much more subtle balance of warming and cooling effects blurred by greater natural variability than previously assumed, then the case for specific action is more difficult to make. In this situation, tackling climate change will slip down the list of priorities. It will become one of the many factors that have to be considered in making the difficult decisions that democratic societies make about environmental and social conditions.

For the time being, however, it has to be argued that the observed global warming is due principally to a build-up of greenhouse gases, and so this is the obvious target for our efforts. This does not mean we should ignore evidence that other factors have been underestimated. There is a danger that having set our sights on a specific target there will be a sense of denial when contrary evidence emerges from our monitoring efforts. Confronted by what looks like contradictory information it is all too easy to sideline it for fear that it will undermine sensitive political decisions. This is a natural response to the pressing case for policy options, which must take full account of the conflicting economic, environmental and social issues involved. This process recognises also that there are other priorities (e.g. education, health and law and order) that may place more urgent demands on public spending. In the circumstances, it does not help to gloss over the limitations in our current understanding of climate change. To do so and argue that there is a simple answer runs the risk of being accused of not having understood the question.

11.5   What can we do about global warming?

These cautious observations about the challenges facing governments in reaching decisions on what action should be done to minimise the threat of climate change look like a cop out. Given the scale of the work that has been done by the IPCC, surely there must be sufficient agreement about what should be done. In particular, the international understandings, which have emerged from major UN-sponsored conferences like those held in Rio de Janeiro in 1992, Berlin in 1995 and Kyoto in 1997, led to the

Kyoto Protocol that defines action to be taken to curb the emission of greenhouse gases. The Protocol was eventually ratified in late 2004, when Russia signed up to its obligations.

The fact that neither the USA nor Australia have signed the protocol and that both of them together with Japan and Canada have no expectation of reaching their targets by 2012 shows we have a long way to go. Although France, Germany and the UK have done much better, principally as a consequence of the composition of their energy supplies (e.g. France's heavy reliance on nuclear power), the prospect of their being able to deliver more demanding targets in the future is less rosy.

It is probably best to leave the basic arguments about how the consumption of fossil fuels will rise during the twenty-first century on one side. Here we will continue work on the AR4 scenario A1B, as an optimistic measure of what might be achieved by a combination of technology, economic growth and specific action to curb emissions. This would mean that the CO_2 level will reach the critical level of 550 ppmv around 2050. Higher growth scenarios are considered in AR4, which represent greater growth in population and economic growth where in the extreme case the CO_2 level by 2100 could exceed 1000 ppmv. At the other end of the spectrum of scenarios is a set that emphasise a convergent world with rapid change in economic structures towards a service and information economy, reductions in material intensity and the introduction of clean technologies. These make relatively little difference out to around 2050, but thereafter may be capable of delivering stable CO_2 levels at around 550 ppmv at the end of the twenty-first century.

An intermediate approach is to consider what strategies might be needed to restrict future CO_2 level so that we do not exceed the widely toted critical warming figure of exceeding 3 °C at some point in the twenty-first century. Given the median sensitivity estimate of a 3 °C warming for a doubling of CO_2 quoted in AR4 this equates the upper limit of CO_2 levels at being 550 ppmv (roughly twice the pre-industrial level). Whether this target is achievable depends not only on the pace of climate change in the next 50 years, but also on economic and political circumstances. One answer lies in the work of Stephen Pacala and Robert Socolow, at Princeton University who argue that we already possess the fundamental scientific, technical, and industrial know-how to solve the carbon and climate problem for the next half-century. A portfolio of technologies now exists to meet the world's energy needs over the next 50 years and to limit atmospheric CO_2 to a trajectory that avoids a doubling of the preindustrial concentration.

The analysis considers a business-as-usual scenario, which postulates that the current emissions levels [equivalent to 7 gigatons of carbon per year (GtCpa)] will rise to 14 GtCpa by 2054. It proposes various technical options that are presented as a set of *wedges* that are capable of reducing these emissions to 7 GtCpa by 2054. These consist of various forms of energy conservation and efficiency, switching to renewables, sequestration of carbon in forests and the soil, nuclear fission and fuel switches. Each of these options would start at zero now and rise to a level such that each could make a comparable contribution to limiting CO_2 emissions to achieve the proposed target. The net result of these actions is estimated to make only limited impact on the rise in CO_2 levels during the first 50 years, but thereafter would further reduce carbon emissions and stabilise levels at 500 ppmv. Failure to take action over the next 50 years would mean that levels rise to 850 ppmv and then remain at this level even with the application of the same set of actions.

In some cases these proposals look technically realistic and either are already part of existing strategies or could be introduced relatively rapidly. The real challenge is whether they can be sold to the electorate. For instance, the case for increased investments in renewable sources of energy is politically acceptable given the benefits of security of supply, which may outweigh the rising costs of energy. Against a background of rising oil prices, and dependence on imports, plus a fear of nuclear power, the electorate has already given tacit approval to price rises and action is in hand. Whether they will continue to accept this approach will depend critically on whether prices are kept at an acceptable level.

In the case of more conventional forms of energy supply, rising prices – the surest way to reduce demand – are highly unpopular, and have politicians tiptoeing through an electoral minefield. In particular, economic action is regarded as a political anathema in respect of transportation, so the situation is far less rosy. Socolow and Pacala proposed a wedge for automobiles; to make a proportionate contribution to stabilizing carbon emissions every car in the world must be driven half as much as today, and to become twice as fuel efficient. The first objective will almost certainly involve rising taxation, road pricing and limited access to cities. The second is more achievable providing you can persuade motorists to give up their obsession for gas-guzzlers and SUVs (4 × 4s in Europe), and drive smaller vehicles. Nevertheless, the chances of the developed nations, and, in particular, the US adopting fiscal measures to curb demand is, to say the least unlikely to happen in the short term. It is interesting that Pacala and Socolow did not single out the airline industry for particular action.

Clearly, progress on the more difficult areas will depend on political will and the capacity of nations to reach real agreements to make proportionate reductions in emissions. In respect of political action, much will centre on the impact of climatic events plus the influence of major opinion formers. In this context, recent examples of the Stern Report (see Section 11.3), and the book and film *An Inconvenient Truth* by Al Gore show how thinking can be developed. It will need hard-hitting analyses such as these to bring the public round to accepting that action is essential.

The Stern Report is interesting in this context as it effectively put flesh on the bare bones of Pacala and Socolow's work. It too takes the view that a realistic strategy should aim at stabilising CO_2 levels at around 500 ppm by around 2050. The range of consequences predicted for the scale of climate change (see Table 11.1) could reduce the world's GDP by 5 to 20% by 2050. If, however, concerted international action were set in hand now it would need an investment 1% of the global GDP on abatement activities over the same period to stabilise global greenhouse-gas levels. The cost of not taking action could be over £3.5 trillion.

How the burden of reducing emissions is apportioned between the developed and developing nations is another tricky political issue. Trading emissions between countries will almost certainly have a part to play, but whether it can become a central plank in future strategies depends on how it impacts upon individual countries. Negotiations to produce workable schemes are bound to continue for many years. They will be coloured by how the climate actually behaves and shifting perceptions about whether other human activities are making important contributions to the observed changes. The going will get tougher as the economically attractive and politically acceptable options are used up. Advanced industrial nations will be willing to reduce CO_2 emissions by sensible energy conservation measures, improved public transport in major urban areas, and closing down ageing coal-fired power stations, but when it comes to the imposition of unpopular taxes on energy consumption or the restriction of the use of private vehicles, and travel, their resolve to meet demanding targets may weaken. At the same time, developing countries, whose per capita energy consumption is much lower, will continue to argue that they should be allowed leeway to achieve economic growth. If these complications are combined with confusing developments about other physical factors contributing to climate change the negotiations will become increasingly fractious and muddied.

All of this means there is huge uncertainty about what action can be taken to minimise the impact of human activities on the climate. For this reason the many scenarios of the future build-up of greenhouse gases in the atmosphere together with the forecasts of how these figures would be reduced by the adoption of certain strategies are not reviewed here. Suffice it to say they cover a wide range. Depending on future economic growth and the action to reduce emissions, the level of CO_2 in the atmosphere could either rise to a level over 1000 ppm by the end of the twenty-first century or be stabilised at 500 ppm by around that time. This would involve anthropogenic CO_2 emissions either more than doubling, or being reduced to barely a third of current levels by the year 2100. The essential feature of these figures is that whatever action is taken now to reduce emissions, the levels of greenhouse gases in the atmosphere are bound to rise, if not peak, in the next century or so.

Depending on what is done and how quickly it is implemented, the rate of growth will be slowed, but the consequences will only become appreciable in the longer term. The crux of the matter is how fast the climate reacts to this inevitable build-up of greenhouse gases. Given that all scenarios are close to the dangerous threshold of 550 ppm by around 2050, it shows what we are up against. It does, however, give us an opportunity to put our house in order. But if the climate warms up more rapidly or reaches some critical level, which precipitates a flip to some new climatic state, then it could already be too late to do anything about it.

In making estimates of the potential scope for action, there is a danger in oversimplifying the options. Almost everything that is proposed is presented in terms of reducing CO_2 emissions. In fact, in considering the impact of human emissions of greenhouse gases it has been accepted practice to refer to the CO_2 equivalence of these emissions with the somewhat confusing term *carbon emissions*. While this term reflects the physical importance of treating the physical impact of the various gases, it does not mean that reducing CO_2 alone is the answer. As has been emphasised in a variety of places, the contribution of CO_2 to the anthropogenic greenhouse-gas radiative forcing burden is about 60% (see Fig. 7.4). So, even if greenhouse-gas forcing is all that matters in terms of future climate change, which, to say the least, is a heck of an assumption, as it will at most address just over half the problem. Even here there are complications depending on the response of other gases, notably water vapour and ozone.

Once we add in the non-linear responses of the climate to both our efforts to alter the anthropogenic impacts, plus the natural variability in the climate, then focusing too much on CO_2 emissions may be a mite myopic. This does not mean the initial emphasis on CO_2 is wrong; it is the most obvious place to start. Where it becomes more questionable is when governments start setting targets for reductions of say 60%, or even 80% by 2050. Long before we get to these draconian levels we will need to be maintaining a highly critical assessment of whether what we are achieving matches the predicted climatic response. This must include improved measures of cloudiness and features of the Earth's surface that alter the albedo.

In this process, unfolding weather events will a central role in political perceptions of what needs to be done. These will vary from country to country and governments will only take more radical action to reduce emissions when this is seen as essential to address pressing needs. So, concerted action will not be taken until there is incontrovertible evidence of damaging changes in the climate, by which time many of these shifts are bound to run their course. This means that improving our knowledge of what is really driving current climate change is the only way of getting a better handle on future developments.

One way of thinking about this difficult issue of uncertainty is to consider the consequences of a sudden and dramatic shift in the climate. For instance, as discussed in Sections 3.7 and 8.3, the circulation in the North Atlantic has in the past switched into different modes. If future changes were to reverse the warming experienced at the end of the Younger Dryas, 12 kya, the impact on winter temperatures across much of the northern hemisphere would be catastrophic. They would drop by several degrees Celsius, especially in northern Europe, and the economic consequences would be vast. Given such an event has not happened for 12 kyr, and is by its very nature unpredictable, it is not realistic for governments to press ahead with costly policies to prevent the climate flipping. All they can do is stick to those policies, which appear best targeted, whether or not the climate is going to throw a wobbly. These are bound to concentrate on actions which make economic sense now; which may well delay global warming; and which could conceivably have the additional benefit of reducing the risk of more catastrophic events occurring, albeit only by a tiny amount.

11.6   The Gaia hypothesis

Having explored the range of explanations of climate change, and what we can do about it, and come up with what can only be described as an equivocal, if not gloomy, set of conclusions, you may well ask whether there is a more convincing way of presenting the arguments. I would insist that this is not the case, but you could respond by saying I would say that wouldn't I. So as a closing presentation or even an epilogue, I will leave you with what could either be regarded as an alternative explanation or simply another way of looking at the evidence. This the controversial *Gaia hypothesis*.

This hypothesis was first proposed by James Lovelock, an independent English scientist and named after the 'mother Earth' goddess of the ancient Greeks. It aims to explain why, unlike other planets in the solar system, the Earth's atmospheric composition and history cannot be described in terms of physics and chemistry, but reflects the strong influence of biology. It explains the survival of life on Earth for nearly 4 billion years by treating life and the global environment as two parts of a single system. In effect, micro-organisms, plants and animals behave in such a way that the Earth's environment becomes adjusted to states optimum to their maintenance. This is not a conscious act on the part of the biosphere but instead that adjustments arise from natural selection. As such it represents a more sophisticated interpretation of the biosphere than has been considered earlier in this book (see Section 3.5), but what is said here does not go as far as endorsing his latest book that talks in anthropomorphic terms of nature will exact *revenge* for the feckless behaviour of humankind.

A good example of the type of process that might contribute to the stability of the global ecosystem is the production of dimethylsulphide (DMS) by phytoplankton. As noted in Section 3.5 this gas is the by-product of the life cycle of marine algae. Its conversion into sulphate particulates may be a major factor in the formation of clouds over the oceans (see Section 10.2). So it is possible that an increase in SSTs together with a rise in CO_2 levels will lead to greater production of algae and hence release more DMS into the atmosphere. If this results in additional cloud formation this will have a cooling effect – a negative feedback mechanism that effectively acts as a climatic 'thermostat'.

It is not necessary to go further into the competing arguments that have swirled around the Gaia hypothesis since it was first published in 1972, or the various physical processes and models that have been developed to explore its physical validity. What matters here is the concept of the biosphere, in responding the types of physical changes that can cause climate change (see Chapter 6), meets the criterion of being *optimum for the maintenance of living things*. While there is a heated debate about what this criterion means, in the case of many of the examples cited in this book it can be used to consider the overall response of the climate system and the biosphere to change. So whether it be the Earth's temperature staying within the narrow range needed to maintain life over billions of years, while the Sun's output has risen by 30% and the continents have moved across the face of the planet, or the ability of life to adapt to the catastrophe of a large bolide impact, it is helpful to consider the total response to these challenges. The long-term changes of the composition of the atmosphere and the waxing and waning of the ice ages are other examples of how we must consider all the issues if we are to form a sensible picture.

None of this makes climate change any easier to understand, but it may help to maintain a balanced attitude to what has happened; what is happening now, and what could happen in the future. Equally important, by taking a suitably wide view we may reinforce a sense of wonderment for the immensity and complexity of the Earth's climate.

QUESTIONS

1. The term 'consensus' used to describe the agreement reached by the IPCC on the impact of human activities on the climate has been described as being anything from a courageous scientific compromise through an uneasy truce to an unfortunate if not politically inspired fudge. On your reading of the report what do you think is the best way to describe the conclusion and why?

2. Do you think that political parties in both developed and developing countries can reconcile the pressures for economic growth with the demands to cut back emissions of greenhouse gases without the support of the opposition parties if the only solution is to reduce growth and employment? Without this cross-party support is it inevitable that political unity will only develop when there is incontrovertible evidence of climate damage?

3. Consider the arguments for and against taking specific action to modify the climate on a regional or global scale (e.g. programmes of afforestation, or diversion of ocean currents). What issues would need to be resolved before such large-scale activities were put in train?

4. Identify possible low-cost ways of minimising the impact of global warming in urban areas, with particular regard to cutting the demand for air-conditioning and reducing air pollution. What are the economic and social barriers to introducing these options that could reduce the amount of solar energy absorbed and cut down vehicle emissions?

5. If, once every million years (see Fig. 8.11) or so, a bolide impact with the Earth causes major climatic damage, how much money do you think we should spend to protect ourselves against such an event, and what would you recommend we spend the money on?

FURTHER READING

A complete reference list is available at the end of the book but the following is a selection of the best books or articles to follow up particular topics within this chapter. Full details of each reference are to be found in the Bibliography.

Houghton (2004): A penetrating and committed analysis of the issues surrounding global warming and the options for reducing its impact, written by the co-chairman of Scientific Assessment Working Group of the IPCC, which is required reading for anyone who wants an accessible presentation of the arguments for action without going through all the IPCC reports.

Lovelock (1988): The best source for an up-to-date analysis of the basic content of the Gaia hypothesis of the stability of the Earth's biosphere.

Glossary

Absolute temperature (K) a temperature scale based on the thermodynamic principle that the lowest possible temperature is absolute zero (0 K) and the ice point is 273.16 K measured in degrees Kelvin (K) which are of the same magnitude as degrees Celsius (°C).

Absorption the process by which incident radiation is taken into a body and retained without reflection or transmission, thereby increasing the internal or kinetic energy of the molecules or atoms composing the absorbing medium.

Aerosols particles, other than water or ice, suspended in the atmosphere. They range in radius from one hundredth to one ten-millionth of a centimetre – or 10^2 to 10^{-3} micrometres (μm), and may be of natural or anthropogenic origin. Aerosols are important as nuclei for the condensation of water droplets and ice crystals, and as participants in various atmospheric chemical reactions. Aerosols resulting from volcanic eruptions can lead to a cooling at the Earth's surface.

Albedo the fraction of the radiation falling upon a non-luminous body that it diffusely reflects, often expressed as a percentage.

Angiosperms flowering plants; first arrived in the Cretaceous.

Biosphere the system of earth and its atmosphere that supports life. In the global carbon cycle, the biosphere serves as a sink (reservoir): carbon is stored and preserved in living organisms (plants and animals) and life-derived organic matter (litter, detritus). The biosphere controls the magnitude of the fluxes of several greenhouse gases, including CO_2 and methane, between the atmosphere, oceans, and land. The terrestrial biosphere includes living biota (plants and animals) and the litter and soil organic matter on land. The marine biosphere includes the flora, fauna and detritus in the oceans.

Black body radiation   the radiation that is emitted by a surface that absorbs all incident radiation at all wavelengths. The temperature of the surface defines the wavelength-dependence of this radiation.

Blocking   a phenomenon, most often association with stationary high-pressure systems in the mid-latitudes of the northern hemisphere, which produces periods of abnormal weather.

Bolide   any solid object from outer space; a term usually restricted to objects larger than common meteorites.

Carbonate sediments   sediments composed of the calcium carbonate ($CaCO_3$) minerals: aragonite and calcite or the calcium-magnesium carbonate ($CaMg(CO_3)_2$) dolomite.

Chlorofluorocarbons (CFCs)   a family of inert, non-toxic and easily liquefied chemicals, implicated in two major environmental problems: ozone-layer depletion and global warming. CFCs are used as coolants in refrigerators and air-conditioners, as propellants in aerosol cans, as solvents, and as blowing-agents that inflate flexible foams. It takes about 15 years for a CFC molecule to drift into the upper atmosphere, where it can last 100 years or more, destroying an estimated 10 000 ozone molecules over that time.

Climate   the long-term statistical average of weather conditions. Global climate represents the long-term behaviour of such parameters as temperature, air pressure, precipitation, soil moisture, runoff, cloudiness, storm activity, winds, and ocean currents, integrated over the full surface of the globe. Regional climate, analogously, is defined by the long-term averages for geographically limited domains on the Earth's surface.

Coccolithophorida   a group of plankton that construct a calcareous shell of round platelets known as coccoliths, which may accumulate on the ocean floor as sediment, ultimately contributing to limestone, such as chalk (see also phytoplankton).

Cold front the boundary line between advancing cold air and a mass of warm air under which the cold air pushes like a wedge.

Continental drift the lateral movement of continents as a result of sea-floor spreading (see also plate tectonics).

Convection a type of heat transfer which occurs in a fluid by the vertical motion of large volumes of the heated material by differential heating (at the bottom of the atmosphere) thus creating, locally, a less dense, more buoyant fluid.

Coriolis force the term used to explain the fact that a moving object detached from the rotating Earth appears to an observer on Earth to be deflected by a force acting at right angles to the direction of motion. Deflection of moving objects in the northern hemisphere is to the right of the path of motion. Deflection in the southern hemisphere is to the left of the path of motion.

Cretaceous (65–144 Ma) an era in which mean global temperatures may have been 10 °C warmer than they are today, and deep-water temperatures may have been 18 °C warmer. The last period of the Mesozoic era, the Cretaceous was marked by the rapid rise and spread of deciduous trees, by shallow seas submerging most of the Earth's present land surface, and by dinosaurs – which became extinct after this period.

Cryosphere the portion of the climate system consisting of the world's ice masses, sea ice, glaciers, and snow deposits. Snow cover on land is largely seasonal and related to atmospheric circulation. Glaciers and ice sheets are tied to global water cycles and variations of sea level, and change over periods from hundreds to millions of years. The ice sheets of Greenland and the Antarctic contain 80% of the existing fresh water on the globe, thereby acting as a long-term reservoir in the hydrological cycle.

Deforestation loss of forest. At least eleven million hectares of tropical forest are lost every year. Although the causes vary by region, one estimate indicates that slash-and-burn agriculture and scavenging for wood-fuel, often in the wake of commercial road building, accounts worldwide for 40–50% of deforestation. Grazing accounts for 10%, commercial agriculture for 10–20%, forestry and plantations for 5–10%, and forest fires for 1–15%.

Depression
a part of the atmosphere where the surface pressure is lower than in surrounding parts – often called a 'low'.

Dendrochronology
the dating of past events and variations in the environment and climate by studying the annual growth rates of trees.

Dendroclimatology
the science of reconstructing past climates from the information stored in tree trunks as the annual radial increments of growth. Wide rings signify favourable growing conditions, absence of disease and pests, and favourable climatic conditions. Narrow rings indicate unfavourable growing conditions or climate. Tree rings record responses to a wider range of climatic variables.

Desertification
Land degradation in arid, semi-arid and dry sub-humid areas resulting from various factors, including climatic variations and human activities.

Ecliptic
the great circle in which the plane containing the centres of the Earth and the Sun cuts the celestial sphere.

Eemian (125–130 kya)
the interglacial optimum period prior to the current one when it was warmer than today. This warmth may have been a consequence of the Earth's orbit giving markedly more radiation during the Northern hemisphere summer.

Electromagnetic radiation
the emission and propagation of electromagnetic energy from a source in the form of electric and magnetic fields, which need no medium to support them and which travel through a vacuum at the velocity of light. This radiation encompasses the entire frequency range from γ-rays to radio waves.

El Niño Southern Oscillation (ENSO)
a quasi-periodic occurrence when large-scale abnormal pressure and sea-surface temperature patterns become established across the tropical Pacific every few years.

Emissivity	the ratio of the emissive power of a surface at a given temperature to that of a black body at the same temperature and the same surroundings.
Eocene (37–58 Mya)	an epoch, part of the Tertiary era. Like the Cretaceous era, the Eocene epoch was significantly warmer than the present day. Seas expanded far beyond their present boundaries; palm trees grew where London and southern Alaska are today.
Eustasy	the worldwide global changes in sea level caused by changes in water volume due to the formation and melting of ice sheets, changes in the temperature of the water, or changes in the volume of ocean basins induced by changing volumes of ocean ridges.
Eustatic	a global change in sea level.
Evapotranspiration	the process of water vapour transfer from vegetated land surfaces into the atmosphere; an essential part of the global hydrologic cycle. Evapotranspiration includes *evaporation* (the change of liquid water, from bodies of water and wet soil, into water vapour) and *transpiration* (in which water is drawn from the soil into plant roots, transported through the plant, and then evaporated from leaves and other plant surfaces into the air).
Feedback mechanism	a process of system dynamics in which a system reacts to amplify or suppress the effect of a force that is acting upon it. For example, in the climate system, warmer temperatures may melt snow and ice cover, revealing the darker land surface underneath. The darker surface absorbs more solar energy, causing further temperature increases, thus melting even more snow and ice cover and so on. This is positive feedback, in which warming reinforces itself. In negative feedback, a force ultimately reduces its own effect. For example, when the Earth's surface grows warmer, more water evaporates, forming more clouds. If the clouds which form are extensive and widely distributed, covering large areas of the surface, they will tend to reflect more solar radiation back into space than the

dark ground underneath would, cooling the Earth's surface and reducing the impact of warmer temperatures.

Foraminifera
(foraminifers, or forams for short)
a group of microscopic marine organisms belonging to the Protozoa; they are planktonic or benthic and are grazers and predators. Their calcareous shells provide a major part of the paleoceanographic and paleoclimatic record.

Fourier transform spectral analysis
the mathematical determination of the amplitude of the harmonic components of a time series and the presentation of these in the form of a power spectrum (see Power spectrum).

Gaia hypothesis
this hypothesis holds that living organisms on Earth (including micro-organisms) actively regulate atmospheric composition and climate, helping provide climate stability in the face of challenges like the increasing luminosity of the sun, or increasing anthropogenic greenhouse-gas emissions.

General Circulation Models (GCMs)
a computational model or representation of the Earth's climate used to forecast changes in climate or weather.

Glacial epochs
periods during the history of the Earth when there were larger ice sheets (continental-size) and mountain glaciers than today. The most recent glacial epoch, the Pleistocene, has encompassed much of the last 1.6 Mya. In overall occurrence, all the glacial epochs that have ever occurred occupy only 5–10% of all geologic time. During major glacial epochs, which seem to recur at intervals of 200 to 250 Mya, great ice sheets form in the high latitudes and spread out to cover as much as 40% of the Earth's land surface.

Glacial rebound
(see isostasy).

Greenhouse effect
an atmospheric process in which the concentration of atmospheric trace gases (greenhouse gases) affects the amount of radiation that escapes directly into space from the lower atmosphere. Short-wave solar radiation can pass through the clear atmosphere relatively unimpeded. But long-wave terrestrial radiation, emitted by the warm surface of the Earth, is

	partially absorbed and then re-emitted by certain trace gases.
Greenhouse gases	the trace gases that contribute to the greenhouse effect. The main greenhouse gases are not the major constituents of the atmosphere – nitrogen and oxygen – but water vapour (the biggest contributor), carbon dioxide, methane, nitrous oxide, and (in recent years) chlorofluorocarbons. Increases in concentrations of the latter four gases have been linked to human activity.
Hadley cell	the basic vertical circulation pattern in the tropics where moist warm air rises near the equator and spreads out north and south and descends at around 20–30° N and S.
Hale cycle	the 22-year cycle in solar activity which is a combination of the 11-year cycle in sunspot number and the reversal of the magnetic polarity of adjacent pairs of sunspots between alternate cycles, which may be a cause of the 20-year cycle that is detected in many climatic records.
Half-life	time in which half of the atoms of a given quantity of radioactive nuclide undergo at least one disintegration.
Holocene	the relatively warm epoch, which started around 10 000 years ago and runs up to present time. It is marked by several short-lived particularly warm periods, the most significant of which, from 6.2–5.3 kya, is called the Holocene optimum.
Hurricane	the name given primarily to tropical cyclones in the West Indies and Gulf of Mexico.
Insolation (from INcoming SOLar radiATION)	the solar radiation received at any particular area of the Earth's surface, which varies from region to region depending on latitude and weather.
Interglacial	warmer periods during **glacial epochs** when the major ice sheets recede to higher latitudes.
Interstadial	a relatively warmer stage within a glacial phase during which the ice advance is temporarily halted.
Intertropical Convergence Zone (ITCZ)	a narrow low-latitude zone in which air masses originating in the northern and southern hemispheres converge and generally produce

cloudy, showery weather. Over the Atlantic and Pacific it is the boundary between the northeast and southeast trade winds. The mean position is somewhat north of the equator but over the continents the range of movement throughout the year is considerable.

Isostasy
the process whereby areas of the crust tend to float in conditions of near equilibrium on the plastic mantle; where ice sheets have melted this process leads to a slow rise in the crust as it returns to equilibrium (glacial rebound).

Isotopes
atoms of a single element (with the same number of protons) that have different masses (because they have a different number of neutrons). Isotopes are labelled with the atomic mass preceding the symbol of the element; ^{18}O, for example, denotes an oxygen isotope with an atomic mass of 18, instead of the mass of 16 which the most abundant form of oxygen has (as ^{16}O). Some isotopes have characteristics making them useful for analysing chemical history; for example, they release electrons or other sub-atomic particles, allowing their presence to be detected, and they decay (gradually changing into another element) at a steady, measurable speed. Carbon-14 (^{14}C) decays into ^{14}N, with a half-life of 5730 years. When detected in samples of sediment or ice, isotopes can be analysed to compile records of past climate characteristics.

Jet stream
strong winds in the upper troposphere whose course is related to the major weather systems in the lower atmosphere and which tend to define the movement of these systems.

Kelvin waves
gravity-inertia waves, which occur in the atmosphere and the oceans, where either the effect of the Coriolis force is negligible (i.e. close to the equator) or where this force is balanced by the pressure gradient. The most important examples are in the equatorial stratosphere and in the thermocline of the equatorial Atlantic and Pacific close to the equator (in both cases the waves propagate eastwards relative to the Earth).

Last Glacial Maximum (21 kya)
the last prolonged period of Ice Age cold climate before the present day.

Lithification
the process of conversion of unconsolidated sediment to a solid rock.

Lithosphere
the outer rigid shell of the Earth.

Little Ice Age (AD 1550–AD 1850)
a period marked by more frequent cold episodes in Europe, North America, and Asia, during which mountain glaciers, especially in the Alps, Norway, Iceland and Alaska expanded substantially.

Magma
naturally occurring molten or partially molten rock formed within the Earth from which igneous rocks crystallize.

Maunder minimum
a period during the seventeenth century when the level of solar activity, as reflected by the number of sunspots, was much lower than in subsequent centuries.

Mean sea level (MSL)
the average height of the sea surface, based on hourly observation of the tide height on the open coast, or in adjacent waters that have free access to the sea. In the United States, MSL is defined as the average height of the sea surface for all stages of the tide over a nineteen-year period.

Micrometre (μm)
10^{-6} metres.

Monsoon
a seasonal reversal of wind, which in the summer season blows onshore, bringing with it heavy, rains, and in winter blows offshore – it is of greatest meteorological importance in southern Asia. The word is believed to be derived from the Arabic word '*mausin*', meaning a season.

Non-linearity
the lack of direct proportionality of the input and output of a physical system.

North Atlantic Oscillation (NAO)
an index of the circulation in the North Atlantic that is measured in terms of the difference in pressure between the Azores and Iceland. In winter this index tends to switch between a strong westerly flow with pressure low to the north and high in the south and a weaker opposite pattern; the former tends to produce above-normal temperatures over much of the northern hemisphere, the latter the reverse.

Nutation
oscillation of the Earth's pole about the mean position. It has a period of about 19 000 years and is superimposed on the precessional movement.

Obliquity of the ecliptic the angle at which celestial equator intersects the ecliptic. At present this angle is slowly decreasing by 0.47 arc seconds a year, due to precession and nutation. It varies between $21°\,53'$ and $24°\,18'$.

Ozone a molecule made up of three atoms of oxygen (O_3). In the stratosphere, it occurs naturally and provides a protective layer shielding the earth from ultraviolet radiation and subsequent harmful health effects on humans and the environment. In the troposphere, it is a major component of photochemical smog.

Pelagic ooze a deep-ocean sediment formed from the hard parts of pelagic organisms and very fine suspended sediment.

Photosynthesis the process by which plants convert carbon dioxide (CO_2) and water (H_2O) of the air into carbohydrates by exposure to light.

Phytoplankton microscopic marine organisms (mostly algae and diatoms) which are responsible for most of the photosynthetic activity in the oceans.

Plate tectonics the interpretation of the Earth's structures and processes (including oceanic trenches, mid-ocean ridges, mountain building, earthquake zones and volcanic belts) in terms of the movement of large plates of the lithosphere acting as rigid slabs floating on a viscous mantle.

Pleistocene (10 kya to 1.6 Mya) the geological period, which together with the Holocene makes up the Quaternary. This epoch was characterised by numerous (at least 17) worldwide changes of climate, cycling between glacial (cool) and interglacial (warmer) periods, with periodicities of 100, 41 and 23 kyr.

Pliocene (1.6–5.3 Mya) a warm epoch with only limited glaciation. The mid-Pliocene was the last time that comparable temperatures existed to those predicted by General Circulation Models to take place within 300 years. Precipitation was greater than at present, including in the arid regions of Middle East, Asia and Northern Africa, where temperatures were lower than at present in summer.

Power spectrum	the presentation of the square of the amplitudes of the harmonics of a time series as a function of the frequency of the harmonics.
Precession of the equinoxes	the westward motion of 50.27 arc seconds per year of the equinoxes, caused mainly by the attraction of the Sun and the Moon on the equatorial bulge of the Earth. The equinoxes thus make one complete revolution of the ecliptic in 25 800 years and the Earth's pole turns in a small circle of radius 23° 27' about the pole of the ecliptic.
Proxy data	any source of information that contains indirect evidence of past changes in the weather (e.g. tree rings, ice cores and ocean sediments).
Quasi-biennial oscillation (QBO)	the alternation of easterly and westerly winds in the equatorial stratosphere with an interval between successive corresponding maxima of 20 to 36 months. Each new regime starts above 30 km and propagates downwards at about one kilometre a month.
Quaternary	the geological period covering the lasts 1.6 Myr or so, which includes the Pleistocene and the Holocene.
Radiative forcing	A change imposed upon the climate system that modifies the radiative balance of that system. The causes of such a change may include changes in the sun, clouds, ice, greenhouse gases, volcanic activity, and other agents. Convention lumps all these together as agents of radiative forcing. Radiative forcing is often specified as the net change in energy flux at the tropospause [watts per square metre ($W\,m^{-2}$)]. Many climate models seek to quantify changes in Earth's temperature, rainfall, and sea level in terms of a specified change in radiative forcing.
Radiatively active trace gases	gases, present in small quantities in the atmosphere that absorb incoming solar radiation or outgoing infrared radiation, thus affecting the vertical temperature profile of the atmosphere. These gases include water vapour, carbon dioxide, methane, nitrous oxide, chlorofluorocarbons, and ozone.

Radiolaria   marine pelagic micro-organisms with a siliceous skeleton.

Radiometer   an instrument that makes quantitative measurements of the amount of electromagnetic radiation falling on it in a specified wavelength interval.

Radiometric date   the age of a rock in years determined by the relative proportions of radioactive isotopes and their decay products present within the rock.

Radio-sonde   a free balloon carrying instruments that transmit measurements of temperature, pressure, and humidity to ground by radiotelegraphy as it rises through the atmosphere.

Rossby wave   in the atmosphere a wave in the general circulation in one of the principal zones of westerly winds, characterised by large wavelength (c. 6000 km), significant amplitude (c. 3000 km) and slow movement, which can be both eastward and westward relative to the Earth. In the ocean, similar waves have a wavelength of an order of a few hundred kilometres and nearly always move westward relative to the Earth.

Sea-floor spreading   the process by which lithospheric plates either side of an ocean ridge grow by addition of new material as the plates either side of the ridge move apart.

Solar radiation   the amount of radiation or energy received from the Sun at any given point. The *solar constant* is a measure of solar radiation ($\approx 1376\,\mathrm{W\,m^{-2}}$), at a point just outside the Earth's atmosphere, located on a surface that is perpendicular to the line of radiation, and measured when the Earth is at its mean orbital distance from the Sun.

Stadial   a period during **glacial epochs** when the ice sheets advanced to lower latitudes.

Stevenson shelter   a standard housing for ground-level meteorological instruments designed to ensure that reliable shade temperatures are measured.

Stratosphere   a region of the upper atmosphere, which extends from the tropopause to about 50 km above the Earth's surface, and where the temperature rises slowly with altitude. The properties of the stratosphere include very little vertical mixing, strong horizontal motions, and low water-vapour content compared to the troposphere.

Sunspots
dark blotches on the Sun indicating increased solar activity.

Thermocline
a region of rapidly changing temperature between the warm upper layer (the epilimnion) and the colder deeper water (the hypolimnion) in the ocean.

Thermohaline circulation
the deep-water circulation of the oceans driven by density contrasts due to variations in salinity and temperature.

Time series
any series of observations of a physical variable that is sampled at set constant time intervals.

Troposphere
the lower atmosphere, from the ground to an altitude of about 8 km at the poles, about 12 km in mid-latitudes, and about 16 km in the tropics. Clouds and weather systems, as experienced by people, take place in the troposphere.

Typhoon
a name of Chinese origin (meaning 'great wind') applied to tropical cyclones that occur in the western Pacific Ocean. They are essentially the same as hurricanes in the Atlantic and cyclones in the Bay of Bengal.

Uniformatorianism
the principle that geological events can be explained by processes observable today (the present is the key to the past). This assumes these processes have not changed during geological time.

Variance
the mean of the sum of squared deviations of a set of observations from the corresponding mean.

Volcanism
the phenomena of volcanic activity. Large volcanic eruptions spew massive amounts of ash into the atmosphere that absorb solar radiation, thus potentially generating a cooling effect on planetary temperatures. At the same time, volcanoes release carbon dioxide and sulfur dioxide, and decrease stratospheric concentrations of ozone.

Weathering	the chemical or physical breakdown of rocks at the surface by atmospheric agents and physical processes.
Younger Dryas (12.9–11.6 kya)	a sudden, abrupt cold episode, which interrupted the sustained warming trend between the Last Glacial Maximum and the Holocene.

Bibliography

Alexander, L. V. *et al.* (2006). Global observed dazes in daily climate extremes of temperature and precipitation. *Journal of Geophysical Research*, **111**.

Alley, R. B. (2000). *The Two-Mile Time Machine: Ice Cores, Abrupt Climate Change and Our Future*. Princeton, NJ, USA: Princeton University Press.

Baillie, M. G. L. (1995). *A Slice through Time: Dendrochronology and Precision Dating*. London, UK: Batsford.

Baldwin, M. P. *et al.* (2001). The Quasi-biennial Oscillation. *Reviews of Geophysics*, **32**, 179–229.

Barber, D. C. *et al.* (1999). Forcing of the cold event of 8,200 years ago by catastrophic drainage of Laurentide lakes. *Nature*, **400**, 344–8.

Barnston, A. G. (1995). Our improving capability in ENSO forecasting. *Weather*, **50**, 419–30.

Barry, R. G. & Chorley, R. J. (2003). *Atmosphere, Weather & Climate*. Eighth edition. London, UK: Methuen.

Bertage (to follow)

Benton, M. J. (1995). Diversification and extinction in the history of life. *Science*. **268**, 52–8.

Bigg, G. R. (2003). *The Oceans and Climate*. Second edition. Cambridge, UK: Cambridge University Press.

Bonan, G. B., Pollard, D. & Thompson, S. L. (1992). Effects of boreal forest vegetation on global climate. *Nature*, **359**, 716–18.

Bradley, R. S. & Jones, P. D. (eds) (1995). *Climate since AD 1500*. London, UK: Routledge.

Briffa, K. R. *et al.* (1990). A 1400-year tree-ring record of summer temperatures in Fennoscandia. *Nature*, **346**, 434–9.

Briffa, K. R. *et al.* (1995). Unusual twentieth-century summer warmth in a 1000-year temperature record from Siberia. *Nature*, **376**, 156–9.

Broecker, W. S. (1994). Massive iceberg discharges as triggers for global climate change. *Nature*, **372**, 421–5.

Broecker, W. S. (1995a). Chaotic climate. *Scientific American*, **267**, No. 11, 44–50.

Broecker, W. S. (1995b). Cooling the tropics. *Nature*, **376**, 212–13.

Brown, G. C., Hawkesworth, C. J. & Wilson, R. C. L. (eds.) (1992). *Understanding the Earth: a New Synthesis*. Cambridge, UK: Cambridge University Press.

Bryant, E. (1997). *Climate Process and Change*. Cambridge, UK: Cambridge University Press.

Burroughs, W. J. (1978). On running means and meteorological cycles. *Weather*, **33**, 101–9.

Burroughs, W. J. (1991). *Watching the World's Weather*. Cambridge, UK: Cambridge University Press.

Burroughs, W. J. (1994). *Weather Cycles: Real or Imaginary?* Cambridge, UK: Cambridge University Press.

Burroughs, W. J. (1997). *Does the Weather Really Matter? The Social Implications of Climate Change*. Cambridge, UK: Cambridge University Press.

Burroughs, W. J. (2001). *Climate Change: A Multidisciplinary Approach;* First Edition. Cambridge, UK: Cambridge University Press.

Burroughs, W. J. (2003). *Weather Cycles: Real or Imaginary?* Second edition. Cambridge, UK: Cambridge University Press.

Burroughs, W. J. (2005a). *Climate Change in Prehistory: The End of the Reign of Chaos*. Cambridge, UK: Cambridge University Press.

Burroughs, W. J. (2005b). *Does the Weather Really Matter?* Cambridge, UK: Cambridge University Press.

Cess, R. D. *et al.* (1991). Interpretation of snow-climate feedback as produced by 17 general circulation models. *Science*, **253**, 888–92.

Cess, R. D. *et al.* (1995). Absorption of solar radiation by clouds: observations versus models. *Science*, **267**, 496–9.

Chahine, M. T. (1992). The hydrological cycle and its influence on the climate. *Nature*, **359**, 373–9.

Chuine, I. *et al.* (2004). Grape ripening as a past climate indicator. *Nature*, **432**, 289–90.

CLIMAP Project Members. (1976). The surface of the ice-age Earth. *Science*, **191**, 1131–7.

CLIMAP. (1981). *Seasonal Reconstruction of the Earth's Surface at the Last Glacial Maximum*. Geological Society of America, Map and Chart Series, Vol. **C36**.

Cook, E. R. *et al.* (1995) The 'segment length curse' in long tree-ring chronology development for palaeoclimatic studies. *The Holocene*, **5**, 229–37.

Courtillot, V. (1999). *Evolutionary Catastrophes: The Science of Mass Extinctions*. Cambridge, UK: Cambridge University Press.

Christy, J. R. & Norris, W. B. (2004). What may we conclude about global tropospheric temperature trends? *Geophysical Research Letters*, **31**.

Dansgaard, W. & Oeschger, H. (1989). In *The Environmental Record in Glaciers and Ice Sheets*, H. Oeschger & C. C. Langway (eds.). Chichester, UK: Wiley, pp. 287–318.

Dansgaard, W. *et al.* (1993). Evidence of general instability of past climate from a 250-kyr ice-core record. *Nature*, **364**, 218–20.

Davis, B. A. S. *et al.* (2003). The temperature of Europe during the Holocene reconstructed from pollen data. *Quaternary Science Reviews*, **20**, 1701–6.

Dawson, A. G. (1992). *Ice Age Earth: Late Quaternary Geology and Climate*. London, UK: Routledge.

Diaz, H. F. & Markgraf, V. (eds.) (2000). *El Niño and the Southern Oscillation*. Cambridge, UK: Cambridge University Press.

Doherty, R. M., Hulme, M. & Jones, C. G. (1999). A gridded reconstruction of land and ocean precipitation for the extended tropics from 1974–1994. *International Journal of Climatology*, **19**, 119–42.

EPICA Community Members. (2004). Eight glacial cycles from an Antarctic ice core. *Nature*, **429**, 623–8.

Fagan, B. (1999). *Floods, Famines and Emperors: El Nino and the Fate of Civilisations*. London, UK: Plimlico.

Fagan, B. (2004). *The Long Summer: How Climate Changed Civilisation*. New York, USA: Basic Books.

Folland, C. K. & Parker, D. E. (1995). Correction of instrumental biases in historical sea surface temperature data. *Quarterly Journal Royal Meteorological Society*, **121**, 319–67.

Frakes, L. A., Francis, J. E. & Syktus, J. I. (1992). *Climate Modes of the Phanerozoic*. Cambridge, UK: Cambridge University Press.

Frich, P. *et al.* (2002). Observed coherent changes in climatic extremes during the second half of the twentieth century. *Climate Research*, **19**, 193–212.

Fritts, H. C. (1976). *Tree Rings and Climate*. London, UK: Academic Press.

Giovanelli, R. (1984). *Secrets of the Sun*. Cambridge UK: Cambridge University Press.

von Grafenstein, U. *et al.* (1999). A mid-European decadal isotope-climate record from 15,500 to 5000 years B.P. *Science*, **284**, 1684–7.

Gray, L. J., Haigh, J. D. & Harrison, R. G. (2005). The influences of solar changes on the Earth's climate. Hadley Centre Technical Note 62 (HCTN62), available from the Hadley Centre website (http://www.met-office.gov.uk/research/hadleycentre/pubs/HCNT/.)

Gray, W. M. (1990). Strong association between West African rainfall and US landfall of intense hurricanes. *Science*, **249**, 1251–6.

Greenland Ice Core Project (GRIP) Members. (1993). Climate instability during the last interglacial period recorded in the GRIP ice core. *Nature*, **364**, 203–7.

Grootes, P. M. *et al.* (1993). Comparison of oxygen isotope records from the GISP 2 and Greenland ice cores. *Nature*, **366**, 552–4.

Grove, J. M. (1988). *The Little Ice Age*. London, UK: Methuen.

Gurney, R. J., Foster, J. L. & Parkinson, C. L. (eds.) (1993). *Atlas of Satellite Observations Related to Global Change*. Cambridge, UK: Cambridge University Press.

Haigh, J. D. (2000). Solar variability and climate. *Weather*, **55**, 399–405.

Hammer, C. U., Clausen, H. B. & Dansgaard, W. (1980). Greenland ice sheet evidence of post-glacial volcanism and its climatic impact. *Nature*, **288**, 230–5.

Hansen, J. E. *et al.* (1995). Satellite and surface temperature data at odds? *Climate Change*, **30**, 103–17.

Harries, J. E. (1990). *Earthwatch: the Climate from Space*. London, UK: Ellis Horwood.

Hastenrath, S. (2002). Dipoles, temperature gradients, and tropical climate anomalies. *Bulletin of the American Meteorological Society*, **83**, 735–8.

Haug, G.H. *et al.* (2003). Climate and the collapse of Maya civilisation. *Science*, **299**, 1721–5.

Hickey, M. & King, C. (1988). *100 Families of Flowering Plants*, Second Edition. Cambridge, UK: Cambridge University Press.

Hostetler, S.W. & Mix, A.C. (1999). Reassessment of ice-age cooling of the tropical ocean and atmosphere. *Nature*, **399**, 673–6.

Houghton, J. (2002). *The Physics of Atmospheres.* Third edition. Cambridge, UK: Cambridge University Press.

Houghton, J. (2004). *Global Warming: the Complete Briefing.* Third edition. Cambridge, UK: Cambridge University Press.

Hulme, M. & Barrow, E. (eds.) (1997). *Climates of the British Isles: Present, Past and Future.* London, UK: Routledge.

Hulme, M. & Jones, P.D. (1991). Temperatures and windiness over the United Kingdom during the winters of 1988/89 and 1989/90 compared with previous years. *Weather*, **46**, 126–36.

Hurrell, J.W. (1995). Decadal trends in the North Atlantic Oscillation: Regional temperatures and precipitation. *Science*, **269**, 676–9.

Hurrell, J.W. (1996). Influence of variations in extratropical wintertime tele-connections on Northern Hemisphere temperature. *Geophysical Research Letters*, **23**, 665–8.

Imbrie, J. & Imbrie, J.Z. (1979). *Ices Ages: Solving the Mystery.* London, UK: Macmillan.

Imbrie, J. & Imbrie, J.Z. (1980). Modelling the climatic response of orbital variations. *Science*, **207**, 943–53.

Imbrie, J. *et al.* (1992). On the structure and origin of major glaciation cycles. 1. Linear responses to Milankovitch forcing. *Paleoceanography*, **7**, 701–38.

Imbrie, J. *et al.* (1993). On the structure and origin of major glaciation cycles. 2. The 100 000-year cycle. *Paleoceanography*, **8**, 699–735.

International Federation of Red Cross and Red Crescent Societies. (1999). *World Disasters Report.*

IPCC. (1990). *Climate Change: The IPCC Scientific Assessment.* J.T. Houghton, G.J. Jenkins & G.G. Ephraums (eds.). Cambridge, UK: Cambridge University Press. 365pp.

IPCC. (1992). *Climate Change 1992: the Supplementary Report to IPCC Scientific Assessment*, J.T. Houghton, B.A. Callander & S.K. Varney (eds.). Cambridge, UK: Cambridge University Press.

IPCC. (1994). *Climate Change 1994: Radiative Forcing of Climate and an Evaluation of the IPCC IS92 Emission Scenarios*, J.T. Houghton, L.G. Meira Filho, J. Bruce, Hoesung Lee, B.A.Callendar, E. Haites, N. Harris, & K. Maskell (eds.). Cambridge, UK: Cambridge University Press.

IPCC, (1995). *Climate Change 1995: The Science of Climate Change*, J.T. Houghton, L.G. Meira Filho, B.A. Callendar, N. Harris, A. Kattenberg & K. Maskell. (eds.). Cambridge, UK: Cambridge University Press.

IPCC, (2001). *Climate Change 2001: The Scientific Basis*, J. T. Houghton, Y. Ding, D. Griggs, M. Noguer, P. J., van der Linden, X., Dai, K., Maskell & C. A Johnson, (eds.). Cambridge, UK: Cambridge University Press.

IPCC, (2007). *Climate Change 2007: The Scientific Basis*, Cambridge, UK: Cambridge University Press.

Jablonski, D. (1997). Progress at the K-T boundary. *Nature*, **387**, 354–5.

Jolliffe, I. T. (1986). *Principal Component Analysis*. New York, USA: Springer-Verlag.

Jones, P. D. *et al.* (1999). *Review of Geophysics*, **37**, 173–99.

Kalnay, E. *et al.* (1996). The NCEP/NCAR 40-year reanalysis project. *Bulletin of the American Meterological Society*, **77**, 437–71.

Karl, T. R. *et al.* (1995). Critical issues for long-term climate monitoring. *Climatic Change*, **31**, 185–221.

Karl, T. R. *et al.* (1996). Indices of climate change for the United States. *Bulletin of the American Meteorological Society*, **77**, 279–92.

Kendall, M. (1976). *Time Series*. London, UK: Charles Griffin.

Kripalani, R. H. & Kulkarni, A. (1997). Climatic impact of the El Niño/La Niña on the Indian monsoon: a new perspective. *Weather*, **52**, 39–46.

Laird, K. R. *et al.* (1996). Greater drought intensity and frequency before AD 1200 in the Northern Great Plains, USA. *Nature*, **384**, 552–4.

LaMarche, V. C. Jr. (1974). Paleoclimatic inferences from long tree-ring records. *Science*, **183**, 1043–8.

Lamb, H. H. (1972). *Climate: Present, Past and Future*. Volume 1. London, UK: Methuen.

Lamb, H. H. (1977). *Climate: Present, Past and Future*. Volume 2. London, UK: Methuen.

Lamb, H. H. (1995). *Climate, History and the Modern World* (2nd edn). London, UK: Routledge.

Lambeck, K., Esat, T. M. & Potter, E.-K. (2003). Links between climate and sea levels for the past three million years. *Nature*, **419**, 199–206.

Landsea, C. W. *et al.* (1994). Seasonal forecasting of Atlantic hurricane activity. *Weather*, **49**, 273–84.

Landsea, C. W. *et al.* (1999). Atlantic basin hurricanes: Indices of climatic changes. *Climatic Change*, **42**, 89–129.

Lean, J. (2000). Evolution of the Sun's Spectral Irradiance since the Maunder Minimum. *Geophysical Research Letters*, **27**, 2425–8.

Le Roy Ladurie, E. & Baulant, M. (1980). Grape harvests from the fifteenth through the nineteenth centuries. *Journal of Interdisciplinary History*, **10**, 839–49.

Li, X. *et al.* (1996). Dominance of mineral dust in aerosol light scattering in the North Atlantic trade winds. *Nature*, **380**, 416–19.

Lisiecki, L. E. & Raymo, M. E. (2005). A Pliocene–Pleistocene stack of 57 globally distributed benthic $\delta^{18}O$ records. *Paleoceanography*, **20**, PA 1003, doi:10.1029/2004PA001071.

Lockwood, M. & Stamper, R. (1999). Long term drift in the coronal source magnetic flux and total solar irradiance. *Geophysical Research Letters*, **26**, 2461–5.

Lovelock, J. E. (1988). *The Ages of Gaia*. Oxford, UK: Oxford University Press.

Lundin, R., Eliasson, L. & Murphree, J. S. (1991) The quiet time aurora. In *Auroral Physics*, C.-I. Meng, M. J. Rycroft & L. A. Frank (eds.). Cambridge, UK: Cambridge University Press.

Luterbacher, J. *et al.* (2004). European seasonal and annual temperature variability, trends, and extremes since 1500. *Science*, **303**, 1499–503.

Mann, M. E. & Jones, P. D. (2003). Global surface temperatures over the past two millennia. *Geophysical Research Letters*, **30**, 1820, doi:10.1029/2003GL017814.

Mangini, A, Spotl, C, & Verdes, P. (2005). Reconstruction of temperature in the Central Alps during the past 2000 yr from a $\delta^{18}O$ stalagmite record. *Earth & Planetary Science Letters*, **235**, 741–51.

Manley, G. (1974). Central England Temperatures: monthly means 1659 to 1973. *Quarterly Journal Royal Meteorlogical Society*, **100**, 389–405.

Markson, R. (1978). Solar modification of atmospheric electrification and possible implications for the Sun-weather relationship. *Nature*, **244**, 197–200.

Marshall, J., *et al.* (2001). North Atlantic climate variability: Phenomena, impacts and mechanisms: A Review. *International Journal of Climatology*, **21**, 1863–98.

Martinson, D. G., *et al.* (1987). Age dating and the orbital theory of the Ice Age:development of a high resolution 0 to 300 000-year chronostratigraphy. *Quaternary Research*, **17**, 1–30.

McIlveen, R. (1992). *Fundamentals of Weather and Climate*. London, UK: Chapman and Hall.

Mitchell, J. F. B. *et al.* (1995). Climate response to increasing levels of greenhouse gases and sulphate aerosols. *Nature*, **376**, 501–4.

Mitchell, J. M., Stockton, C. W. & Meko, D. M. (1979). Evidence of a 22-year rhythm of drought in the Western United States related to the Hale Solar Cycle since the 17th century. In *Solar-Terrestrial Influences on Weather and Climate*, B. M. McCormac & T. A. Seliga, (eds.). Dordrecht, The Netherlands: Reidel Publishing Co.

Mitchell, J. M. (1990) Climatic variability: past, present & future. *Climatic Change*, **16**, 231–46.

Mix, A. C., Bard, E. & Schneider, R. (2001). Environmental processes of the ice age: land, oceans, glaciers (EPILOG). *Quaternary Science Reviews*, **20**, 627–57.

Moberg, A. *et al.* (2005). Highly variable Northern Hemisphere temperatures reconstructed from low and high-resolution proxy data. *Nature*, **433**, 613–17.

Musk, L. F. (1988). *Weather Systems*. Cambridge, UK: Cambridge University Press.

Palmer, T. (1993). A nonlinear dynamical perspective on climate change. *Weather*, **48**, 314–25.

North Greenland Ice Core Project members. (2004). High-resolution record of Northern Hemisphere climate extending into the last interglacial period. *Nature*, **431**, 147–55.

Open University (2001). *Ocean Circulation*. London: Butterworth Heinemann (in association with The Open University, Milton Keynes).

Paillard, D. (1998). The timing of Pleistocene glaciations from a simple multiple-state model. *Nature*, **391**, 378–81.

Parker, D. E., Legg, T. P. & Folland, C. K. (1992). A new daily Central England temperature series, 1772–1991. *International Journal of Climatology*. **12**, 317–42.

Parry, M. & Duncan, R. (eds.). (1995). *The Economic Implications of Climate Change in Britain*. London, UK: Earthscan.

Pecker, J. C. & Runcorn, S. K. (eds.). (1990). *The Earth's Climate and Variability of the Sun over Recent Millennia: Geophysical, Astronomical and Archaeological Aspects*. London, UK: The Royal Society.

Petit, J. R. *et al.* (1999). Climate and atmospheric history of the past 420,000 years from the Vostok ice core, Antarctica. *Nature*, **399**, 429–36.

Pfister, C. (1995). Monthly temperature and precipitation in central Europe 1525–1979: quantifying documentary evidence on weather and its effects. In *Climate Since AD 1500*, R. S. Bradley & P. D. Jones (eds.) (Chapter 6). London, UK: Routledge. [Data on temperature and precipitation indices available on disk from the National Geophysical Data Center, Boulder, Co 80309, USA.]

Philander, S. G. H. (1983). El Niño Southern Oscillation. *Nature*, **302**, 295–301.

Pimental, D. *et al.* (1995). Environmental and economic costs of soil erosion and conservation benefits. *Science*, **267**, 1117–23.

Piekle, R. A. & Landsea, C. W. (1998). Normalized hurricane damage in the United States: 1925–95. *Weather and Forecasting*. **13**, 621–31.

Rampino, M. R. & Self, S. (1992). Volcanic winter and accelerated glaciation following the Toba super-eruption. *Nature*, **359**, 50–2.

Raup, D. M. (1991). *Extinction: Bad Genes or Bad Luck?* New York, UK: W. W. Norton.

Richerson, P. J., Boyd, R. & Bettinger, R. L. (2001). Was agriculture impossible during the Pleistocene but mandatory during the Holocene? A climate change hypothesis. *American Antiquity*, **66**, 1–50.

Robin, G. de Q. (1983). *The Climatic Record in Polar Ice Sheets*, Cambridge, UK: Cambridge University Press.

Rosenzweig, C. & Parry, M. L. (1994). Potential impact of climate change on world food supply. *Nature*, **367**, 133–8.

Sherratt, A. (1980). *The Cambridge Encyclopedia of Archaeology*. Cambridge, UK: Cambridge University Press.

Sherwood, S., Lanzante, J. & Meyen, C. (2004). Radiosonde daytime biases and late-20th century warming. *Science*, **309**, 1556–9.

Shoemaker, E. M. (1983). Asteroid and comet bombardment of the Earth. *Annual Review of Earth and Planetary Science*, **11**, 461–94.

Smith, D. G. (1982). *Cambridge Encyclopaedia of Earth Sciences*. Cambridge, UK: Cambridge University Press.

Stommel, H. & Stommel, E. (1979). The year without a summer. *Scientific American*, **240**, June, 134–40.

Stothers, R. B. (1984). Mystery cloud of AD 536. *Nature*, **307**, 344–5.

Stouffer, R. J., Manabe, S. & Vinnilcov, K. Ya. (1994). Model assessment of the role of natural variability in recent global warming. *Nature*, **367**, 634–6.

Strzepek, K. M. & Smith, J. B. (eds.) (1995). *As Climate Changes: International Impacts and Implications*. Cambridge, UK: Cambridge University Press.

Taylor, K. C. *et al.* (1993). The "flickering switch" of late Pleistocene climate change. *Nature*, **361**, 432–6.

von Storch, F. & Zwiers F. W. (2002) *Statistical Analysis in Climate Research*. Cambridge, UK: Cambridge University Press.

Tengen, I., Lacis, A. A. & Fung, I. (1996). The influence on climate forcing of mineral aerosols from disturbed soils. *Nature*, **380**, 419–22.

Tett, S. F. B. *et al.* (1999). Causes of twentieth century climate change. *Nature*, **399**, 569–72.

Thomas, D. S. G. & Middleton, N. J. (1994). *Desertification: Exploding the Myth*. Chichester UK: Wiley.

Toumi, R., Bekki, S. & Law, K. (1995). Indirect influence of ozone depletion on climate forcing by clouds. *Nature*, **372**, 348–51.

Trenberth, K. E. (ed) (1992). *Climate System Modelling*. Cambridge, UK: Cambridge University Press.

Tucker, C. J., Dregne, H. E. & Newcomb, W. W. (1991). Expansion and contraction of the Sahara Desert from 1980 to 1990. *Science*, **253**, 299–301.

Van Andel, T. H. (1994). *New Views on an Old Planet: a History of Global Change*. Cambridge, UK: Cambridge University Press.

Weaver, A. J. & Hughes, T. M. C. (1994). Rapid interglacial climate fluctuations driven by North Atlantic ocean circulation. *Nature*, **367**, 447–50.

Weaver, A. J., Sarachik, E. S. & Marotzke, J. (1991). Freshwater flux forcing of decadal and interdecadal oceanic variability. *Nature*, **353**, 836–8.

Whitlock, C. & Bartlein, P. J. (1997). Vegetation and climate change in northwest America during the last 125 kyr. *Nature*, **388**, 57–61.

Wigley, T. M. L., Ingram, M. J. & Farmer, G. (1981). *Climate and History: Studies in Past Climates and their Impact on Man*. Cambridge, UK: Cambridge University Press, UK.

Wigley, T. M. L., Lough, J. M. & Jones, P. D. (1984). Spatial patterns of precipitation in England and Wales and a revised homogeneous England and Wales precipitation series. *Journal of Climatology*, **4**, 1–25.

Willson, R. C. & Hudson, H. S. (1991). The Sun's luminosity over a complete cycle. *Nature*, **351**, 42–4.

Wiscombe, W. J. (1995). An absorbing mystery. *Nature*, **376**, 466–7.

Zachos, J. *et al.* (2001). Trends, rhythms and aberrations in global climate 65 Ma to present. *Science*, **292**, 686–93.

Index